W9-BND-147

MANAGING OUR
NATURAL RESOURCES

Fourth Edition

**Delmar is proud to
support FFA activities**

Managing Our Natural Resources

Fourth Edition

William G. Camp
Virginia Polytechnic Institute and State University
Blacksburg, Virginia

Thomas B. Daugherty
Maconaquah High School
Bunker Hill, Indiana

Heidi M. Martin
Fauquier High School
Warrenton, Virginia

Susan Aksamit
Clintwood High School
Clintwood, Virginia

DELMAR
THOMSON LEARNING

Australia Canada Mexico Singapore Spain United Kingdom United States

Managing Our Natural Resources, 4th Edition
by William G. Camp and Thomas B. Daugherty

Business Unit Director:
Susan L. Simpfenderfer

Executive Editor:
Marlene McHugh Pratt

Acquisitions Editor:
Zina M. Lawrence

Development Editor:
Andrea Edwards Myers

Editorial Assistant:
Elizabeth Gallagher

Executive Production Manager:
Wendy A. Troeger

Production Manager:
Carolyn Miller

Production Editor:
Kathryn B. Kucharek

Executive Marketing Manager:
Donna J. Lewis

Channel Manager:
Nigar Hale

Cover Design:
Dutton & Sherman Design

For permission to use material from this text or product, contact us by
Tel   (800) 730-2214
Fax   (800) 730-2215
www.thomsonrights.com

Library of Congress Cataloging-in-Publication Data
Camp, William G.
   Managing our natural resources / William G. Camp, Thomas B. Daugherty.—4th ed.
      p. cm.
   Includes index.
   ISBN 0-7668-1554-4
      1. Natural resources—United States—Management—Juvenile literature. 2. Natural resources—Management—Vocational guidance—United States—Juvenile literature. [1. Natural resources—Management. 2. Conservation of natural resources. 3. Occupations.] I. Title.
HC103.7.C33 2000
333.7'0973—dc21   00-064523

NOTICE TO THE READER

Publisher does not warrant or guarantee any of the products described herein or perform any independent analysis in connection with any of the product information contained herein. Publisher does not assume, and expressly disclaims, any obligation to obtain and include information other than that provided to it by the manufacturer.

The reader is expressly warned to consider and adopt all safety precautions that might be indicated by the activities herein and to avoid all potential hazards. By following the instructions contained herein, the reader willingly assumes all risks in connection with such instructions.

The publisher makes no representation or warranties of any kind, including but not limited to, the warranties of fitness for particular purpose or merchantability, nor are any such representations implied with respect to the material set forth herein, and the publisher takes no responsibility with respect to such material. The publisher shall not be liable for any special, consequential, or exemplary damages resulting, in whole or part, from the readers use of, or reliance upon, this material.

Contents

PREFACE

There are many different ways to look at the management of our natural resources. Three different perspectives that must be considered are those of the preservationist, the conservationist, and the exploiter.

If we look at the world as does a preservationist, nature is something that should be left intact as much as possible. From that perspective, humankind is a great destroyer and pillager, and we have no right to destroy the land or tear the earth apart. To a preservationist, managing nature is a foreign concept because nature is something to be left alone. The preservationist would stop the construction of a pipeline needed by our economy to move Alaskan oil, if he or she believed it would make the landscape ugly or endanger the migratory habits of caribou herds.

If we take the viewpoint of the exploiter, natural resources are merely a source of wealth and power. They represent something that is there for the taking and for our own use. From that perspective, nature is something to be reshaped for our economic benefit, with little regard for other creatures in the ecosystem. The exploiter would build a dam to supply electricity even if, by doing so, the habitat of a small, rare fish would be destroyed.

The conservationist viewpoint is somewhere between those two. From that perspective, nature provides resources that should be used carefully for our collective economic benefit. The conservationist would encourage the harvesting of forests to produce wood and paper as long as the trees are replaced by other trees having economic value, but not necessarily the same kind of tree harvested originally. The conservationist would allow the strip mining of coal as long as the land is restored to a balanced, natural state when the mining is completed and as long as no meaningful collateral environmental damage occurred during the process.

Which viewpoint is right and which is wrong? There is no simple answer to such a question. If there were a right answer, there would not be such divergent viewpoints; and consensus among the different factions would be possible. What we can say for certain is that the world is not a very large place, and there are many people. As the number of people on Earth continues to increase, the pressure applied to the environment also increases. It is clear that we cannot follow the exploiter's approach forever but, by the same token, a world human population of over 6 billion cannot exist without affecting the shape and condition of the environment.

It is not the purpose of this book to convince you that any one of these three divergent perspectives is the right one. The preservationist argument will be made that such creatures as the bald eagle and the sea otter have legitimate rights to exist on Earth. At the same time, it is difficult to argue convincingly that the people of a village in India should be forced to live in constant danger of being killed and eaten by tigers to safeguard the tiger population. Without the exploiters, our nation would never have grown into the powerful and rich society it is today. It was conservationists who gave us kudzu (a vine) to prevent road bank erosion, and now much of the south is overgrown with the rapidly spreading plant. It is obvious that members of all three groups are both right and wrong in many ways.

It is the purpose of this book to present a balanced viewpoint of the place of humans in the world as long-term residents. Discussions will be presented that deal with soil formation, erosion, reclamation, and conservation; water use and improvement; endangered species of wildlife; hunting game animals; fishing; safety in boating, hiking, and other forms of outdoor recreation; conservation farming; land-use planning; construction practices that minimize the impact of exploitation on the environment; energy resources use, abuse, conservation, and alternatives; mineral use and recycling; and career opportunities in each of those diverse fields.

Managing our natural resources is a very broad topic, and a book that attempts to deal with it cannot go into great depth in any of the areas considered. This book should be used as a survey of many broad areas rather than as a definitive treatment of any one area of study.

Managing Our Natural Resources was written because we perceived a need for a book that took a broad look at the whole panorama of preservation, exploitation, and conservation of natural resources. There are many books on soils, for instance, but such books do not treat soil management as it re-

lates to wildlife management and fisheries development. This book attempts to take just such an approach.

From Tom Daugherty: I dedicate my portion of this book with love to my wife Jennifer Daugherty and to my children Heather, Jeremy, and Laura.

From Bill Camp: I dedicate my efforts on this book to my wife Betty and to my four super grandchildren Robbie and Kayla Thorpe and Emmie and Lewis Camp.

Our thanks also go to the editorial staff of Delmar/Thomson Learning, especially Judith Nelson, Zina Lawrence, and Andrea Myers for the encouragement and motivation so constantly supplied. Finally, we would also like to express special recognition to Ms. Heidi Martin, an agricultural education teacher in Hanover County Virginia for writing the Case Studies that add so much to this edition of the book, and to Ms. Susan Aksamit, an agricultural education teacher in Clintwood, Virginia for writing the first draft of Chapter 10.

ACKNOWLEDGMENTS

The authors and Delmar wish to express special appreciation to the following reviewers who provided their input and assistance in preparing this text:

M. Ray Gillis
Ponce De Leon High School
Ponce De Leon, Florida

Perry Richard
Harmony Grove Schools
Camden, Arkansas

Douglas Prevette
North Iredell High School
Olin, North Carolina

Section I
Introduction

1

OUR NATURAL RESOURCES

OBJECTIVES

After reading this chapter, you should be able to

- define and discuss the concept of natural resources
- list and describe the major categories of natural resources in America
- explain what makes something a natural resource

TERMS TO KNOW		
natural resource	vertebrate	minerals
topsoil	noncommercial forest	recreational resources
usable water	mature forest	

Endless, boundless, inexhaustible gifts—we once looked upon the world's natural resources that way. Indeed, when there were fewer people to use the resources, when our technology did not allow us to change the face of the earth so dramatically, that appeared true. But now machines allow one person to do things that armies of workers could not do before. According to the United Nations, the world's population exceeded 6 billion about October 12, 1999. Their description of that event was:

> During this final year of the twentieth century, a child will be born, bringing the world's population to 6 billion. No one knows when or where the baby will arrive. It could be a girl or a boy, the child of a millionaire or—far more likely—the child of a family living on less than a dollar a day.

Yet, many continue to treat nature's gifts as endless, boundless, and inexhaustible.

This cannot continue. It is to help you appreciate this dilemma that we dedicate this book.

WHAT IS A NATURAL RESOURCE?

One authority defines a **natural resource** as any form of energy that can be used by humans. Others would tell us that natural resources are objects people use. In an ecological sense, everything that touches upon our lives is a natural resource.

For the purposes of this book, we will use the ecological approach. Natural resources can be defined as all those things with which people come in contact and can be used to perform any useful function. This includes all energy forms that can be harnessed by human ingenuity. It includes objects, creatures, and materials that can be moved, shaped, built upon, built with, or manipulated for any useful purpose. It includes those things that inspire, relax, or strengthen humans as individuals or groups.

This obviously covers too many areas for any book, so we must limit our discussion. We will look at only those natural resources used on a large scale and in an organized way. These include our soil, water, fish and wildlife, forests, metals and minerals, fossil fuels, other major energy sources, and recreational resources.

This leads us to a working definition of natural resources: *Natural resources are objects, materials, creatures, or energy found in nature that can be used by humans.*

Usefulness Changes

Many things affect our definition of usefulness. Religion affects Hindu attitudes toward cattle, for instance. Custom affects most Western attitudes toward dogs as a food source. Technology and science affect our use of outer space as a communication medium.

In particular, the usefulness of many things changes over time as our science and technology improve. For instance, the natural resources humans have used to provide light have changed many times. For thousands of years, humans burned wood to provide light at night. Later, people learned that wood torches could be dipped into animal fat and made to burn longer, so fat from animals became a source of light. Still later, people learned that whale oil could be used to burn in lamps, and whales became a resource used to produce light.

Over a century ago, we learned that petroleum could be refined to produce products that could be burned to produce light, and a new natural resource was born. In this century, falling water and nuclear energy have been harnessed to produce electricity to provide light (among other things). None of these

things was new. Animal fat, whale oil, petroleum, falling water, and nuclear energy have existed all along; but their usefulness has changed because changes in our technology have meant that we could use things that have existed all along in new ways.

Many of the things we consider as resources now were not natural resources at all in earlier times. Nuclear energy, gasoline, most metal alloys, electricity from falling water, deep groundwater—these and many more were not natural resources when Europeans first came to America. They were not natural resources because they were not useful to humans.

Our situation changed too. When the first permanent English settlement in America was established in 1607 on the James River in what is now Virginia, the settlers faced many problems. One of the most serious was trees. Certainly, a few trees were needed for building and for burning, but once those needs were met, forests were a liability. They seemed to go on forever. In that sense, forests were a hindrance rather than a resource.

With this working definition of natural resources in mind, let us continue. The remainder of this chapter will briefly examine the major categories of natural resources.

SOIL RESOURCES

Land Area

The United States has a total land area of 3,675,545 square miles in its 50 states. That equates to 2.26 billion acres. The surface of our country ranges from 282 feet below sea level in Death Valley to 20,320 feet above sea level on Mt. McKinley. This vast area is covered by many things: rocks, sand, water, organic matter, parent material, subsoil, and soil. It is from the **topsoil**, the uppermost layer of soil, that we must get almost all of our food and natural fibers; it is also where we, for the most part, live, work, and play.

Of the land in this country in 1607, there seemed to be no practical limit. There was more than it seemed possible to use. Even 100 years later, there was more land than we could settle. Today that is no longer true.

In the 2000s, almost one-third of our land area is not suitable for farming. Another 8 percent is covered by cities, factories, homes, highways, and other artificial structures. The remaining 60 percent is useful for food and fiber production. Of that, only about 385 million acres, or 17 percent of the total, is usable for crop production. Even less is good farmland, and much is of only marginal value.

The soil's major enemy has been erosion. In the years since our nation began, we have lost one-third of our topsoil to erosion. Only one-fourth of our cropland is undamaged by this menace. Another problem we are beginning to face is the conversion of agricultural land to urban or other uses. Once an acre of prime corn-belt land is covered by concrete or asphalt, it is hard to grow corn there. (See Figure 1-1.)

Urban expansion, industrialization, highway construction, and other alternative uses for our land surface are becoming more and more important. This is not necessarily bad. Land is an important natural resource for many purposes—food and fiber production is only one purpose. Unfortunately, this expansion tends too often to occur on our best land

FIGURE 1-1 Prime farmland is often taken up by urban expansion. Both are valuable uses.

for farm production. Land-use planning is more important now than ever. We must, as a nation, establish our priorities for land use. We must then allocate our land and soil resources based on those priorities. It would seem that, with the world population explosion, food production must rank very high in those priorities. Nevertheless, manufacture of other products, transportation, processing, and distribution are also critical to our way of life.

Managing our soil and land resources is a complex problem. Hard decisions need to be made now and in the future. (See Figure 1-2.)

WATER RESOURCES

"Water, water everywhere" was the wail of the ancient mariner.[1] Indeed, this resource is abundant. Why worry about managing our water resources when 70 percent of the earth's surface is covered by it?

The ancient mariner continued "nor any drop to drink." There is a great difference between water and **usable water**. Remember

---

[1] Coleridge, Samuel Taylor, *The Rhyme of the Ancient Mariner.*

FIGURE 1-2 Farming and urbanization compete for the same land. Prime farmland is relatively limited, and urbanization often takes the very best land. This farm is located outside Richmond, Virginia. *(Photo by Tim McCabe, courtesy U.S. Department of Agriculture)*

our definition of natural resource: Water is a natural resource only when it can be put to use by people. The oceans provide us with marine products and a medium in which ships can travel. They contain minerals and metals, and tide flows can be harnessed for electrical generation. The ocean floors contain vast deposits of oil. We could go on and on, but these uses are minor compared to our total uses for water.

Most of our usable water is always on its way back to the sea. We can increase its usefulness by slowing its journey. We can contain it for other uses and take its energy for our own purposes. We can drink, wash, clean, cool, grow our food, and manufacture with it. To do these things, we must manage the water.

Water was an early source of power in this country. Running water carried logs, floated boats, and turned water wheels. By 1900, less than 4 percent of America's power still came from this source. Still, water is needed in even greater quantities today. Every day Americans use 300 billion gallons of water. Of this, about 60 billion gallons are temporarily removed from the water cycle. The remaining 240 billion gallons are returned directly to the hydrologic cycle (discussed in Chapter 13), but much of that is either heated or damaged by pollutants.

Another facet of our water resource management problem is the control of excess runoff. At least 37 states have average annual runoffs of 10 inches or more, yet much of this area has low annual rainfall totals—as little as 20 inches in several of the states. Thus, much of the water that falls on the land simply flows directly to the sea—it remains part of the hydrologic cycle without being put to use by people. This is a potential natural resource we may be able to tap with effective water management.

There is not a shortage of water in this country. We have plenty of fresh water; but try telling that to a deep-well farmer in Arizona! Yes, there is plenty of water; but it may not be where it is needed. Moreover, it may not be of a usable quality when it is where it is needed. The problems are control, quality, and distribution.

FISH AND WILDLIFE RESOURCES

Wildlife is the natural resource that resembles humans most. Though not thought of as a resource by many, wildlife species definitely play an important role in our lives. In the conservationist sense, fish and wildlife are defined as nondomesticated animals, either game or nongame. Broadly interpreted, however, it can also embrace uncultivated plant life. The key is "wildness."

Fish and animal wildlife are considered to be renewable, but this is only true while a species is alive and reproducing. Since colonial times, 48 **vertebrate** species have been exterminated in our nation and another 228 were regarded as rare and endangered in 1999. Thus, in one sense, they are not renewable, and they certainly are not inexhaustible.

Although wildlife is not as important for food as it was when the country was young, it still is of value to us. The pleasure that wild animals, fish, and birds afford us; the meat still produced from them; and their instinctive insect-destroying ability (valued at over $1 billion per year) are all assets that they contribute. More than $1 billion is also brought in each year through hunting and fishing.

Until recently, the structure of wildlife conservation has dictated a program especially for hunters and sports fishers. However, lately a new aspect of fish and wildlife management has begun to emerge. This aspect deals with satisfying the demands of the nonhunting and nonfishing public for the pleasure afforded by seeing wild creatures in nature. Thus, parks and preserves are becoming more dedicated to this idea. They are concentrating on the return of "nature" to such recreation areas. Aesthetic values encourage the preservation of ecosystems in their natural states.

As long as the population of a species of fish or wildlife remains fairly stable, there is little concern for its long-term survival. For such animals, hunting and fishing are acceptable forms of management. When the population of a species starts to fall too low, it may become "threatened." Threatened species are those that appear to be declining in numbers toward the point that the species' survival may be in danger. If the population continues to fall, it may be declared "endangered" or "rare." When that happens, exceptional efforts are often required to help the species recover. Chapters 23 and 24 provide more in-depth discussion of rare, endangered, and even extinct species.

FOREST RESOURCES

In 1607, at least half of our land area was forested. This amounted to over a billion acres. Almost 70 percent of that area is still in forest. Of the 738 million acres of forest in the United States today, about one-third is **noncommercial forest**. This means it is generally not usable for forest production. The other two-thirds are usable for commercial forestry. (See Figure 1-3.)

Of the 30 percent of our original forest that is now gone, almost all would be considered commercially usable if we still had it. As in nonfarm uses of farmland, this nonforest use of forest land is not necessarily bad. Our society

FIGURE 1-3 Since colonial days, our forested land in the United States has decreased by nearly one-third; yet we grow more wood each year than ever before. This historic reconstruction of a colonial fort is at Yorktown, Virginia.

could not function without highways built over some good farmland. By the same token, people cannot be fed without clearing the forests for farming.

Since colonial times, U.S. forests have produced about 2,700 billion board feet of timber. Each decade we take more wood from our forests than the decade before, yet each year the trees in our forests grow more wood than we harvest. Our 760 million acres produce more wood each year than our over 1 billion acres did in 1607. Forest management works!

Mature Forests

When you think of the term *forest*, what do you think of? Is it a pine plantation with rows of trees all the same height, or do you think of a dark, cool place with tall, beautiful trees? Most people would think of the latter.

When a forest tree begins to grow, it starts slowly. It may gain a foot a year in height, but even after a few years the tree doesn't have much wood. As it gets older, its crown increases in size and it produces wood much faster. As it becomes even older, its growth slows, and it gains less in height each year. Like people, a tree grows slowly, then faster, and then matures and doesn't grow very fast any more. Eventually the tree dies.

When a forest is left alone for centuries, it becomes a **mature forest**. This means that its canopy is dominated by mature, slow-growing trees. A mature forest is beautiful, dark, and cool; but it produces very little wood and provides a home for relatively few birds and animals. When we need to produce wood, we need a growing forest—a young forest. When this country was settled, the colonists found mostly mature forests. Today, we have few mature forests in our commercially usable forest land.

Forest management techniques and improved varieties of trees have helped greatly. Don't feel sad to see trees being cut. After all, they are not a natural resource until they become useful. More importantly, by good management, trees are renewable. This means we can cut a millon trees and still have more than before.

OTHER RESOURCES

Energy Sources

Most of our energy comes either directly or indirectly from the sun, the earth's most vital natural resource. Capturing the energy that begins in the sun has always been our greatest challenge.

Over the centuries we have used the force of moving air, the wind, to move our goods and to pull water from the earth. Windmills once dotted the American landscape. Sailing ships crossed the oceans and great lakes of the world. We cannot compare the quantity of wind in the past to that today. Neither can we conserve nor waste wind. We can only use it whenever possible and try to keep it clean. With the energy crisis of the 1970s, there was a growing interest in the use of wind power. Then the energy "glut" of the 1980s lowered the interest in wind power.

Another great energy source is coal. It is probably the most widely distributed storehouse of the sun's energy. Coal was first discovered in America in 1673, in what is now Illinois. In 1824 about 81,000 tons of coal were mined in this country. By 1947 that figure had grown to 676 million tons. According to U.S. Department of Energy statistics, production of coal in this country increased slightly during the decade of the 1990s from just less than 1 million tons in 1991 to 1.1 million tons in 1998. A total of 35 billion tons has been produced in the United States since mining began. That represents only 1 to 2 percent of the total available. Authorities estimate reserves of 2.75 trillion tons of U.S. coal. Coal is a one-time resource—but it is very plentiful. That means we have been finding coal much faster than we have been using it. There is no doubt that known coal deposits would last for centuries. Moreover, there is little doubt that we have enough coal still undiscovered to last through the next thousand years. As we have already seen, usefulness changes over time. It seems likely that the Earth's supply of coal will last long after coal as a source of heat energy has been replaced by less expensive and cleaner natural resources that we may not even know about yet.

Another sun-charged energy resource is oil. A science textbook for high school students in the late 1950s said that the world's reserve of oil would only last about 14 years at the rate of use then in effect. Today we use much more oil, but we have a greater store of known oil than we did 50 years ago. Because of improved oil finding and recovery techniques, the world's known oil reserves have grown steadily over the years. Yet oil is also a one-time resource, so it must be carefully conserved.

In 1960, world production of crude oil totaled 20.96 million barrels per day. By 1989, that had increased to 59.46 million barrels per day. Known world crude oil reserves totaled almost 135 billion metric tons in 1990. In 1990, the known crude oil reserves, just in the Middle East countries of Saudi Arabia, Iraq, The United Arab Emirates, Kuwait, and Iran, totaled over 70 billion metric tons. In contrast, the known crude oil reserves in the United States in 1990 were only 3.6 billion metric tons. A metric ton equals 1,000 kilograms or about 2,240 pounds.

Until recently, natural gas was considered a waste product from oil fields. Now, however, it has been recognized for what it is—a clean, efficient fuel. Proven reserves total 260 trillion cubic feet. Currently, natural gas is being discovered faster than it is being used.

Minerals

Minerals are yet another type of natural resource. The most widely used are the metals, such as iron, copper, aluminum, magnesium, lead, zinc, tin, and several others. These metals, though not as vital as some of our other resources, are still important and need conservation. Improved mining and processing methods are doing a great deal to help conservation. To get some idea of the known reserves of some minerals, see Table 1-1.

Geologists generally agree that our undiscovered mineral reserves far exceed what we have already found, and the possibilities for further discovery are tremendous.

These estimates do not take into consideration future technological advances or possibilities of mining lower-grade ore.

TABLE 1-1 Known World Reserves of Selected Metals*

Mineral	Years of Proven Reserve
Aluminum	222
Copper	33
Iron	161
Lead	18
Mercury	43
Nickel	51
Tin	168
Zinc	20

* Estimates are based on current proven reserve, usage rates, existing technology, and current prices.

Recreational Resources

Certainly as America has become richer, our people have found more time for recreation. We have moved from an agrarian to an industrial to a technological world. These changes place pressures on our time and on our mental capacities. People need to relax, to enjoy life, to enjoy nature.

Popular **recreational resources** in this country include forests, lakes, beaches, mountains, parks, game animals, and fish. As the United States moves further into its third century, we must closely guard the quality of our ecosystem. We are not alone on this planet, and we do not own exclusive rights to live here. Beyond this philosophical reason for carefully managing our environment, we need the relaxing and inspirational values we gain from unspoiled nature. We need clean water in which to swim. We need unlittered trails to hike. We need clean parks in which to picnic and camp.

SUMMARY

Our natural resources can be defined as objects, materials, creatures, or energy found in nature that can be put to use by humankind. Some of these resources are in fairly short supply. Others, such as coal, are plentiful. Some, such as clean drinking water, can be easily spoiled. And some, like oil, can be used only once and then are gone.

The most important characteristic of a natural resource is its usefulness. Those things that are useful to us today may no longer be useful in the future. Things in our past that had no usefulness can be very valuable resources today. Custom and technology greatly affect the usefulness of a natural resource.

The United States is a vast country rich in natural resources. However, many of our important resources are available in only limited quantities. Managing those natural resources for the long-range benefit of our people is the challenge facing all of us.

DISCUSSION QUESTIONS

1. Nature's resources once seemed limitless. Why is this no longer true?

2. What is a natural resource?

3. How does the usefulness of a natural resource change over time? What factors affect usefulness most?

4. What is the land area of the United States? How much of that is suitable for farming? How much is suitable for crop production?

5. If there is so much water in the world, why is there a water shortage problem in this country?

6. How many species of wild animals, birds, and fish have become extinct in this country since colonial times?

7. What was the forested area in this country 300 years ago? What is it today? How can it be that our smaller forest area produces more wood today?

8. What is the direct or indirect source of most of our energy resources? Give some examples of indirect energy from this source.

9. What are some of our key mineral resources and what are our known reserves?

10. What are two very good arguments for carefully maintaining our natural resources for recreation?

SUGGESTED ACTIVITIES

1. Inventory the natural resources in your community. Include soil resources; water resources; forest resources; fish and wildlife resources; metals, minerals, and energy resources; and outdoor recreation resources. Remember, a natural resource is something that is useful to people. Also remember that there are many kinds of usefulness.

2. Prepare a report for discussion in class on any one type of natural resource. Each class member should select a different natural resource to discuss.

2

A HISTORY OF CONSERVATION IN THE UNITED STATES

OBJECTIVES

After reading this chapter, you should be able to

- contrast exploitation, conservation, and preservation as they relate to natural resources management

- outline the history of conservation in the United States

- describe the role of the federal government in conservation

TERMS TO KNOW		
exploitation	migratory waterfowl	soil conservation district
conservation	"duck" stamp	ASCS
New York Sporting Club	CCC	
market hunter	NRCS	

The history of this country has been one of **exploitation**, or using up, of our natural resources. Our industrial greatness was built upon our forests, water, iron, coal, oil, and other natural resources. Our agricultural greatness was built on our soil and water resources. When people were few and resources were seemingly without limit, there seemed to be little need to be careful. As a result, little thought was given to the future, but that was another time and another century. The future is here.

As we will see in Chapter 3, some natural resources are exploited; some are changed without being used up (developed). When we are exploiting or developing our natural resources carefully, so that they will last as long

as possible, we are practicing conservation. **Conservation**, as we use the term in this book, refers to the careful use of our natural resources to provide as much usefulness as possible to people both now and in the future.

In the fields of forestry, fish and wildlife management, soil and water management, energy, and mineral resources management we have begun to turn the tide. Wise management of our natural resources is beginning to replace shortsighted exploitation. We cannot afford to let our progress in the last quarter of the twentieth century give us a false sense of security.

The challenges facing us in the future are greater than ever before. With a world population that now is doubling every few decades, we must continue and even accelerate our country's recent progress in natural resources management.

With that in mind, let us look into the past. This chapter briefly examines the history of the conservation movement in this country. The areas of forest management, soil and water conservation, and wildlife management are examined. We will not look at the energy or mineral areas. Large-scale conservation efforts in those fields have been fairly recent developments.

WILDLIFE MANAGEMENT

The Early Years

Early accounts of America's wildlife speak in awed terms. The settlers in the Virginia colony and the Pilgrims further north wrote of deer, hares, and fowl "in incredible numbers." At least one writer estimated the number of passenger pigeons as "millions of millions." There is no reason to doubt that America was one of the world's greatest storehouses of wildlife. Yet, these early settlers probably badly overestimated the true picture in the eastern part of the country. Of course, they had no way of knowing about the vast game herds in the Great Plains at the time.

When the European settlers came, the colonies were covered largely by mature forests. These forests were broken only by streams, marshes, natural meadows, and clearings created by Native Americans, beavers, or lightning-caused fires.

Virgin forests and grasslands could support vast flocks of passenger pigeons. The streams could support millions of beaver. The marshes could support great flocks of waterfowl.

Deer, turkey, quail, rabbits, and many other important game animals need clearings to

exist in large numbers. The animals were plentiful along the coast, streams, and around the great marshes, but they could not have been plentiful in the great inland mature forests. The early settlers, by choice, also lived near the coasts and along the streams. Thus, they drew a mistaken picture of the true populations of most important game animals.

This mistaken impression may help explain the rapid depletion of early game populations. Hunters could depopulate deer herds along the coast. When they went further inland looking for more deer, they found the hunting to be less successful.

Because deer herds preferred certain types of land, they were easy to find. Because deer were good sources of meat and hides, they were hunted mercilessly. By 1639, regulations on deer hunting were imposed by the town charter of Newport, Rhode Island. Such minor attempts at conservation grew until 1698 when two colonies, Connecticut and Massachusetts, imposed limits on the deer hunt. By then, the deer herds were almost gone in those areas.

In 1844, the **New York Sporting Club** was formed. A few years later it became the New York State Game Protective Society. This group was made up of a group of men who hunted primarily for sport. It sought to promote restrictions against market hunters. The first state-administered game and fish commission was created in Massachusetts in 1865. Soon, all of the states had similar agencies. These early fish and wildlife commissioners were instrumental in getting laws passed for the conservation of game species.

Market Hunters

When there were only about a million Native Americans in what is now the United States, there was no problem in terms of wild game populations. They hunted for food and hides. They took game and fish as they were needed. There was no reason for wholesale slaughter. When the settlers came, that changed. Europeans prized beaver hides, and colonists needed meat from game animals and birds. The Native Americans found they could live better with less uncertainty by supplying those furs and meat, and Europeans saw the opportunity to take this source of wealth for themselves. Hunting and trapping fur and meat animals became big business overnight. People who killed birds and animals to sell their feathers, furs, and meat became known as **market hunters**.

Market hunters found that their quarry could be taken all year long. Passenger pigeons could be killed most easily during nesting season. Migratory waterfowl could be found most easily on their breeding grounds. Such shortsighted slaughter led to the extinction of the passenger pigeon. The same thing almost happened to the buffalo (actually American bison). By 1890, there were only 500,000 deer in North America, and those were hidden in dense swamps along the southern coastlines.

We all agree that such practices were distasteful and shortsighted. In most cases, however, the market hunters caused less real damage than they were credited with. The ox, the plow, herds of livestock, and fire drastically changed the face of America. It is certain that market hunters needlessly damaged our wildlife resource. It is also certain that our clearing of the land for farming changed the wildlife habitat in extreme ways.

The Road Back

After the turn of the last century, the first federal act dealing with wildlife was passed. The Lacey Act of 1900 made the interstate transportation of game taken against state law a federal crime. This legislation spelled the doom of large-scale market hunting. It also went a long way toward saving a number of fish and wildlife species from destruction.

In 1916, the United States and Great Britain signed the Migratory Bird Treaty. In 1918, the federal Migratory Bird Treaty Act was passed, making the treaty effective. This was the first effective legislation for the protection of **migratory waterfowl**, in this case waterfowl that breed in Canada and fly across the United States each year.

Changing patterns of land use also came to the aid of America's disappearing game animals and birds. All across the eastern half of the country, farm acreages began to decline beginning in the early 1900s. This meant that many pastures and fields were no longer being farmed. On such land, scrub growth and new forests began to spring up. This made ideal habitat for wildlife. By 1900, there were practically no deer in most of the eastern states. By the 1920s, deer populations had returned to all those states. This dramatic recovery can be attributed both to federal regulations and to the conversion of farmland back into forests.

In 1933, Aldo Leopold, a professor of game management at the University of Wisconsin, published *Game Management*. This landmark book still forms the basis for much of what we do today in wildlife management.

In 1934, the Duck Stamp Act was passed. This law required waterfowl hunters to purchase a $1 stamp. The resulting sale of **"duck" stamps** produced $600,000 in its first year. With an increase to $3, the duck stamps have since raised up to $6 million a year. The money generated by this act has been used to finance numerous projects to protect and expand North American waterfowl populations.

The United States Fish and Wildlife Service of the Department of Interior was established on June 30, 1940. Today every state operates its own fish and wildlife agency. The sale of hunting and fishing permits generates many millions of dollars annually.

Today, partly as a result of better game management laws and enforcement, deer populations in the United States are at their highest in this century. That is also true of many other game animals. Wolves and coyotes are reappearing in parts of the United States where they had all but vanished. Alligators are becoming so plentiful that they have become a nuisance in parts of Florida and other southeastern states. You need only drive along the highways of the country to see evidence of this change in the form of animals that are being killed in ever-increasing numbers in traffic accidents.

Without serious argument, all must agree that the large and healthy game animal, bird, and fish populations in the United States today are a result of conservationist efforts. And most of those efforts have been led and funded over the last century by hunters and sports fishers. It is ironic, but nevertheless true, that hunters and sports fishers have made possible much of America's abundance in wildlife.

FOREST MANAGEMENT

Probably the earliest recorded timber shortage occurred in China about 5,000 years ago. The Egyptians experienced a shortage of timber about 4,000 years ago. The Romans had to import wood from their conquered lands before the birth of Christ. By A.D. 1000, central Europeans were running out of wood, and strict regulations were needed to preserve timber.

In 1626, Plymouth Colony passed America's first ordinance controlling the sale of timber. By 1650, several of the colonies had passed laws against burning of the forests. These attempts, however, were the exception and not the rule in our early history. More often than not, our forests were seen as endless, as a gift from God for us to exploit.

In the late 1700s and early 1800s, the U.S. forest preservation effort centered around saving live oaks for use in building ships. In fact, the efforts were for military purposes and not for forest conservation. With the advent of ironclad ships during the Civil War, this interest ended. Even though a tree species was being protected, forest conservation was not the motive. It had been merely an attempt to prevent the live oaks from being cut for nonmilitary uses.

The forests of the United States produced about a billion board feet of lumber in 1840. That grew to 35 billion board feet in 1869 and 46 billion board feet in 1906. During the period of settlement of this country, land grants were made to individuals and companies. The forests were harvested to pay for other enterprises, or they were simply cleared and burned to make way for farming. Forest use for developing our nation was essential, but such wasteful exploitation was certainly a great tragedy.

The American Forestry Association was organized in 1875 to promote timber culture and forestry. A year later a forestry agent was appointed to the United States Department of Agriculture (USDA). The forestry office became the Division of Forestry of the USDA in 1881.

A decade later, in 1891, Congress authorized the creation of forest reserves from public-owned land. By 1900, 33 million acres of forest reserves had been set aside in the western states. These areas were controlled under the Department of the Interior.

Gifford Pinchot became head of the USDA's Forestry Division in 1898. Under his leadership, and with President Theodore Roosevelt's backing, the Division of Forestry was upgraded to bureau status and became the United States Forest Service in 1905. Also in 1905, the forest reserves were transferred from the Department of Interior to Pinchot's control in the Forest Service. The reserves were renamed national forests at that time. Pinchot is generally considered one of the leading influences in the development of our system of national forests.

Together, Pinchot and Roosevelt greatly expanded the national forest system. The United States Forest Service's national forest system covered 182 million acres in 1983. During the administration of former President Reagan, large parcels of national forestland were sold to private landowners.

The Weeks Law of 1911 gave the President authority to purchase forest lands for river watershed protection. This officially linked forestry with soil and water conservation as well as navigable waterway transportation and flood control. Certainly, this connection was far-reaching in its effects. Also under this legislation, forest fire prevention and control measures were authorized.

World War I brought an expanded federal role in the forestry business. Timber was needed for the war, and many thousands of soldiers were used to harvest, process, and ship it.

Another great impetus to the federal role in forestry came with the Great Depression. Much of the famous Civilian Conservation Corps (**CCC**) was involved in forestry work. As a result of these involvements during the war and the Depression, many people, who would later remain in the field, were trained and gained experience in forestry. In addition, much work of immediate as well as long-range value was done in our national forests under this and other programs.

World War II brought another expansion in the need for timber. Following the war,

expansion in the housing industry increased America's need for wood.

Perhaps as significant in recent years has been the upsurge of private forestry. The giant forest and forest-products industry has recognized now that the cut-and-move methods of the past are no longer adequate. Intensive forest management has replaced the wasteful earlier methods. Chapters 18 and 19 deal with this topic in some detail.

Early in our history, Americans cut and burned or cut and used forests with no regard for the future. It was a wasteful approach. Today we waste very little of our forest resources. In general, it is safe to say that today we produce more wood in this country each year than we use. Our forests are our greatest renewable resource. Yet, we are not safe for the future. Removal of forestlands for farm, highway, residential, and industrial uses poses a great threat to our forest industry. Careful management is the key to the future in forestry.

SOIL CONSERVATION

In the beginning of this country, almost every worker was either a full- or part-time farmer. Educated and foresighted farmers from earliest colonial times saw the effects of erosion. They watched clean rainfall become muddy runoff. They saw the hillsides become sterile and the valleys become either rich with topsoil or clogged with debris.

Jared Eliot (1685–1763) was one of the first to experiment and write about soil erosion and drainage in America. He and other early colonists recognized the developing problem. Yet, in general, farmers failed to heed their warnings.

This mentality of farming for today and leaving tomorrow to take care of itself grew for a reason. After all, America was the land of plenty. It stretched to the mountains and beyond. It went so far and was so big that our ancestors thought of it as without practical limit. Thus, it was easier to clear new land and abandon "worn out" farms than to take care of the land.

In general, the three centuries after the settlement of Jamestown saw this pattern of soil abuse continue. Wherever the land was tillable it was cleared. The cleared land was used for agricultural production without regard to its productive potential. Hillsides in Georgia and Pennsylvania were row cropped. Grasslands in the Southwest were overgrazed.

Scattered attempts at soil conservation were made. Hillside "ditching" campaigns were tried, and individual farmers established good farming practices on their own land. These attempts were the exception, not the rule. Between 1607 and the mid-1800s, first tobacco, then corn, and then cotton were our main farm crops. All of these are very hard on the soil. Since the 1600s, we have lost at least one-third of our precious topsoil to erosion.

In general, early attempts at soil conservation in this country were quite limited. For a long time, soil conservation merely meant the prevention of sheet and gully erosion. Soil surveys conducted by the Bureau of Soils of the United States Department of Agriculture began to show the results of erosion by the early 1900s.

Dr. Hugh H. Bennett was an early advocate of soil conservation. His work in soil surveys in Virginia, North Carolina, and South Carolina convinced him that something must be done. In 1928, he and W. R. Chapline published the USDA's first soil conservation bulletin, *Soil Erosion—A National Menace.*

As a result of Dr. Bennett's efforts, Congress established a series of soil erosion research stations under his supervision. Dr. Bennett later

concluded that enough soil was being washed away from American fields to load a string of freight trains that would stretch around the earth eighteen times at the equator! Congress and the American people were convinced.

In 1933, the Soil Erosion Service was established in the Department of the Interior. Tied in with the economic recovery from the Great Depression, the CCC directed much of its effort toward erosion control. In 1935, the Soil Erosion Service was moved to the Department of Agriculture. In that same year, it became the Soil Conservation Service (SCS). Today, this agency is known as the Natural Resources Conservation Service (**NRCS**).

Experience showed that farmers were more likely to accept conservation practices if they were directly involved in planning and decision making. This gave rise to the concept of the soil conservation association. In 1937, the Secretary of Agriculture mandated that all SCS work on private land be done only through such associations. In that way, the farmers themselves would be responsible for group planning and decisions. SCS personnel would serve as advisors and technicians. In 1937, President Franklin Roosevelt proposed that the states enact laws to establish **soil conservation districts**. Each district would consist of an association of local farmers, business persons, and others interested in the conservation of local soil and water resources. (See Figure 2-1.)

Also in 1936, the federal government instituted a program of grants to farmers. These grants, or incentive payments, were to assist in the cost of soil conservation and soil building practices—some of which could be very expensive. Construction of terraces, drainage systems, waterways, and farm ponds were very greatly expanded by this effort. These, along with other programs, now fall under the USDA's Agricultural Stabilization and Conservation

FIGURE 2-1 Since 1937, USDA Soil Conservation Districts have been an important part in the fight to manage soil erosion, flooding, and water resources.

Service (**ASCS**). In effect, the NRCS helps the farmer develop a plan for soil and water conservation. The ASCS helps in financing the practices.

As a result of these successes, other federal agencies began to enter the field of soil and water conservation: the Bureau of Indian Affairs, the Department of Defense, the Bureau of Land Reclamation, and the Bureau of Land Management. Americans' interest in this field has led to the establishment of the National Association of Soil and Water Conservation Districts and the Soil Conservation Society of America.

Much has been done since Dr. Bennett conducted a soil survey in Louisa County, Virginia, in 1905. Yet, much still needs to be done. Two-thirds of U.S. cropland needs additional soil conservation practices. Three-fourths of our pasture and rangeland is in need of improvements. Over half of our forestland needs improved practices. Even on those areas where soil and water management is good, efforts must be continuous.

Nature does not provide us with thousand-acre cornfields or with quarter-acre gardens. Humans must control nature to make those things happen. Whenever we change the surface of the land for our own purposes, we must be prepared to deal with undesirable side effects. Soil erosion is the main side effect of food and fiber production. It is a continuous process. After all, our soil is our most important non-renewable resource.

WATER MANAGEMENT

Transportation and Flood Control

In the early years of our country, water was the determining factor of where people would live, work, and play. Settlers built homes only where there was adequate water from lakes, rivers, streams, springs, or wells. To a lesser extent rainwater, captured and stored in cisterns, met some of this need. Cities could be built only where water could be supplied. Because land was plentiful, this was no problem. Water was used also as a means of waste disposal in many areas. When a city built a sewage system, it typically emptied directly into the river.

Early interest in water management in this country came more out of the forestry movement than as a genuine concern over water itself. Forest conservationists emphasized the benefits of forests in regulating stream flow, preventing silting of waterways, and preventing flooding. In 1882, the Commonwealth of Massachusetts authorized its cities to purchase "municipal forests" to protect their watersheds. The American Association for the Advancement of Science made a general effort to advocate conservation of resources during the

1890s. Among its interests were stream flow, water supplies, and watershed maintenance.

Federal legislation dealing with forest management used water management as part of its justification. Unfortunately, authorities could not agree on the effects forests had on the hydrologic (water) cycle (see Chapter 13) even as late as the 1920s.

In 1879, the Mississippi River Commission was set up to help the states improve the river as a waterway. The Rivers and Harbors Acts of 1917 and 1927 expanded the federal role in establishing and maintaining navigable waterways in this country. A century prior to that, the famous Erie Canal had been completed. This connected the Great Lakes with the Hudson River in 1825.

Thus, early emphasis in this country was not on water conservation as such. It was on water as a channel for transportation and water as a by-product of forestry. By the late 1920s, the federal government began to assume some responsibility for flood control. The second major concern in water management arose—prevention of flooding.

It became clear that flooding was a function of two other problems. Loss of good forest cover led to greater runoff and, thus, more flooding. Soil erosion led to clogging of waterways and silt deposits on streets, residences, and highways. It also caused the filling in of lakes, rivers, and streams. This, in turn, led to more flooding. Thus, the main emphasis on water management still came from foresters and soil conservationists.

The Flood Control Act of 1936 authorized the Soil Conservation Service to develop and implement plans for upstream soil and water conservation in order to reduce sedimentation and flooding. This and following work led to the 1954 Watershed Protection and Flood Prevention Act, which transferred responsibility

for decision making to state and local organizations. In general, most watershed management projects today are handled through the local soil and water conservation districts described earlier in this chapter.

Throughout the 1990s, a series of record-setting weather events have resulted in increased emphasis on the need for flood control efforts in this country and worldwide. Massive rains in the Midwest in 1993 resulted in the most severe flooding in at least a century in the Mississippi Valley. Entire towns simply ceased to exist as a result of that flooding episode. Figure 2-2 shows a dam collapsing in the 1993 floods.

Again in late 1999 a series of hurricanes followed one after the other in eastern North Carolina and Virginia. The heavy rains caused the ground to become saturated. Then in September, yet another hurricane, this one named Floyd, came ashore along the Eastern Seaboard dumping as much as 2 feet of rain in a single day in some places. The floods resulting from Hurricane Floyd were the worst in the history

FIGURE 2-2 The flooding of the Mississippi Valley in early 1993 was the most severe in this century for that region. *(Courtesy U.S. Army Corps of Engineers)*

of that part of the country. Figure 2-3 shows some of the flooding resulting from that storm in North Carolina.

Over the last 50 years, the emphasis has greatly broadened. Prior to 1930, our water management efforts had been on water transportation and flood prevention. With expanding population, industry, and irrigation farming, the quantity of water used by Americans has grown drastically. Also, with a growing concern for the quality of life in this country, Americans have demanded cleaner water. Advances in our understanding of the human body and of medicine have shown the importance of cleaner water supplies. As a result, wastewater that once was dumped directly into rivers is now treated and disinfected before reintroduction into the water cycle. Federal legislation in the 1960s and 1970s emphasized these health concerns.

Another problem becoming critical is in the distribution of our water supply. Lowered water tables in the western states have resulted largely from deep-well irrigation. This is a problem that must be addressed immediately.

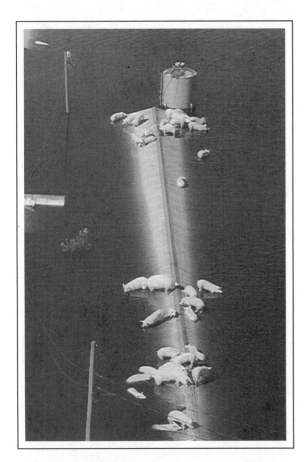

FIGURE 2-3 Hurricane Floyd caused massive destruction in the fall of 1999. These hogs were attempting to survive the rising floodwaters near Trenton, North Carolina. *(Photo courtesy the News and Observer Publishing Company, Raleigh, NC. Photo taken by Mel Nathanson)*

SUMMARY

Limitless bounty led to open exploitation that produced damage, waste, and led to problems. Recognition of problems led to concern for the future. That, in turn, led to study, research, planning, and action. The result has been the conservation movement in this country.

The original theme of conservation was to prevent damage. Today the natural resources management efforts in this country look not only to the wise use of our resources but also to the goal of redressing the damage our thoughtless exploitations of the past have caused. The very quality of life is being improved today as a result of the conservation movement. We should never forget the contributions of great American conservationists like Gifford Pinchot, Aldo Leopold, and Hugh Bennett. They, and thousands like them, have guided us in the right direction; but their work, indeed all of our past efforts, are not adequate for the future. Our nation and our world must write an even greater history of conservation as we embark upon the twenty-first century.

DISCUSSION QUESTIONS

1. Why did Americans have such wasteful practices in using our natural resources in the past?

2. In what ways were they right? How were they wrong?

3. What were market hunters? Why were they so unpopular among American sports persons?

4. If sport hunters and sports fishers had not fought market hunters, what would have happened to our fish and game animal populations?

5. Who pays for most wildlife conservation in this country?

6. What was the Weeks Law of 1911? Why was it important?

7. What is a soil and water conservation district? How does it work?

8. How has the federal government helped local farmers and other landowners work to solve their soil and water conservation problems?

9. Why is soil and water conservation a federal concern?

10. Early water management efforts in America centered around three needs. What were they?

SUGGESTED ACTIVITIES

1. **(a)** Interview several persons who have lived in your community at least 60 years. How have the hunting and fishing changed over 50 years? Are there more or fewer forested areas? What have been the worst floods? Have they seen any severe erosion or sedimentation problems develop?

   (b) If you live in a large city, interview a relative or friend who is at least 60 years old and who was raised in a rural area. Ask him or her the same questions about the area where he or she grew up compared to today.

2. Try to find some old (pre-1920) books or booklets on conservation in your library and read them. Try to determine how our perspectives on conservation have changed. Have the goals of conservationists changed?

3 CONCEPTS IN NATURAL RESOURCES MANAGEMENT

OBJECTIVES

After reading this chapter, you should be able to

- explain the differences between nonexhaustible, renewable, and exhaustible natural resources
- discuss the concept of balance in natural ecosystems
- discuss the role of food chains in maintaining balanced ecosystems
- discuss the role of ecology in human efforts at natural resources management

TERMS TO KNOW		
environment	environmentalist	biome
nonexhaustible	amoral	biogeography
renewable	food chain	population level
exhaustible	producer	carrying capacity
ecology	transformer	preservation
ecosystem	decomposer	multiple use

People have always lived in an **environment**. That environment (soil, water, plants, animals, energy, minerals, atmosphere) has provided everything for our survival. If any part of our requirements had not been met, then we would have ceased to exist. At one time, our natural resources could be used without fear. There were fewer people, and our technology did not allow for rapid and massive use of natural resources: Trees were cut by saws and axes, and coal was harvested by hand. Those days are gone now.

People still live in an environment, but now we have the ability to use our resources on unbelievably large scales. We can take water from a water table hundreds of feet underground, and with the aid of modern technology we can level mountains, change the flow of rivers, and take energy from the atom. Sounds impressive, doesn't it?

Unfortunately, technology and our growing population present a problem. We find that we must, in the future, live in the same environment—our "spaceship earth." And it is not getting appreciably larger. We have discovered that many of nature's resources are not going to last forever.

Wise use of our natural resources is not a new idea; but today, more than ever, it must become a common goal. Managing our natural resources for the future as well as the present must become a priority for all of us. This chapter explains a number of important ideas in natural resources conservation and management.

THE NATURE OF RESOURCES

In a very real sense, everything in our environment could be considered a natural resource. Rocks may be used as gravel for our roads, facings for our buildings, or material for our statues. Wind, falling water, still air, resting water, minerals, insects—virtually everything around us can be considered a resource. When someone takes an object and uses it to perform work or change other parts of the environment, that object has become a resource. Even before its use, the potential for use makes the object a natural resource.

Those things that have become or show promise of becoming important to us are the natural resources we are concerned about in this book. Those natural resources may take many forms. More importantly, they may be capable of going on forever. On the other hand, they may be very limited. They may be usable over and over, or they may be gone forever with a single use. Let us look at these a little closer.

Nonexhaustible Resources

Natural resources that can last forever regardless of human activities are **nonexhaustible**. They renew themselves continuously. This does not mean that such resources are not limited. It also does not mean that human misuse cannot damage such resources—it certainly can.

A good example is surface water. If you take a gallon of water from a river, another gallon will replace it. If you dam a stream, the water will simply go elsewhere. If we damage a watershed so that its rainfall does not soak into the ground, the rainfall will simply go elsewhere. Little that we do will affect the total amount of water that comes to earth in the form of precipitation.

Water supplies may be very limited. Chapter 13 deals with our water supply and its users. At the beginning of that chapter the hydrologic cycle is explained. It is because of this cycle that water is a nonexhaustible resource. (See Figure 3-1.) Chapter 14 points out the problems we face in damaging our water supply by pollution. Nevertheless, our water supply is nonexhaustible, for all practical purposes.

Another example is air. We use air to breathe, to grow plants, to fly airplanes, to power windmills, to dry food and clothes. We can damage the air with pollution. We could even make it unusable, as many environmentalists would argue, but we cannot use it up. It is nonexhaustible, for all practical purposes.

FIGURE 3-1 Maintaining an adequate supply of clean water may be one of the greatest challenges to our society in the future. It is the location that is the main problem—not the amount of water.

Renewable Resources

Natural resources that can be replaced by human efforts are considered **renewable**. Simply because a resource is renewable does not mean it will never be used up. On the other hand, it is possible to use such resources and yet have as much left afterward as before that use.

One example is the forest. (Chapter 18 discusses forests and forest products.) We use more wood today than ever before, yet we produce more wood each year than we use in this country. The types of wood have changed. We no longer harvest as many large hardwoods as we once did. Yet we have no foreseeable shortage of wood or wood products. This is true because of the advances made in forestry, both in woodland management and in genetics.

Another example is our fish and wildlife population. In our nation's past, there have been times of great waste. Huge droves of passenger pigeons were destroyed, and the popular food bird became extinct. Great herds of bison, or buffalo, were killed for their hides and meat, and this great American natural resource neared extinction. Yet, with techniques of game management such as those discussed in Chapter 24, their numbers rebounded. Fish populations, too, respond readily to fisheries' management techniques. (See Chapters 25 and 26.)

Exhaustible Resources

Many of our natural resources exist in finite quantities. Those limited resources that cannot be replaced or reproduced are known as nonrenewable or **exhaustible**. In the case of exhaustible resources, we cannot manage them for renewal. They do not renew themselves; once they are gone, they are gone—forever. We can conserve our exhaustible resources. We can learn how to use less. We may try to find more of it. We may even be able to recycle some of them; but once the resource is gone, we simply have to do without it.

Even though an exhaustible resource exists in a finite (limited) supply, that does not mean it is necessarily a limited resource. Many exhaustible resources exist in such huge amounts that they are practically nonexhaustible. For instance, there is so much coal on the planet that, though it is exhaustible, there is no practical limit to coal. There is so much iron ore that, though iron is an exhaustible resource, there is no practical limit to the metal.

One very important exhaustible resource is oil. We constantly hear of the "energy crisis." There is only so much oil in the ground, and when we have removed all we can find, it is gone. We will just have to develop other sources of energy.

Another example is our mineral resources. We use lead, cobalt, zinc, and other minerals to make our goods. We depend on these mineral resources for our way of life. We must manage those resources to make them last as long as possible.

Soil probably fits into this category. Chapter 4 explains that soil is constantly being formed by nature. We can improve existing soil, make it more fertile, move water to it, and supply missing minerals. We can even make soil substitutes in small quantities, but we cannot really make soil. Only nature can do that. Why, then, is soil not a nonexhaustible resource? It is exhaustible because nature makes soil so slowly. True, a soil destroyed by improper use will probably be replaced—in 500,000 years. As far as we are concerned, that is not renewal that is useful to us. Thus, soil is a nonrenewable, or exhaustible resource.

AN ECOLOGICAL OUTLOOK

Humans, trees, water, animals, fish, grasses, sunlight—all these and millions more fit into an overall system. This system, with all its interactions and interdependencies, makes up our ecosystem.

Ecology is the branch of science dealing with the complex relationships among living things and their environment. An ecological system, or **ecosystem**, is any partially self-contained environmental and living system. A lake might be thought of as an ecosystem. A forest, a large valley, or a desert might be considered an ecosystem.

In a very real sense, we exist in an ecosystem. We depend on our environment for life itself. For many thousands of years, that was no problem; but today our numbers are increasing. Our technology is becoming tremendously powerful. In ancient times, our use of natural resources had little effect on our ecosystem; but in the past few centuries, we have had an ever-increasing impact on that ecosystem.

Perceptions in Ecology

It is important to note that ecology is a science. It can be defined as the study of the interactions of organisms and their environment. Environmentalism, which is simply a strong concern for the environment, is not the same thing at all. Environmentalism is based on emotion, values, beliefs, and politics. An **environmentalist** is a political activist with a special interest in some aspect of the environment. Environmentalists concern themselves with right and wrong as they perceive it; with good and bad, again as they perceive it; and with morality, as they perceive it.

As a science, ecology is based on observation and objective interpretation of data. An ecologist is a scientist. In the role of scientists, ecologists do not attempt to make decisions based on moral interpretations (value judgments). We say that science is **amoral**, which simply means that value judgments about good and bad should not be a part of science.

To illustrate the distinction, consider this situation: A person in a fishing boat catches a large shark just off a beach used by swimmers. Should he or she release the shark because the fish has a right to live? Should he kill the shark out of concern for the human swimmers? That sort of judgment is one of values. It is a moral question not a question for science. An ecologist (while acting as a scientist) would simply look at the shark as a large predator and an important part of the food web. To the ecologist, the question of killing the shark or releasing it would probably center around the shark's place in the ecosystem and whether there were too many or too few sharks for the food web to support. The role of science in this situation would be amoral—non-value oriented.

This does not mean a scientist cannot also be an environmentalist. When a scientist stops doing science and starts advocating environmentalist positions on political issues, he or she is not talking as a scientist but rather as an environmentalist—although, perhaps a very well-informed environmentalist.

The Science of Ecosystems

An ecosystem is a given set of organisms, organic residues, physical and chemical components, and conditions (i.e., light, temperature, etc.) that interact and transfer energy and matter in form and location. Ecosystems consist of biotic (living) subsystems as well as abiotic (nonliving) subsystems. An example of a biotic subsystem is the relationship among the plant and animal members of a food web or **food chain**. An example of an abiotic system is the water of a lake and the chemicals that dissolve from the atmosphere and land that affect the water's acidity level.

In its most basic sense, an ecosystem is an energy system. Every part of an ecosystem interacts with the other parts of the system and depends on them. Fish in a lake use oxygen from the water that is dissolved from the atmosphere. The plants in the lake use light from the sun and minerals from the lake bottom to grow. All the processes in an ecosystem depend on energy. In fact, we can say accurately that nothing happens in an ecosystem without the flow of energy.

To have a complete ecosystem, there must be three components present. There must be **producers**, basically green plants that produce new food (sugar) by means of photosynthesis. There must be **transformers** that can take that primary source of food, incorporate other chemicals and energy forms, and change it into more complex organic compounds, foods, and tissue. Finally, there must be **decomposers** that break the organic materials back down into their constituents for reuse in the ecosystem. One organism can be a member of more than one of the components. For instance, a green plant can be both producer and transformer. A fungus can be both transformer and decomposer, as are animals.

There are four fundamental biotic processes that go on in ecosystems: synthesis, photosynthesis, respiration, and decomposition. The basic process is that of biological synthesis. Biological synthesis refers to any change in the composition, shape, size, or structure of the plants or animals in the ecosystem. Examples are the use of nutrients, minerals, and water to produce growth and reproduction in plants and animals.

A second fundamental process is that of photosynthesis. In photosynthesis, plants convert water and carbon dioxide into sugar. The process requires the presence of the catalyst chlorophyll, which generally gives the green color to healthy plants. It also requires energy from the sun—energy that is incorporated into the sugar molecules. We have all heard that sugar is a "high energy" food. That is true because energy from the sun is stored in it. Photosynthesis is the original source of almost all foods in the ecosystem. Photosynthesis also produces oxygen, which benefits all animals, including humans.

A third fundamental process in ecosystems is that of respiration. Respiration is a process that takes place within the individual cells of plants and animals. It involves the breaking down of foods into their components along with the release of energy. An example is the digestion of sugar into water and carbon dioxide, with the release of the stored energy being made available for use by the plant or animal cells.

A fourth fundamental process in ecosystems is that of decomposition, which is the process by which organic matter (plant or animal tissue) is reduced to organic compounds. Only by means of decomposition can the chemicals in plant and animal bodies be ready to recycle into the biotic subsystems in the ecosystem.

An ecosystem can be defined in many different ways. In one sense, a terrarium in your classroom is an ecosystem. In another sense, the classroom itself makes up an ecosystem and the terrarium is simply a part of that ecosystem. In yet another sense, your whole school is part of an ecosystem that could be defined in geographic terms.

That brings us to what could be called the "Ultimate Concept in Ecology." Everything on earth is part of one or more ecosystems. In each system, if you do something to one part, it affects some or all the other parts of the system. The effects are often unpredictable and may be very extreme.

The emerging science of Chaos recognizes the "Butterfly Theory," which states that the flapping of the wings of a butterfly in one part of the world might ultimately affect in some way the weather in another part of the world. A more realistic example is that burning high-sulfur coal in the United States promotes acid rain in Canada.

Ecological Succession

At any given time in a particular ecosystem, there is probably a wide variety of living things. No ecosystem is ever completely and permanently stable. For that reason, there is no real "balance of nature." All ecosystems are dynamic, that is, constantly changing. One species of plant or animal is replaced by another as conditions change and as the ecosystem matures. The replacement of one species by another in an ecosystem is ecological succession.

Perhaps an example would make this clear. Consider a pond that has been constructed by a landowner. Initially, algae will be the only plant in that pond. Later, water lilies and cattails will appear. After many years, the pond will fill up with sediment and become dry land again. Then the algae and water plants will be replaced by grasses and small shrubs. After a few more years, cedars and small trees will dominate the area. Later still, those will be replaced by taller trees until, eventually, the area

will become forested, perhaps with tall hardwood trees.

The term climax species implies eventual stability in ecosystems. If an ecosystem were to become completely stable, as in the example just given, the species of plants that would dominate the system would be known as the climax species. Of course, as we have seen, there is no permanent climax species because no ecosystem is ever truly stable forever. In the example of the pond that fills in and becomes a hardwood forest, there will someday be a fire, a disease, a logger, an ice age, or something else that changes the conditions of the ecosystem, and the whole process will start over.

An important concept in ecology is that of the **biome**. A biome can be thought of as the biotic sub-system (living organisms) in an extensive ecosystem. When we speak of biomes, we describe them in terms of their geography and of their climax vegetation. The major biomes are the Arctic, the Antarctic, tundra, deserts, coniferous forests, deciduous forests, grasslands, the freshwater biome, and the marine biome. The study of the distribution and residents of the world's biomes is called **biogeography**.

Balance of Nature

We have all read or heard about the balance of nature, but is nature really balanced? If nature were perfectly balanced, there would be little change. As a gallon of water flowed to the ocean, another gallon would evaporate and enter the hydrologic cycle. As one rabbit died, another would be born. Rivers would not change their course. Ice ages would never occur. Clearly this is not the case. There is, however, a sort of balance.

There is no such thing as a true balance of nature. In reality, the forces of nature are constantly counteracting each other. The result is a constant change of nature. Change is different from balance. Balance implies no change. The reality is that change in the environment is both continuous and natural. What is important is that the changes in nature be gradual to allow time for organisms such as humans to adapt to the changes. Gradual change is what we are really talking about when we refer to a balance of nature. Please hold that idea in mind whenever the term *balance of nature* is used in this book.

Some plant and animal species develop. Other ones become extinct. Fires, insects, or diseases destroy a forest. Grasses and brush replace the forest, only to be replaced eventually by a new forest.

As stated earlier, people change things. We clear forests, plow fields, drain swamps, and build cities and highways. Such massive changes in the ecosystems affect every living thing, including humans. It is because of this problem that books such as this one are written.

Managing our natural resources wisely means many things. One thing that it means is controlling nature so that we can use its resources without destroying its balance, or at least without permanently upsetting that balance.

Food Webs

All living things make up parts of both food chains and food webs. In its simplest form, a food chain is a sequence of organisms, each of which provides a source of nutrients for the next organism in the chain. A food web is a set of overlapping food chains. Let us examine a couple of simple food chains and then see how they could overlap to form a simple food web.

A grass seed germinates. It uses nutrients from the remains of dead plants and animals to grow. Later, the grass blades and seeds are eaten by grasshoppers and other insects. A family of mice make meals of the grasshoppers. The mice also feed on acorns that fall from a nearby oak tree. Still later, one of the mice is eaten by a hawk, and a second mouse is eaten by a fox.

This situation describes several very simplified food chains: (1) oak-acorn-mouse-fox, (2) grass-insect-mouse-fox, (3) oak-acorn-mouse-hawk, and (4) grass-insect-mouse-hawk. Yet, all four food chains are related because they are connected by the family of mice. (See Figure 3-2.) This series of interconnected food chains makes up an example of a very simplified food web.

In fact, the food chains just described were not quite complete. The hawk and fox, in their own turn will die and the nutrients from their bodies will return to the food chain to nourish the next generation of grasses and trees. Moreover, not all food chains contain predators at the top, like the hawk or fox. The next example illustrates a food web without a predator.

An acorn falls from a great oak tree. It germinates and begins to grow. Its roots take in the nutrients from the decaying bodies of dead plants and animals of the forest floor. They absorb water and minerals from the soil. In the meantime, the great oak that produced the acorn eventually weakens and dies. The growing tree now has more sunlight and less competition for soil moisture and nutrients. It matures and produces many acorns. Most of the acorns are eaten by insects and rodents, but a few germinate and begin to grow. Eventually, the tree will be damaged or weakened by age. Insects and disease may finish it off; when it dies, its nutrients will return to

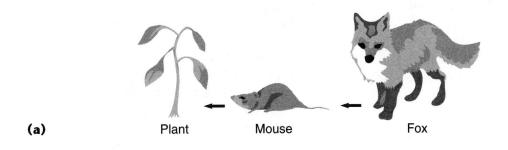

(a) Plant   Mouse   Fox

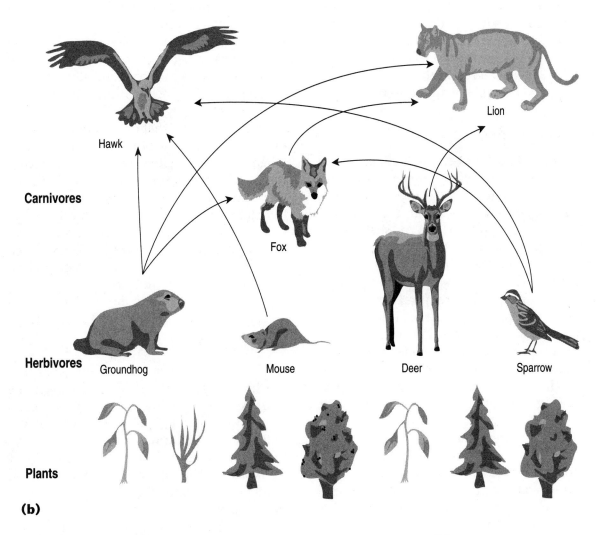

(b)

FIGURE 3-2 Food chains (a) can interact to make up complex food webs (b).

the soil. Its moisture will reenter the hydrologic cycle.

Two of these examples represent fairly simple food chains. In the first, the chain is grass-insect-mouse-fox-grass. In the second, the chain is grass-insect-mouse-hawk-grass.

The third situation is harder to visualize but is more realistic than the two food chains. When two or more food chains overlap with each other, the result is referred to as a food web. In the two examples given, the mouse might also eat some of the insects from the oak tree's food chain. That would mean, in effect, that the wolf would be feeding from both the grasses and from the oak tree. This is a simplified example of a food web.

Carrying Capacity

A **population level** can be defined as the number of a given species of plant or animal in a given area at a particular point in time. **Carrying capacity** refers to the ability of an ecosystem to provide food and shelter for a given population level. Population levels cannot long exceed the carrying capacity of the ecosystem.

Carrying capacities are affected by the food chain of the species in question. Here is an example: Quail eat insects and plant seeds. A covey of quail must have an adequate supply of food to thrive and reproduce—even to survive. The population level of quail in a given area cannot exceed the carrying capacity for that area.

Population levels are also affected by water availability, shelter, and predators. All these factors, as well as diseases and parasites, help keep populations of each species at an acceptable level in the "balance" of nature. Chapter 24 deals with this concept in more detail. When a population exceeds its ecosystem's

carrying capacity, diseases, predators, or starvation reduce the population level. That is the way of nature.

In a previous section, we discussed the concept of actions in one part of an ecosystem affecting other parts of the system. The most profound effect by humans on the ecosystem has been from the advent of agricultural production of food and fiber. Farming has drastically increased the carrying capacity of the world for humans by increasing the amount of food that can be produced on the land. Intensive crop and livestock production have enhanced that aspect of the ecosystem. At the same time, we cannot so drastically alter one aspect of the ecosystem without also affecting other parts of the system.

We are learning that we cannot long ignore or bypass ecological principles. For instance, when we harvest crops, we remove much of the organic matter that would normally be returned to the soil by decomposition. Over the long term, this has a drastic effect on the ecosystem. Thus, modern conservation practices such as incorporating crop residues and green manure crops into the soil are direct applications of ecological principles.

HUMAN POPULATION

Earlier in this chapter, it was pointed out that the human population has grown over the years. At about the time of the birth of Christ, the world's human population was about 300 million. That figure was a great increase over the estimated 10 million at the beginning of the new stone age (about 6000 B.C.), and the rate of increase has jumped in the past few centuries to unbelievable levels. (See Table 3-1.)

Those figures are startling. As late as A.D. 1800, there were fewer than 1 billion living

TABLE 3-1 Estimated and Projected World Population Over 8,000 Years

Year	Population (in billions)
6000 B.C.	0.01
A.D. 1	0.30
1000	0.31
1250	0.40
1500	0.50
1750	0.79
1800	0.98
1850	1.26
1900	1.65
1910	1.75
1920	1.86
1930	2.07
1940	2.30
1950	2.52
1960	3.02
1970	3.70
1980	4.44
1990	5.27
2000	6.06
2010	6.79
2020	7.50
2030	8.11
2040	8.58
2050	8.91
2100	11.20
2200, stabilized at or near 11.60 billion	

Sources: United Nations, World Population Prospects: The 1998 Revision (United Nations, New York) and the World Resources Institute.

humans. The number had grown to 1.6 billion by 1900. It had skyrocketed to 4.3 billion in 1979 and surpassed over 6 billion in 1999. It is even more astonishing to see this information on a graph. (See Figure 3-3.)

Surely the question of our population level and the world's human carrying capacity must be an important one. If we are to continue to feed, shelter, and clothe ourselves into the next

century, we must plan now. We must begin to better manage our natural resources. We must learn how to improve our natural resource management skills and tools.

There is no immediate danger of worldwide starvation or death from disease. We are not running out of natural resources, but the challenge is before us. Managing our natural resources carefully is more important now than ever before.

RESOURCE USE

All living things depend on their ecosystem for survival. With modern transportation and communication, the world is our ecosystem. We must derive everything we need to have from that system, and it is simply not possible to sustain the number of people alive today without altering that ecosystem. Our very survival depends upon our use of natural resources.

On the other hand, the abuse or misuse of the resources nature provides cannot be allowed. Our society must have food in large quantities. Such large amounts of food can only be produced by modern, large-scale farming, but the farmer must not allow the soil to be destroyed by erosion. It is possible to produce agricultural products with little damage to the soil. Soil erosion and its control are discussed in Chapters 5, 6, and 7.

Our homes, farms, and factories produce vast amounts of waste; in the past, this was simply poured into rivers. As the amount of waste increased, water pollution has become a serious world problem. Chapter 14 deals with this topic.

Metals and minerals can be recycled and used more than once. We can seek out new

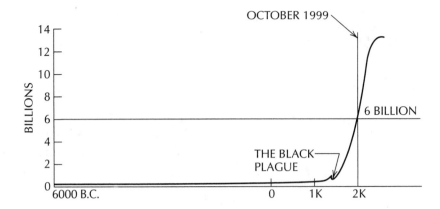

FIGURE 3-3 World population over 8,000 years, forecast through A.D. 2250.

substitutes and new supplies of these exhaustible resources.

In the end, we have no choice but to use our natural resources as fully and wisely as we can. In working for today, we must keep an eye on the future.

CONSERVATION

Chapter 2 deals extensively with the conservation movement in this country. In essence, conservationists believe in using nature to produce the maximum long-range benefit for people.

Conservation does not set aside resources simply to have them. A forest is not something to be prized simply for its own sake. It is a natural resource to be managed for wood production, a guard against erosion, and a sponge to soak up rainfall for a thirsty world. It is a place to hunt, hike, camp, picnic, or study.

To the conservationist, a deer is a natural resource. It provides pleasure for the naturalist. It reproduces and provides food and recreation for the hunter. Just as the forest is managed partially to harvest trees, a deer population is managed partially to harvest deer.

Conservation brings to mind sentimental thoughts of nature and beauty, but the true test for conservation is whether society benefits from its undertakings. Conservation is oriented toward practical use for today and for the future.

PRESERVATION

Some things are worth having and guarding just for their own sake. The Liberty Bell is a national treasure. Yosemite, the Grand Canyon, and the American bald eagle are other examples. Such heirlooms of our heritage should be preserved. Why? Simply because we, as a people, value them. No other reason is needed.

The National Park Service was established in 1916, and by 1970 it administered some 30 million acres. An additional 7.4 million acres were included in state parks throughout the United States.

At the same time, we must balance our desire for **preservation** with our needs as a people. Preservation is a part of conservation, but only a small part. In seeking to preserve a natural area, such as millions of acres in Alaska, we must ask several questions. Can we use such resources without destroying them, as we can with forests? Can we afford to set aside such resources? Which is more important, economic growth or the preservation of nature? With this last question, the growth of the human population surely must be considered. Is it important to preserve a tiny fish from extinction? After all, once it is gone, it is gone forever from the earth. Or is it more important to keep workers employed as a result of a dam being built?

Such questions are being asked in this country each year. Conservationists and preservationists would probably answer from different perspectives. We all live in the same ecosystem. We must learn to work together. We need a new national attitude toward managing our natural resources.

FIGURE 3-4 This multiple-use area serves as wildlife habitat, conserves soil and water, and is used for human recreation and education.

MULTIPLE USE

A concept originating with foresters seems to be gaining popularity among all conservationists. That concept is **multiple use**.

A forest can be used for recreation as well as for the production of wood. A lake can be used as a water reservoir, for fishing, and as a flood control measure. Windbreaks can be seeded with plants that produce wildlife food and so produce improved hunting. Winter cover crops can be grains that are useful to migrating birds.

In essence, the concept of multiple use is a very productive one. It encourages us to plan natural resource management activities to produce more than one benefit. It allows us to combine soil and water management, forest management, mineral and energy management, and recreation—all at the same time. (See Figure 3-4.)

A NATIONAL ATTITUDE

Stewart Udall, Secretary of the Interior during the mid-1960s, had this to say:

The conservation goal of America's third century as a nation must be the development and protection of a quality environment which serves both the demands of nature for ecological balance and the demands of man for social and psychological balance.

The landscapes and cityscapes that comprise the face of our continent present a partial statement about the state of our civilization, in much the same manner as the cut of a man's clothes tells something of the man himself. As the republic nears its 200th birthday, the cut of the countryside bespeaks ambivalence.

The need for a new national attitude toward our environment has grown until today it is an absolute necessity for human survival. Technology has stretched and magnified our natural resource potential in many areas. It has also supplied a harassed people with an infinite number of pain-killers and tranquilizers. But it cannot provide us with one square inch of additional planetary surface, nor do more than gloss over the mounting environmental insults to humanity.

It becomes increasingly apparent that runaway population, noise and psychological pressures of too-close living will eventually run us out of space and nervous energy even if food and minerals and fuels were never to flag.

Any new national attitude toward natural resources must take into consid- eration at least two factors—the quality of human life we seek to establish, and the specific meld of environmental ingredients that make up that quality.

SUMMARY

Humans live in an environment. We take everything we need for our survival from that environment and reshape or move or process parts of our environment to suit our needs. For thousands of centuries, our ancestors could do this with little worry about the environment. They did not have to be too concerned about running out of natural resources. Yet, today these are two of the most pressing concerns of humanity—maintaining a livable environment and managing our limited natural resources.

Natural resources can be classified as non-exhaustible, renewable, and exhaustible. Nonexhaustible resources are those that nature provides and renews continuously, like wind and water. This does not imply that they are without limit at any given moment in time.

Renewable resources can be replenished or reproduced by the efforts of people. Examples are wildlife, forests, and grasslands. Exhaustible resources are those that can be used but never renewed. Examples are coal, oil, and lead. Once coal is burned, it is gone forever. Lead can be used and then recycled over and over; but once a quantity of lead is gone, it cannot be renewed. Some exhaustible resources are in short supply; others are so plentiful that conservation is not a major concern yet.

All life exists in an ecological system. Organisms are interrelated by many things. One important relationship among all plants and animals is illustrated by the concepts of food chains and food webs. The ability of a given ecosystem to feed and shelter a given plant or animal population level is the carrying capacity for the species. Once the system's carrying capacity is exceeded, the excess part of that population must die. Mass disease and starvation of overpopulated deer herds are well known.

The human population level on this planet has grown so rapidly in the last century that there is cause for concern. We may number over 11 billion by the year 2050! Managing our natural resources to care for such vast numbers of people will be your challenge.

DISCUSSION QUESTIONS

1. Define nonexhaustible natural resources. Give examples.

2. Define renewable natural resources. Give examples.

3. Define exhaustible natural resources. Give examples.

4. What is an ecosystem?

5. What is humankind's ecosystem? Why?

6. Is there an accurate balance in nature? What would an accurate balance in nature mean?

7. What are some ways that nature is balanced?

8. What is a food chain? Give an example.

9. What is carrying capacity?

10. Describe the trends of human population growth over the past 8,000 years.

11. Describe the difference between conservation and preservation.

SUGGESTED ACTIVITIES

1. Make a list of some of the nonexhaustible, renewable, and exhaustible natural resources you have used today.

2. Make a bulletin board to show the differences between nonexhaustible, renewable, and exhaustible resources.

3. Organize a conservationist-versus-preservationist debate about the best future for a national forest or park near your community.

EYES IN THE NIGHT: THE DEBATE OVER WOLF REINTRODUCTION IN NORTH AMERICA

Wolf. Fierce predator. Killer. *Wolf.* Magnificent animal. Beauty. Our world is dominated by images. What do you imagine when you hear the word *wolf*? The answer to that question may be determined by where you live, what your parents do for a living, or even the first movie you saw about a wolf. Today, we are faced with the decision of what to do about the wolf. Once inhabiting most of North America, humans have forced some wolves North and others to extinction.

HISTORY

When settlers first came to the shores of North America, they found several things. They found a land inhabited by an exotic people that was rich in resources and in wolves. Fearful of the four-legged predator, people began to methodically exterminate wolves from the lower 48 states. It began as the mere protection of their livestock. However, people and wolves often face conflict when their territories cross. The wolves lost the fight against humans. Today, they may be coming back. (See Figure I-A.)

Once hunted as deadly predators, wolves are now the romantic hero symbolizing the tragedy of the Exploitation era in American Natural Resource use. Wolves were reintro-

FIGURE I-A A Gray Wolf may be seen as a menacing predator or a magnificent symbol of nature. Your perception of this animal all depends on the viewpoint you take. *(Photo courtesy U.S. Department of the Interior, Fish and Wildlife Service)*

duced into Yellowstone Park some years ago. In the late 1990s a group of local ranchers sued in Federal Court to have the wolves removed. According to a National Wildlife Federation poll published in 1998, 64 percent of the American public opposes removing the wolves; 84 percent oppose removing them by killing them; and 68 percent favor Congressional action to allow the wolves to stay. With such a high desire among the public for reintroduction, it seems that the outcry against reintroduction would be low. Such logic, however, fails to recognize the fact that most Americans do not live in or around the suggested introduction sites. Those people who are directly affected often make up the 30 percent who do not want wolves. This grass-roots outcry is not just against the wolf. It rails against what many feel is just another way for the government to extract more power from its citizens. Thus, more than 70 years after the last wolf was removed from Yellowstone Park in 1926, Americans are embroiled in a raging debate over the need for reintroduction.

FACT AND FICTION

People favor or oppose wolf reintroduction on many grounds. Some organizations are split over where they stand on this issue because of the details brought up on these grounds. Notice that the Sierra Club does not favor wolf reintroduction in Idaho because of genetic grounds. The Sierra Club has often championed the causes of wildlife.

The following table outlines the views of both sides of this controversy:

Grounds	Favor	Oppose
Biological	• Wolves serve the necessary role of predator. • Wolves have shown an excellent success rate when introduced to the wild.	• As a predator, wolves kill ungulates,[1] including cattle. • Captive animals often have a bad record of success. • Wolves are surplus killers (they often kill animals with no intention of using them for food).
Economic	• Wolves help maintain healthy ungulate populations and should therefore be beneficial to states that want to increase trophy animals for hunters.	• If wolves kill ungulates, then some states, like Alaska, may limit the number of available hunting permits. • In 1995 the Department of the Interior spent $6 million on 14 reintroduced wolves, which comes out to $207,000 per animal. That does not include agency costs for agencies involved in the program but not in the DOI.
Political	• Many people consider reintroduction to be the governmental apology for the Exploitation Era.	• Some fear that this is just another form of government interference.
Ethical	• Wolves were here first and have the right to exist in their former world.	• Isn't wolf reintroduction a way to satisfy the guilt of the American people? Reintroduction is a last ditch effort that is too late to do wolves any good.
Genetic (The biggest argument is whether or not natural populations still exist at introduction sites.)	• There are no wild populations in the introduction sites. Wolves from Canada are the closest populations from which to start new groups.	• There are wild populations in the introduction sites. In Idaho, the Sierra Club filed suit against the Interior Department on the basis that the ESA[2] was being violated by the presence of Canadian wolves.

[1] An ungulate is a mammal having hooves.
[2] ESA, Endangered Species Act.

You decide: Should wolves be reintroduced in the wilds of the western United States? On second thought, wolves were native to all of North America. Should they be reintroduced in YOUR community?

Section II
Soil and Land Resources

SOIL CHARACTERISTICS

OBJECTIVES

After reading this chapter, you should be able to

- outline the processes involved in soil formation

- describe a mature soil profile

- discuss the eight land capability classes

- define soil series and explain how those differ from land capability classes

TERMS TO KNOW		
soil	original tissue	erosion
organic matter	humus	topsoil and subsoil thickness
parent material	soil horizons	
loess deposit	slope	land capability class
alluvial deposit	texture	soil classification
marine deposit	soil drainage	soil survey
weathering	flood hazard	soil survey report

L ife on this planet is very complex. We depend on many things for survival: air, water, shelter, food, and security. Certainly, no single natural resource is more important to that survival than our land, and the most valuable and productive part of the land is its surface layer—what we call **soil**. This chapter describes the nature and some of the important characteristics of the soil. It is only by understanding the nature of the soil that we can best manage it for both the present and the future.

WHAT IS SOIL?

Defining soil is not as simple as it would seem. Ask a hundred people to define the term *soil* and you will probably get a hundred different definitions. For the purposes of this book, we consider soil to be the layer of natural materials on the earth's surface containing both organic and inorganic materials and capable of supporting plant life. The material covers much of the earth's surface in a thin layer. It may be covered by water, such as the sea bottom, or it may be exposed to the atmosphere.

This thin layer of life-sustaining soil supports our buildings, catches and stores much of our water, and provides us with our food, fiber, and forest products. It was formed over many centuries. It is so durable it can last as long as our species survives, yet it is so fragile that it can be destroyed almost overnight. It is so complex that it cannot be replaced once it is destroyed.

Soil contains four main components (see Figure 4-1):

- inorganic material
- organic matter
- water
- air

An ideal soil contains about 50 percent solid material and 50 percent pore space, by volume. Again, in an ideal soil, about one-half of the pore space contains water and about one-half is filled with air. The inorganic material of soil consists of weathered mineral and rock particles ranging in size from submicro-scopic to those readily visible to the human eye. The soil's **organic matter** is made up of both dead plant and animal materials in varying stages of decay. Although the soil may contain large rocks, those are not normally considered to be an actual soil component. By the same token, the soil layer normally contains many living plants and animals. These also are not usually considered to be a part of the soil itself until they die and begin to decay.

The percentages of each of the four main soil components vary from one soil type to another. They can even vary in the same soil type depending on the kind of vegetation, amount of mechanical compaction, and amount of soil water present. During dry weather, for instance, soil air may greatly exceed 25 percent by volume. After a long period of row crop production with removal of crop residues, the organic matter percentage may decrease somewhat. After a green manure crop is turned under, that organic content may be unusually high.

SOIL FORMATION

Soil is formed very slowly. It results from natural forces acting on the mineral and rock portions of the earth's surface. The process of soil formation has been going on since the earth first formed. In fact, the powder on the surface of the moon is generally referred to as lunar soil. This implies that "soil" formation may be a normal part of the development of any planet.

Parent Material

In general, soil **parent materials** are those materials underlying the soil and from which the soil was formed. There are five general categories of soil parent materials:

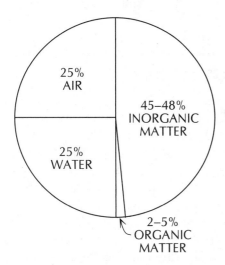

FIGURE 4-1 Physical components of a typical, well-drained loamy soil.

- minerals and rocks
- glacial deposits
- loess deposits
- alluvial and marine deposits
- organic deposits

Minerals

Minerals are solid, inorganic, chemically uniform substances occurring naturally in the earth. Some common minerals for soil formation are feldspars, micas, silica, iron oxides, and calcium carbonates.

Rocks

Whereas minerals are chemically uniform, rocks are not. Rocks, simply aggregates of mineral, are usually classified into three general groups.

Igneous rocks are formed by the cooling of molten materials pushed up to the earth's surface. Lava and magma are two types of igneous rocks.

Sedimentary rocks are those formed by the solidification of sediment. We have all seen sandbars and mud flats along running streams. We have also seen rocks that look like layers of sand stuck together. Chances are that such rocks are sedimentary rocks. They may have formed ages ago when sediment settled along a stream bed or on the ocean floor. Almost three-fourths of the earth's surface is covered by sedimentary rocks.

Metamorphic rocks are simply igneous or sedimentary rocks that have been reformed because of great heat or pressure.

Glacial Deposits

During the four great Ice Ages, glaciers moved across vast areas of the northern hemisphere. (See Figure 4-2.)

As they moved, the glaciers scooped up massive amounts of surface material. They

Glacial Period	Time
Wisconsin	10,000–100,000 B.C.
Illinois	250,000–350,000 B.C.
Kansas	600,000–750,000 B.C.
Nebraska	900,000–1,000,000 B.C.

FIGURE 4-2 The four great Ice Ages.

ground, pushed, piled, gouged, and eventually left behind great deposits of rocks, parent materials, and already formed soil materials. Much of the midwestern United States is covered by soils formed from glacial deposits.

Loess Deposits

Loess deposits are generally thought of as wind-blown silt. Much of the soil in the eastern Mississippi Valley are loess soils.

Alluvial Deposits and Marine Deposits

Both of these categories refer to waterborne sediments. **Alluvial deposits** were left by moving fresh water. **Marine deposits** were formed on ancient ocean floors. As water moves rapidly downhill, it picks up soil particles. As it reaches more level areas, it slows down and tends to "fan out." It is in the level fans that alluvial deposits are common. These alluvial fans are often in the form of floodplains or deltas.

Organic Deposits

In swampy and marshy areas, plant life may be very lush. As the plants die, the vegetation falls into the water where it decays very slowly. Over the years, this partially decayed material begins to build up. It eventually gets thick enough to support plant life itself—it becomes muck or a peat soil. Peat soils are made up of recognizable plant materials. Muck soils are more completely decayed so that plant parts are no longer recognizable.

In the case of a lake that eventually fills with such materials, the peat or muck deposits may become quite thick. This becomes a form of soil in itself.

WEATHERING

When minerals and rocks are exposed to the weather, they begin to break into smaller and smaller pieces. This is called **weathering**. Have you ever placed a cold glass into hot water? Did it crack? Although this may be an extreme example of heating and cooling, similar actions take place in nature every day. The top of a rock can be very warm in the sun while the bottom is still very cold. Such heating and cooling causes rocks to crack into smaller and smaller pieces.

Some minerals are water-soluble, which means they dissolve when exposed to water. When a rock that contains some water-soluble mineral is exposed to moisture, parts of it dissolve. Caves are usually formed by this type of natural weathering action. Other chemical processes also take place that will not be discussed in this book.

Have you ever seen a tree or shrub growing in the cracks of a large rock? The force exerted by the growing roots of a plant can become great. You have probably seen sidewalks or streets that have been cracked by tree roots. Once a crack forms in a large rock and soil starts to form in the crack, plants will not be far behind. The growing plants thus speed up the weathering process.

Another natural force that is important in weathering has to do with ice. As water changes into ice, it expands. One reason we put antifreeze in our cars is to prevent the water from freezing and cracking the engine. Bottles of liquids will sometimes shatter if they are left in the freezer until they are solid. By the same token, if a crack forms in a large rock it may sometimes fill with water. If it freezes there, the expanding ice can literally break the rock into pieces.

There are other means by which rocks are broken into smaller pieces. Sand may be blown against a large rock by high winds. That causes both the sand particles and large rock to weather. As glaciers move rocks, the rocks grind against each other. As water moves soil particles and gravel, the pieces are ground together into smaller pieces.

Thus, the major weathering forces are the following:

- temperature changes
- water action
- plant roots
- ice expansion
- mechanical grinding

The weathering process goes on continuously—day and night, year after year, century after century. Even as soil is being washed away from a field, the parent material beneath the soil is weathering. The problem is one of balance. If soil erosion takes place faster than soil formation, the result can be a destroyed field. A badly managed topsoil can be destroyed in a few years, and it may take nature thousands of years to repair the damage through weathering and soil formation.

SOIL ORGANIC MATTER

In most soils, the proportion of organic matter is relatively small (2 to 5 percent). Its importance in the formation and productivity of the soil is much higher than this small percentage would suggest.

Soil organic matter consists of decaying plant and animal parts. As plants and animals die, their tissues are attacked by microorganisms: fungi, bacteria, and others. The organic matter may be in two basic forms. **Original tissue** is that portion of the organic matter that remains recognizable. Twigs and leaves covering a forest floor are good examples of this. The other category is known as **humus**. Soil humus is organic matter that is decomposed to the point where it is unrecognizable. The brown color of some topsoil is a result of its humus content.

The soil's organic matter serves many important functions:

- It affects the soil structure by serving as a cementing agent.

- It returns plant nutrients to the soil, most notably phosphorous, sulfur, and nitrogen.

- It helps store soil moisture.

- It makes soil more tillable for farming.

- It provides food (energy) for soil microorganisms, which makes the soil capable of plant production.

CHARACTERIZING SOILS

The Soil Profile

Most soils have three visibly distinct layers or **soil horizons**. Often, each horizon may have subhorizons. These horizons are called the A horizon (topsoil or surface soil), the B horizon (subsoil), and the C horizon (parent material). (See Figure 4-3.)

A1
A2

B1
B2

B3

C1

C2

Topsoil
(A HORIZON)

Subsoil
(B HORIZON)

Parent Material
(C HORIZON)

FIGURE 4-3 Soil profile showing A, B, and C horizons.

We have already discussed parent material, the C horizon. Except for the peat and muck soils, this parent material consists of weathered rocks and minerals. It is usually quite like the rock from which it came. It is soft enough to dig in or to crumble. It may be a structureless mass. It may contain many larger pieces of partially weathered rocks. There will probably be very few plant roots in this part of the soil profile.

The B horizon, or subsoil, is more thoroughly weathered than the parent material. It is often a different color than the parent material or the surface soil. It usually contains little organic matter. It may be much finer and harder packed than either the A or C horizons. Many plant roots appear in this part of the soil profile.

The A horizon, or topsoil, lies at the surface. It is characterized by a higher content of organic matter than the other horizons. It may have a grayish, brownish, or blackish color. The topsoil color results from the humus content. (Remember what humus is from an earlier part of this chapter.) Topsoils with more humus are usually darker. As the soil's organic matter content is depleted, its topsoil color becomes lighter. In Chapter 6, we will discuss the use of crop residue management to control this organic matter loss.

The topsoil is also the most productive part of the soil. It is here that most biological activity takes place. It is also here that most of the plant nutrients are available to plant roots. Thus, as you might expect, the A horizon is most valuable in crop production. In Chapters 6 and 7, we will discuss the importance of conserving and managing our topsoil.

Soil Physical Properties

The ability of a piece of land to produce agricultural crops is determined by a number of soil properties.

Slope is the single most important factor in determining the productive potential of the soil. Slope may be defined as the angle of the soil surface from horizontal. It is expressed as a percentage of rise or fall in a given horizontal distance. (See Figure 4-4.)

Slope affects the productive potential of the soil in numerous ways: It influences rainfall runoff rates; it relates directly to the danger of soil erosion; it affects the use of farm machinery; and the size and shape of fields may be determined by the requirements for contour farming.

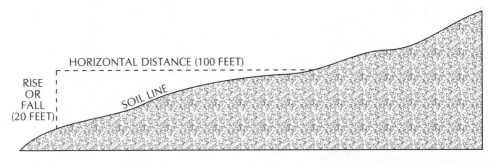

FIGURE 4-4 The slope of a soil surface can be determined by dividing the rise or fall by the horizontal distance. Here, slope = (20/100) × 100 = 20%.

Texture refers to the proportions of sand, slit, and clay in the soil. Coarse-textured soils are those with high proportions of sand. They tend to drain more rapidly and, in extreme cases, may tend toward droughtiness. Fine-textured soils are those with high proportions of clay. They tend to be sticky when wet and absorb surface moisture more slowly than other soils.

Soil drainage reflects the natural ability of the soil to allow water to flow through it. Well-drained soils allow excess soil water to move fairly quickly out of the plant-growing regions of the soil layers. Poorly drained soil holds excess soil water in the upper layers of the soil. Excessive wetness reduces the soil's productive potential by damaging plant root systems. Soils with poor drainage probably will have gray-colored or mottled subsoil or even topsoil.

Flood hazard refers to the likelihood that a given field will receive flood damage. A field in a frequent flood plain does not have a good productive potential. Even though it may be capable of producing a good crop, such a field's long-range potential is lowered by the probability that the crop will be flooded in any given year.

Erosion, as a soil property, refers to the degree the soil has already been damaged. Erosion may range from none to severe. A field that has been used for crop production, yet suffered no or only slight erosion damage, can be used safely for further production. On the other hand, a soil that has suffered severe erosion should be used for agricultural production very carefully. Its potential may be limited to permanent pasture or forest production. The appearance of large gullies or the loss of almost all of the original topsoil indicates severe erosion.

Topsoil and subsoil thickness refers to the depth of those layers that are available for plant root production. A very thin topsoil may limit crop production potential. A combination of a thin topsoil and a thin subsoil will severely limit crop production.

Land Capability Classes

Land capability classes categorize the productive potential of the soil. The classification system begins with Class I land, which is the very best land for agricultural production. The least useful land, from an agricultural standpoint, is Class VIII land. In general, Classes I through IV can be used for row crop production. Classes V through VIII are not suitable for row crop production for various reasons. (See Figure 4-5.)

Class I land is very good, productive land. It can be safely cultivated with minimal erosion-control measure. It is nearly level, well drained, deep, medium textured, not subject to erosion or flooding, and easily cultivated. Very little land is Class I even in the most productive farming areas. In hilly or mountainous regions, sizable areas of Class I land are very rare.

Class II land is good land for all types of farming. It has some limitations. It may have a gentle slope, it may suffer from the effects of past erosion, and it may have less than an ideal soil depth. It may also have some drainage problems, but these can be corrected by installation of drainage systems. This class of land requires careful soil management to prevent erosion damage.

Class III land is moderately good land. It can be cultivated and farmed regularly, but it has some important limitations. It may have a moderate slope and, thus, be highly susceptible to erosion. It may have already received severe erosion damage; it may have problems with drainage or have a shallow rooting zone

FIGURE 4-5 There are eight land capability classes. Can you find all eight classes in this picture?

that causes droughtiness. Some level wet lands may be Class III if they can be improved by drainage systems. Class III land should be farmed with great care to control erosion.

Class IV land has severe limitations but can be cultivated with careful soil management practices. This class of land may have a strong slope and be subject to severe erosion. It may already suffer from severe past erosion. It may be a very shallow, very dry, or very wet soil. If wet, even extensive drainage systems may still allow occasional waterlogging.

Class V land is nearly level but has some soil property making it unsuitable for cultivation. It has little erosion hazard, but it may be in a frequent floodplain. It may be very wet, very dry, or very rocky. Land in swampy areas is frequently Class V. Soil of this type may be

quite suitable for pasture, wildlife habitat, or forest production.

Class VI land has serious limitations. It may have very rocky or very shallow usable surface soil. Land of this class can be used for tree production, for permanent pasture, or for wildlife habitat.

Class VII land has severely limiting properties. It may be very steep, or it may be very severely eroded with large gullies. Very coarse soil combined with strong slopes may cause land to be rated Class VII. Establishing pasture may be impractical, but native grasses can be used for controlled grazing. In addition, this land can be used for forest production, wildlife, and recreation.

Class VIII land has one or more extreme limitations. The limitations may be a rock outcropping or an area with almost solid surface rock. River washes, streambeds, lake bottoms, sand dunes, and other nearly barren areas are usually Class VIII. This type of land can be preserved in its natural state for recreation and wildlife, but it has little agricultural value.

Soil Classification

Land capability classification, discussed in the previous section, is a management and conservation tool. Another very important land-use planning tool is the **soil classification** system. The Natural Resource Conservation Service (NRCS) of the USDA uses the soil taxonomy system described here. (See Figure 4-6.)

The first unit of classification is the order. All soils fit into one of the ten orders in the system. Each order is very broad in scope and is based on the formation process that resulted in the soil. Each order is broken down into suborders, each suborder into great groups, then subgroups, then families. The final step in classifying a soil is to determine its series.

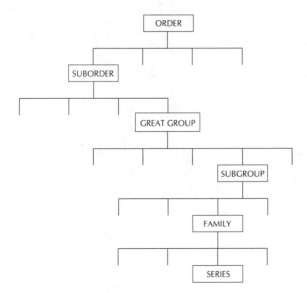

FIGURE 4-6 Levels in the soil taxonomy system.

There are about 10,500 soil series known in the United States. Each soil series is composed of soils that are basically alike in all major characteristics.

The characteristics used to determine a soil's series are

- horizon order and sequence
- horizon thickness or development
- texture of each horizon
- organic-matter content
- pH of each horizon
- parent materials
- depth to hard rock
- pan horizons present
- soil color
- soil structure
- type of clay present
- other factors

Soil Survey

The NRCS has classified much of the soil in this country. This classification process is known as a **soil survey**. The results of a soil survey for a particular location are published in what is known as a **soil survey report**. The report for a given geographic area will include many items of information. The most important item in the report is the soil survey map. (See Figure 4-7.)

The soil survey report, including the maps, can be used as a land-use planning tool. When soil scientists combine the land capability class with the soil series, we have an excellent picture of the characteristics of a particular site. Its ability to produce plant growth and resist erosion are both known. Land-use planning is greatly improved.

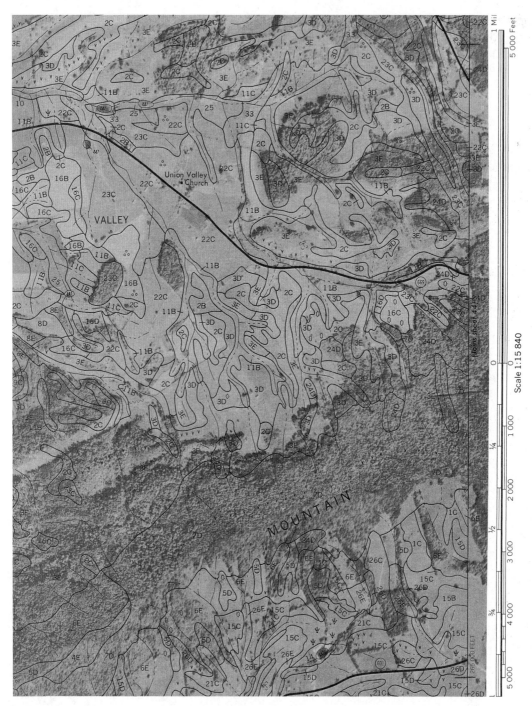

FIGURE 4-7 A soil map from a soil survey can be made from an aerial photograph. This soil map shows an area near the Brush Mountain Range in Virginia. *(Courtesy of the U.S. Department of Agriculture, NRCS)*

SUMMARY

The earth's naturally occurring rocks and minerals are broken down into parent material by weathering forces. Parent materials combine with organic matter through further weathering into soil.

Soils may be classified based on the land's capability to produce plants while resisting erosion. There are eight land capability classes. Class I is the best land, Class VIII is the worst from the standpoint of food production. Classes I through IV are cultivatable. Classes V through VIII are not generally considered to be tillable.

Another soil classification system places a soil into one of almost 10,500 soil series. A soil series has similar chemical and physical properties. All soils of a single series are very similar in color, texture, pH, structure, and many other physical and chemical characteristics.

Once a site's capability class and soil series are known, its best uses can be fairly well determined. Wise land-use planning should be based on these considerations as well as on political and economic considerations. The next two chapters deal with soil management on agricultural and nonagricultural lands.

DISCUSSION QUESTIONS

1. Define soil.
2. Define weathering. List the major weathering forces.
3. How do parent materials differ from rocks and minerals?
4. What are the main categories of parent material? Define each.
5. Differentiate between original tissue and humus. Which gives topsoil its color?
6. List five ways organic-matter content affects the soil.
7. What are the general soil horizons? Describe each.
8. List and define six important physical properties of the soil.
9. A field has a rise of 3 feet for every 50 feet of horizontal run. What is its slope?
10. How does a soil series differ from a land capability class?

SUGGESTED ACTIVITIES

1. Make a collection of rocks from your local area. Try to determine which are igneous, sedimentary, and metamorphic. Try to find at least one of each. A book on rocks and minerals from the library would be helpful.
2. Dig a pit in the soil or clear off a road bank so that you can see the soil horizons.
3. Collect samples of soil with varying amounts of sand, silt, and clay. Moisten the samples and feel them. Can you tell the differences? Try to form ribbons of soil by pressing the samples between your thumb and index finger. Can you see the differences? A very coarse textured soil will not even form a stable ball, but a very fine textured soil will stick together firmly and allow you to form a fairly long ribbon with your fingers.

5

SOIL EROSION

OBJECTIVES

After reading this chapter, you should be able to

- differentiate between natural soil erosion and soil erosion caused by humans
- list the main causes of accelerated soil erosion
- list and define the major types of soil erosion

TERMS TO KNOW		
geological erosion	runoff erosion	rill erosion
accelerated erosion	wind erosion	gully erosion
splash erosion	sheet erosion	sandblow

When the first men and women arrived in the area that is now known as America, they found a rich land. The soil was largely covered by forests, grasses, and other native vegetation. Wildlife was plentiful—fish, birds, deer, bison, and much more. As long as these first Americans could exist as hunters and food gatherers, that was enough. As societies became more complex and the human population grew on this continent, native, wild food soon became inadequate. Primitive farming became necessary. When the Europeans came, they began to clear more and more of the rich land for farming. That society simply could not exist from hunting and food gathering. When this took place, an unfortunate side effect resulted. For untold ages, the topsoil had been forming (Chapter 4 discusses this process). Now, suddenly, the topsoil was being destroyed much more quickly than it was being replaced. This chapter discusses the process by which this took place and is still taking place—erosion.

WHEN IS EROSION A PROBLEM?

Some Erosion Is Normal

Soil erosion is a normal process. It takes place at all times on all soils. Natural erosion is known as **geological erosion**. This is the natural process. Even when the soil is covered by virgin forests, some erosion takes place.

As Chapter 4 discussed, soil is made up of minerals and rock particles and organic matter. Many minerals and rocks are water soluble to some degree. Organic matter is subject to decay and may be partially water soluble. As rainfall strikes the earth, part of it soaks in and moves toward the water table. As it moves, it dissolves these water-soluble materials. This is a form of erosion.

Also, fire is a part of nature. Although most wildfires in this country are caused by humans, many are not. Removal of the soil's plant covering by fire has the same effect whether the fire was caused by nature or not.

Thus, some erosion is always a normal part of the soil formation process. People did not invent soil erosion. This geological erosion is a very slow process. It is usually much slower than the process of soil formation. So, topsoils and subsoils have developed and deepened over the ages.

Accelerated Erosion

The problem arises when the land is cleared to farm and build homes, highways, and other structures. As long as the soil is covered by plants, it is protected from excess erosion. The leaves break the fall of raindrops. The stems and leaves slow down the speed of runoff and wind. The roots bind the soil particles together.

Clearing the natural vegetation away removes this protection. The result is a speedup of the normal rate of soil erosion—**accelerated erosion**. Left unprotected, a topsoil that took many thousands of years to form can be washed or blown away in just a few years.

When the European settlers first came to America, there were few of them and the land was limitless—or so it seemed. Land was cleared to farm, and they used it with little regard for the future until erosion had destroyed its productivity. As more settlers came, they moved further west and cleared more land; after all, there was plenty of land. As these settlers moved westward, they eventually cleared millions of acres of land. They allowed their cattle and sheep to overgraze more millions of acres

of natural grassland. The effects of clearing and overgrazing were much alike.

This clearing and grazing were essential for society. Our nation, our people, our life as we know it could not have survived without the land being farmed. Farming has always been the backbone of America. It has made possible all the other things our society has done, both good and bad. It is sad that the land had to suffer because we didn't know enough or care enough to control the accelerated erosion.

Once the land's natural plant cover is removed, the accelerated erosion begins. Such erosion results from two basic forces: moving water and moving air. The rest of this chapter considers these two forces and discusses the types of erosion as well as the side effects of erosion.

Water-Caused Erosion

Water causes soil erosion in two basic ways, both resulting from movement. The first action of water on the soil is that of falling raindrops, or **splash erosion**. A raindrop strikes the earth at about 20 miles per hour. If you think about the effect of one drop of water hitting the ground, that does not sound too bad. When you stop to consider how many drops of rain fall from just one summer thunderstorm, you can begin to imagine the total result of falling water.

What happens when a raindrop strikes the earth? If it hits a plant leaf, its force is broken. It breaks into smaller droplets, which fall or drip harmlessly to the ground. If it hits bare earth, that force is absorbed by the soil particles. The result is that the surface of the soil is torn apart. Soil particles are separated and thrown about. Some soil particles are suspended in the water and carried away in the runoff. Other small particles filter into the soil's surface layer, plugging the soil's air spaces, or pores. (See Chapter 4.) This makes the soil surface less permeable and causes even more of the falling rain to flow away as runoff.

All of us have seen examples of splash erosion. Have you ever seen soil particles splashed up into the sides of buildings or cars after a heavy rain? (See Figure 5-1.) Have you ever walked in the rain and found sand or mud all over your ankles and shoes? These are examples of the effect of raindrop splash—splash erosion.

The second and more destructive action of water on soil is that caused by runoff. As surface water begins to move downhill, it picks up soil particles and carries them along. The particles may be dislodged by splash effect, or they may simply be washed loose by the moving water. Either way, the faster the water flows the more destructive it becomes. The damage of **runoff erosion** is threefold. First, an existing soil structure is damaged. It becomes less productive. Second, the soil surface may be reshaped by gullies, so it becomes less workable. Third, the soil carried away eventually will be deposited somewhere. Sometimes silt deposits are good. The Nile Valley and the Mississippi Delta are examples of the benefits of silt deposits. This is not the usual case. Most silt deposits are harmful. Clogged waterways, fish kills, filled lakes, muddy streams, and mud flats are the usual results of such deposits.

Wind-Caused Erosion

Wind, like water, has two basic actions that cause erosion. First, moving air tends to dry the soil and dislodge soil particles. Second, the wind moves those soil particles away from their original locations. Particularly in dry areas and where the land is relatively flat with few trees to slow surface winds, **wind erosion** can be a serious problem.

The first action is that of the wind itself. As the wind moves along the surface of land, the evaporation of soil moisture is accelerated. If there are no trees or other surface vegetation to slow the surface wind, this process is even faster. Once the surface of the soil is dry, individual soil particles, except in heavy clay soils, do not cling together as tightly. On bare soil, these particles are easily picked up by strong surface winds. The result is that finer sand and silt particles, which are very important in a healthy and productive soil, may be carried away in a cloud of dust. Thus, a field may find much of its productive capacity simply "gone

FIGURE 5-1 Splash erosion results when raindrops strike exposed earth.

with the wind." Moreover, a plowed field is more susceptible than simply bare soil to drying and blowing.

The second action of wind is the effect of moving soil particles. Once dislodged by the wind, smaller soil particles may be carried great distances by the wind. Dust storms in the plains and southwestern states are extreme examples of wind erosion. Such storms deposit dust in car and truck engines, on homes and factories, and irritate the lungs and eyes of humans and animals alike. The damage done by dust storms can be very great indeed.

Larger particles of sand and certain clay granules may not be carried away at all. If this happens, the soil surface becomes very sandy or covered by a layer of clay. Either situation is detrimental to crop production. If these larger particles are picked up by the wind, they probably will be deposited when they reach some sort of obstacle. Drifting sand dunes at the beach or in the desert are extreme examples of this effect.

RESULTS OF EROSION

Sheet Erosion

Have you ever looked at a freshly plowed field? Was it all the same color—brown or black or maybe deep red? Perhaps it had areas that were lighter in color, maybe yellow or lighter red. If so, the field has probably suffered from the effects of sheet erosion.

Sheet erosion is the gradual and fairly uniform removal of the surface soil. It may occur so slowly that the farmer is not even aware it is taking place. Sheet erosion is caused by splash erosion coupled with a slow runoff of soil-laden water. It results in a gradual reduction of the topsoil thickness. The effect of that

soil loss is to lower the soil's productivity. More fertilizer will be required to get the same output. Eventually, even that won't help, and the field simply will no longer produce as much output per acre.

As sheet erosion becomes more advanced, you will notice colored areas appearing in the plowed field. Those color changes take place when the topsoil becomes very thin. The plow or disk cuts through the topsoil and turns up part of the subsoil. Thus, discolored areas in a field probably indicate that sheet erosion is well advanced; damage has already been done. (See Figure 5-2.)

Rill Erosion

Rill erosion is a more rapid and more visible type of erosion. Rills are simply small streamlets cut into the soil surface by running water. As soon as the rate of rainfall or snowmelt exceeds the soil's ability to take in water, rainfall runoff begins. As the water moves downhill, it washes away soil particles. If the runoff is fast

FIGURE 5-2 Sheet erosion can occur evenly across a broad area. The light areas in this field show where the topsoil has been removed by sheet erosion leaving subsoil at the surface.

enough, it will scratch out tiny furrows or rills. The rills may be very small, smaller even than a pencil, or they may become much larger and more noticeable. During plowing, the rills are smoothed over and the erosion damage is hidden. Regardless, whenever rills appear, erosion has taken place already and the soil has been damaged. (See Figure 5-3.)

Left alone, these rills will continue to wash out and cause further damage. If the soil's slope is steep enough or long enough, the rills may become very large. When a rill becomes too large, it becomes a gully.

Gully Erosion

It would seem that **gully erosion** would be the most serious type. In a sense it is. After all, large gullies are an obvious and quite severe problem. If a gully becomes too large, it will be impossible to get tractors and other farm implements across it. In another sense it is less serious than rill and sheet erosion. These latter forms of erosion are just as harmful but less obvious than gully erosion. With just sheet or rill erosion taking place, almost the entire top-

soil of a field may be lost before the farmer realizes there is a problem. When a gully begins to form, the problem is obvious. (See Figure 5-4.)

Sandblows

Sandblows cause two forms of damage. The first is soil removal. The second is sand deposits or drifts.

Soil removal usually takes a form much like the sheet erosion of the previous section. When the soil's vegetative cover is removed, the surface becomes dry and loose. Then the wind simply picks up soil particles and blows them away. The effect on the soil is obvious and extreme. As the topsoil on a field blows away, the field becomes less productive. It retains less of the rainwater that falls on it. It produces less vegetation, and the result is that even more topsoil is blown away the following year. Also, as in sheet erosion, this form of soil

FIGURE 5-3 Rills result when small streamlets of surface runoff water cut into the soil surface.

FIGURE 5-4 Gully erosion can result from uncorrected rill erosion.

removal may be very damaging but almost invisible. The damage may be very severe before it becomes obvious.

Soil deposits, or sand drifts, are more obvious. They result when the wind is forced to change directions abruptly. An example would be at a fencerow or at the side of a building. (See Figure 5-5.)

When the wind is particularly strong at the soil level, it can be a real problem. The longer an open field is, the stronger the surface wind is likely to be. An interesting paradox comes into play because of those two facts. The most productive farming lands are usually flat and open. In flat, open farm lands, farmers use large machinery. Large machines are most efficient in large fields—the larger and more flat the field, the better. *But* the larger and more flat the bare field, the more likely it will be affected by wind erosion.

The previous section talked about the soil loss and sand deposits caused by wind erosion. Another major problem caused by wind erosion is airborne soil or dust.

Did you ever notice the layer of "dust" that accumulates on furniture every day? Part of it may be pollen or other particles moved by the air, but much of it is airborne soil particles. Airborne soil ranges from the light dust we see on our furniture to great dust storms.

FIGURE 5-5 Sandblows may or may not cause immediately noticeable erosion, but this soil was obviously moved by the wind.

True, when these soil particles are removed, the topsoil is damaged. When they are deposited, if drifts form, further damage is done. No form of erosion is more frightening or awesome than a great dust storm. No result of erosion is more constant and obvious to everyone than the daily deposit of dust on our furniture, cars, homes, and bodies. Both of these—the killer dust storm and the bothersome light dust—are visible and very real effects of blowing soil caused by wind erosion.

One of the most severe national disasters in U.S. history occurred in the 1930s. A long period of overfarming and overgrazing was followed by a multiyear drought in the Great Plains region. Most notably in Oklahoma, Arkansas, and the other states in that region, millions of acres of land were devastated by drought and blowing dust. (See Figure 5-6.) Hundreds of thousands of farmers and other workers whose livelihood relied on farmers were forced to leave their homes to seek work elsewhere. John Steinbeck's famous book *The Grapes of Wrath* and the movie of the same name provide a very graphic description of the environmental and human tragedy that wind-caused erosion produced in the Dust Bowl.

FIGURE 5-6 This huge cloud of dust seems to be chasing the automobile in the foreground during the Great Dust Bowl of the 1930s in the United States. *(Photo courtesy U.S. National Archives)*

SUMMARY

Erosion of the soil is a natural process. It is constantly taking place alongside the soil formation process. This natural erosion is known as geological erosion. Soil erosion becomes a problem when the soil or its vegetative cover is disturbed. When this occurs, accelerated erosion is likely to take place. Accelerated erosion is caused by either moving water or moving air. Water-caused erosion may take the form of sheet, rill, or gully erosion. Wind-caused erosion results in the removal of surface soil from one location and its deposit elsewhere. Soil particles suspended in moving air take the form of dust. This dust may be as severe as great dust storms or as mild as the light layer of dust that falls on your furniture each day. Side effects of water-caused erosion include water pollution, fish kills, filling in of lakes and streams, and sediment deposits. Side effects of wind-caused erosion include scoring of structures, dust damage to machinery, and sand drifts.

DISCUSSION QUESTIONS

1. Define geological erosion.
2. Define accelerated erosion.
3. How do humans cause accelerated erosion?
4. What actions of water cause erosion?
5. What actions of wind cause erosion?
6. List and discuss three types of water-caused erosion.
7. List and discuss two results of wind-caused erosion.
8. What are some side effects of water-caused erosion?
9. What are some side effects of wind-caused erosion?

SUGGESTED ACTIVITIES

1. Find examples of sheet, rill, and gully erosion in your community. Photograph the examples and make a school bulletin board about "Erosion Near Home."

2. Find a wall on a building or a fence where splashing water has thrown soil particles onto the sides. How high do the particles go?

3. Make a bulletin board for the school showing examples of erosion from your community.

6 CONTROLLING EROSION ON THE FARM

OBJECTIVES

After reading this chapter, you should be able to

- explain how land capability classes relate to wise soil use
- describe the main vegetation methods farmers use to control water-caused soil erosion
- describe the main mechanical methods farmers use to control water-caused soil erosion
- describe the control measures farmers use to control wind-caused soil erosion
- explain why soil erosion control should be important to every farmer

TERMS TO KNOW

vegetative erosion control	crop rotation	contour farming
cover crop	strip cropping	terrace
green manure crop	grassed waterway	windbreak
	conservation tillage	shelterbelt

As Chapter 5 pointed out, soil erosion is a natural process. Natural or geological erosion takes place all the time. Thus, soil erosion cannot be completely stopped or prevented. There are, however, a number of steps that can be taken to control accelerated erosion and to keep it to a minimum.

LAND CAPABILITY AND LAND USE

In Chapter 4, we discussed the classification of land based on its capability for safe use. There are eight land capability classes. Class I land is the very best for agricultural production—level, fertile, and relatively safe from erosion. Class VIII is the most extreme land at the other

end of the spectrum. It is very rough, rocky, covered with deep sand, or under water. Why is this an important classification system? What difference could knowing the land's capability class make?

Actually knowing the capability class of a piece of land may not be necessary for a farmer or for any other user, but using the land wisely is important. The use of the land *within its capability* is the first and most important step in land use.

In general, land in Classes I, II, III, and IV is considered suitable for cultivation. Land Classes V through VIII are not considered to be usable for cultivation. A set of general guidelines for the most intensive safe agricultural use for each land class is as follows.

Class I land can be used safely for any kind of farm production. Because it is not generally subject to erosion or flooding, it can be cultivated every year with little damage. Of course, soil fertility, compaction, and other considerations apply, but erosion is not a problem. (See Figure 6-1.)

Class II land is somewhat less safe for farm use than Class I. It probably has a gentle slope and is subject to some erosion. (See Figure 6-2.) Cultivation of Class II land can be done with little danger of excessive soil erosion. Some fairly simple erosion-control techniques are called for, however. Contour plowing, crop residue management, minimum tillage, crop rotations, and runoff-control measures are probably adequate. These and other measures are discussed later in this chapter.

Class III land has more severe limitations and requirements. It probably has a moderate slope and is subject to severe erosion. It can be

FIGURE 6-2 Class II land is nearly level with only minor problems.

used for occasional cultivation, but it requires intensive erosion-control measures. Terracing, careful and regular crop rotations, cover cropping, minimum tillage, and runoff-control are probably the minimum. If Class III land is not sloping, it may have serious drainage problems or droughtiness. These latter problems are not erosion related, and so their solutions are not discussed in this chapter.

Class IV land is either too steep, sandy, rocky, or wet for safe and regular cultivation. (See Figure 6-3.) If it is Class IV land because of steepness, then it is subject to very severe erosion. If that is the case, it should be cultivated only occasionally. It should be kept covered by vegetation almost continuously. No-till planting is best; but if cultivation is to be done, then terracing, combined with strip cropping, rotation systems, residue management, cover crops, and runoff management, will be needed.

FIGURE 6-1 Class I land is the very best agricultural land.

FIGURE 6-3 Class IV land can be cultivated only with great difficulty.

Class V land may be too wet, too sandy, or too rocky for cultivation. It, like Class I land, is level or nearly level; but its other limitations make it suitable only for permanent pasture or forest production. It is not subject to erosion in most cases.

Class VI land is probably a lot like Class III or IV except that it has some very limiting factors. Its shallow topsoil, extreme rockiness, or severe erosion make it unsuitable for cultivation. Its steepness makes it subject to severe erosion if the vegetative cover is removed. Thus, it can be safely used for permanent pasture once a good vegetative cover is established. Overgrazing should be avoided. Forest production is also a good use for this land.

Class VII land is so rough that its use for grazing or even forest production may be limited. (See Figure 6-4.) If Class VII land has an existing vegetative cover, it should be protected. If not, the land should probably be planted with trees. On Class VII forest land, clear cutting of large areas should probably be

FIGURE 6-4 Class VII land has very severe limitations. Even forest production may be limited.

FIGURE 6-5 Any existing vegetation should be protected on Class VIII land. This sand dune near the beach is very susceptible to both wind- and water-caused erosion.

avoided. If this land is clear cut, the harvesting residue should be left to cover the soil until new seedlings are reestablished.

Class VIII land is so rough that any existing vegetation should be protected. (See Figure 6-5.) This land is completely unsuited for farm production in the traditional sense.

CONTROL MEASURES FOR WATER-CAUSED EROSION

As we learned in Chapter 5, water causes soil erosion mostly through two actions—the impact of falling raindrops (splash) and the cutting action of flowing water (runoff). Basically, erosion-control techniques seek to minimize the damage caused by those two actions. In general, there are two categories of control measures: vegetative and mechanical. As you would probably guess, erosion is best controlled when vegetative and mechanical measures are used together. The remainder of this chapter discusses specific techniques that can be used to control erosion when land is being used within its capability.

VEGETATIVE CONTROL OF WATER-CAUSED EROSION

Cover Crops

During the growing season, cultivated land is afforded some protection from water erosion by the growing crop. This protection may not be very good, but it is better than none at all. Between the time the crop is harvested and the beginning of the next season, even this protection may be gone. Various forms of **vegetative erosion control** can be used. On fields where erosion danger may be severe, close-growing, so-called **cover crops** may provide much protection from erosion during this time. In addition, such cover crops may improve the soil's organic-matter content as well as its structure and tilth. If the cover crop is plowed under at the beginning of the next regular production cycle, it is called a **green manure crop**. Close-growing grasses and clovers make good cover crops. For many years, cover crops have been included in standard crop rotation systems. Another use for cover crops is to protect the soil between trees and vines in orchards or vineyards.

Crop Rotation

One of the traditional methods helping to control soil erosion has been **crop rotation** systems. A crop rotation is an orderly and repeated sequence of different crops grown on the same field. A common 3-year crop rotation in parts of the Midwest might be corn-soybeans-wheat. This sequence would be repeated every 3 years. Another system might be a 4-year rotation of corn-cotton-oats-hay (or pasture). These are only two of the many possible and, in fact, common crop rotations that have been used over the years.

Advantages of crop rotation over continuous row cropping are many. Here are a few:

■ Soil tilth (degree to which is crumblike and friable) is better maintained.

■ Water absorption into the soil is improved.

■ Erosion danger is lessened.

■ Soil organic-matter content is increased.

■ If a legume is included, nitrogen fixation helps supply that costly plant nutrient.

■ Soil tillability is improved.

■ Moisture-storing capacity of the soil is improved.

Certainly, short-term economics must be taken into account in planning the cropping system. After all, the farmer must feed and clothe his or her family. The bills must be paid, and the next crop must be put in; but the longer-term results of these decisions must also be considered. If growing 20 years of continuous corn on Class III, sloping land, would completely destroy the topsoil by erosion, that cropping system would be a bad decision.

Strip Cropping

Strip cropping is the production of alternating bands of different crops. It is used as a technique to control both water- and wind-caused erosion. On sloping land, the strips are laid out either (1) on the contour or (2) generally across the prevailing slope. If terraces (which are discussed later) are used, the strips may alternate between terraces or along a terrace line. A typical strip-cropping scheme would alternate row crop, hay, row crop, hay. It may be possible to arrange the strips to provide for the crop rotation system. For instance, with a 3-year rotation of corn-soybeans-hay, the strips of hay could make up one-third of the field area while corn or soybeans make up the other two-thirds. Or one-third of each could be grown. The strips could then be alternated annually.

Another variation is permanent strips of cover crop, pasture, or hay alternated with row-cropped strips with or without a crop rotation system.

FIGURE 6-6 Strip cropping can help to reduce erosion on cultivated land. *(Courtesy of the U.S. Department of Agriculture, NRCS)*

The effect of strip cropping is to provide bands of heavier vegetation alternated with row-cropped bands. The heavier vegetation increases the rate of water absorption, holds the soil in place better, cuts down on gullying, and improves soil tilth in the band. It also serves to slow down the running water so its cutting power is lessened. (See Figure 6-6.)

Grassed Waterways

Grassed waterways are widely used as an erosion-control technique throughout the world. As rainfall or snowmelt produces more surface water than the soil can absorb at one time, runoff occurs. As this runoff flows downhill, it begins to collect in larger and larger quantities.

Clearly, the collecting water must be led safely out of the field, or its uncontrolled movement could lead to the very rapid formation of gullies. Left alone, runoff will drain into a field's natural draws or drainageways. Normally, the best place to put grassed waterways is in those natural drainageways. A grassed waterway is simply a drainageway permanently covered by vegetation.

In a terraced or contour row-cropped field, the rows tend to conduct collecting runoff in some specific direction. The longer the row, the greater the quantity of water the row must conduct away. At some point, the water must be released from the end of the row or terrace. It is typically released into a grassed waterway. If the rows or terraces are particularly long, they probably will be broken by one or more such waterways. (See Figure 6-7.)

FIGURE 6-7 Grassed waterways are drainageways permanently covered with vegetation. This is a small farm near Blacksburg, Virginia.

Grassed waterways may be constructed by the farmer using standard cultivation equipment. The waterway may be saucer shaped, trapezoidal, or V bottomed. (See Figure 6-8.) The most common type is the saucer-shaped waterway.

Once the waterway has been formed, it must be quickly protected by a close-growing sod. Perennial grasses such as fescue, bermuda, or bluegrass are typically used. It is important to get as heavy a sod as possible as soon as possible. It is also critical that the waterway be protected from erosion until the sod is established. This means the seedbed should be heavily seeded, from two to three times the normal rate. Once the seeding is done, the seed-

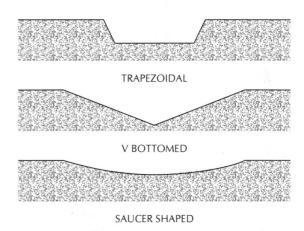

FIGURE 6-8 Three common shapes for waterways: trapezoidal, V bottomed, and saucer shaped.

bed may be lightly mulched. A slightly heavier than normal fertilization rate should be used in the waterway. Remember, the objective is not to maximize profit from the grass in the waterway. The objective is to get a quick, lush growth of grass to produce a firm sod.

Once the sod is produced, the grass may be used for hay. If this is done, the grass can be allowed to grow to normal mowing heights before harvesting. If the grass is not to be harvested for hay, it should be kept fairly short. This allows for more rapid water movement and also encourages a denser sod. The grass should be kept well fertilized and in good health. Very importantly, once the sod is established, the waterway is not cultivated along with the rest of the field. Plows should be lifted and disks should be lifted or straightened when crossing the waterway. If the sod is damaged, it should be repaired immediately.

MECHANICAL CONTROL OF WATER-CAUSED EROSION

Conservation Tillage

"Tillage of the soil" is a phrase that is often taken to be synonymous with farming. Indeed, it has always been necessary to "till" the soil in order to produce crops as we think of them. Over the centuries this tillage has taken many forms. The earliest was the use of a hand-held stick to gouge the earth so that a seed could be planted. This evolved into the use of a forked branch and gave rise to the eventual development of the hoe. (See Figure 6-9.) This type of implement, when combined with animal power, developed into the plow. Early, crude wooden plows led to the development of animal-drawn moldboard plows in more recent history. This, in turn, combined with

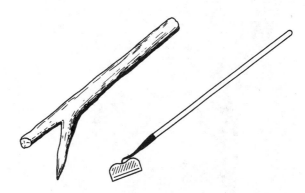

FIGURE 6-9 Manual cultivation implements.

the advent of tractors to produce the array of cultivating equipment available today. (See Figure 6-10.)

The purposes of tillage have always been the same: to prepare the soil to receive the seeds and to assist the crop in its competition against weeds for light, moisture, and nutrients. As first animal power and then tractor power developed, this came to mean the complete plowing under of the vegetation covering the soil as a step in seedbed preparation. The better our equipment became, the more complete was this process.

Conservation tillage is any tillage system that is economically practical for crop production and aids in soil and water conservation. It may take many forms, but it replaces the older methods of clean tillage that left bare soil exposed during the nongrowing season.

Conservation tillage techniques differ widely in different parts of the country. Practices appropriate on sandy soils are probably not appropriate for clay soils. Practices that work on corn may be totally inappropriate on cotton or sorghum. Thus, no attempt will be made here to completely cover this complex subject. Essentially, all these systems rely on

FIGURE 6-10 Modern cultivating equipment.

one common thing. They all depend on the crop residue from the previous crop to protect the soil from erosion until the new crop is well established.

This can mean that a corn or soybean crop is drilled in a field of winter grain that has been harvested but not plowed. It can mean that a field of corn stubble is disked in the fall rather than plowed under. Common techniques are known as minimum tillage, no till, zero till, stubble plant, chisel plowing, lot planting, and strip tilling.

Disadvantages associated with conservation tillage have been an increase in weed, insect, and disease problems. Decaying crop residues tie up soil nitrogen so that heavier fertilization may be required. Increased herbicide use is required to kill weeds that would have been killed by conventional tillage. Also, existing equipment may not be well suited to these practices, necessitating the purchase of new implements.

On the other hand, the advantages are fairly clear as well. Less energy and time are required as the amount of tillage is decreased. Soil moisture levels increase with the increased soil organic-matter level and with the mulch-

ing effect of surface crop residues. Most important, from the standpoint of this chapter, soil erosion is better controlled. (See Table 6-1 and Table 6-2.)

Contour Farming

Contour farming essentially means plowing "around" the hill instead of up and down it. Each row acts much like a small dam to stop runoff from moving straight downhill. The furrow then serves the function of a small ditch to conduct the runoff across the slope to a grassed waterway or tree line where it can be safely controlled.

The most efficient type of row cropping is in large, flat fields with long, straight rows. That way the tractor turns less often and the equipment can cover the field more quickly. However, not all farmers have large, level fields with no danger of erosion. Whenever there is even a gentle slope, the rows should be run generally across the slope to control the speed of the runoff.

As the slope gets steeper, this becomes even more important. Moreover, as the slope be-

TABLE 6-1 Soil Tillage Method and Soil Erosion in Selected Heavy Rains

Tillage Method	Year	Rainfall (inches)	Runoff (inches)	Soil Loss (tons/acre)
Conventional clean plowing	1973	3.9	2.0	15.44
	1975	3.6	2.3	6.02
Conservation (no-till) plowing	1975	4.3	2.0	0.07
	1977	4.0	1.1	0.03
	1989	6.2	1.0	0.01
	1990	5.0	< 0.1	< 0.01

Source: *Journal of Soil and Water Conservation*, May–June, 1992
These field measurements were taken in the same area in the Midwest, planted to row crops under different tillage methods over a 17-year period.

TABLE 6-2 Soil Preparation and Soil Loss[1,2]

Tillage system and row direction	Percentage slope	Runoff as percentage of rain	Soil loss in tons/acre/year
Plow/Clean Till (Sloping Rows)	6.6	80	22.65
Plow/Clean Till (Contour Rows)	5.8	42	3.22
No Till (Contour Rows)	20.7	—	0.03

[1] Runoff and soil loss is on sloping and contour row fields in corn watersheds (Cochocton, Ohio). About 5.3 inches of rain fell within a 7-hour period.
[2] Source: Virginia Cooperative Extension Service.

comes irregular, it is necessary to go to contour plowing instead of plowing in straight rows.

The contour rows are not laid out exactly straight across the slope. They should run downhill very slightly. This allows the runoff to be conducted in the desired direction. In addition, the rows cannot handle too much runoff at a time, so they should empty at fairly regular intervals into pastures, tree lines, or grassed waterways.

Terraces

Contour plowing can control only so much runoff at a given time. Once the rainfall or snowmelt is more rapid than the rows can handle, water spills over the top rows on the hill. Each lower row then has not only its own excess water but that of all the rows above it. You can see clearly that this situation can become very destructive in a hurry.

Terraces are larger surface channels constructed on the contour with a controlled rate of fall. They are designed to accept the runoff and conduct it across the slope to some protected area. Terraces are needed on fields whenever the slope exceeds about 2 percent and where the slope is over a few hundred feet long.

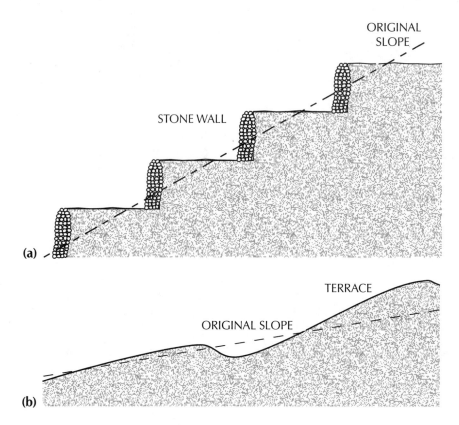

FIGURE 6-11 Terracing extremes: (a) steep slope, (b) gentle slope.

Originally, terracing developed in the Far East where dense populations required the use of all available land for food production. Even very steep hillsides could be farmed using terracing techniques. (See Figure 6-11.)

In this country, such extreme methods have not been necessary. To American farmers, a terrace might look more like Figure 6-11b.

Terraces such as the one shown can be constructed using the conventional moldboard plow or with other earth-moving equipment. The size, length, slope, and distance between terraces are all important considerations. The terracing system should be designed specifically for each field. Assistance in designing and constructing terraces is available from the Natural Resources Conservation Service, because attempting such an undertaking should not be done without seeking help. Terracing is generally not required when no-till farming is being used.

CONTROLLING WIND-CAUSED EROSION

Windbreaks

We usually think of a **windbreak** as a row of trees and/or shrubs planted across the prevailing wind direction. It serves to disrupt the surface movement of the wind. Windbreaks are normally planted to shelter the farm build-

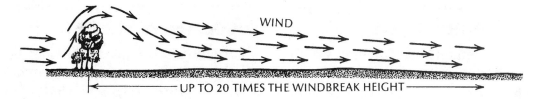

FIGURE 6-12 Windbreaks reduce the effects of high wind on the land surface.

ings and residences from high winds. (See Figure 6-12.)

Windbreaks reduce the cost of heating the home during winter and the cost of cooling the home during summer. The shade provided by the trees is cool and pleasant, and the evaporation of water through the leaves has an additional cooling effect. Windbreaks also help keep the soil moisture from evaporating as rapidly as it ordinarily would.

Shelterbelts

Shelterbelts are similar to windbreaks but are planted to provide protection for crops and livestock. Shelterbelts, like windbreaks, may consist of trees, trees combined with shrubs, or just shrubs. Both the height and density of the shelterbelt determine its effectiveness.

The surface speed of the wind is affected for great distances from the shelterbelt. Some protection is offered downwind at least twenty times the height of the shelterbelt. (See Figure 6-12.)

Conservation Tillage

Just as in controlling water-caused erosion, conservation tillage techniques are valuable in controlling wind-caused erosion. Crop residue left on the field helps conserve soil moisture, reduces surface windspeed, and holds loose soil particles in place.

There is one exception, however, where wind erosion is not a serious concern and soil moisture is critical. Dry land farmers sometimes cultivate the surface of a field into a loose, sandy layer. This has been done to conserve soil moisture better than surface litter. This practice should be reserved for areas with low wind-erosion danger.

Contour Farming

In a very real sense, it is possible to row crop on the contour for wind much like for water. The rows are plowed across the prevailing winds. Each row of the crop then acts as a miniature shelterbelt. The effect on wind patterns is the same as in Figure 6-12.

SUMMARY

The most important step in soil erosion control is to use the land according to its capability. This consideration should come first. Once the decision has been made to cultivate a field, a number of erosion-control practices can be used.

Vegetative controls on water-caused erosion include cover crops, crop rotation systems, strip cropping, and grassed waterways. Clearly, other techniques are possible; but these represent the most important ones.

Mechanical controls for water-caused erosion include, but are not limited to, conservation tillage, contour farming, and terraces. Control techniques for wind-caused erosion include windbreaks, shelterbelts, conservation tillage, and contour farming.

DISCUSSION QUESTIONS

1. What is the most important thing we can do to control soil erosion in farming?

2. Which classes of land are considered to be cultivatable? Which are not? Why?

3. What is a cover crop? How is it used in erosion control?

4. How do crop rotation systems help in erosion control?

5. Define strip cropping.

6. Grassed waterways can have three shapes. Draw and label the three shapes.

7. What is conservation tillage?

8. Contour farming and terracing usually go together. Why?

9. How are windbreaks and shelterbelts different? How are they alike?

10. How can contour farming help in controlling wind-caused erosion?

SUGGESTED ACTIVITIES

1. Construct two model hillsides in flat boxes. On one, construct terraces and contour furrows. On the other, run furrows straight up and down the slope. Sprinkle water on both to see if there is a difference in the resulting erosion.

2. Find a bare soil surface on a sloping area such as a road bank. Smooth two similar areas by raking them. Cover one with straw and leave the other exposed. After a rain, look at both areas to see if the vegetative cover was effective in controlling erosion.

3. Invite your local soil conservation agent or NRCS officer into your class to discuss soil and water conservation in your community.

7

NONFARM EROSION CONTROL

OBJECTIVES

After reading this chapter, you should be able to

- describe the major sources of nonfarm soil erosion

- explain why nonfarm landowners should accept responsibility for soil erosion control on their land

- explain the relationship between mining or construction and erosion

- describe some important techniques used in controlling nonfarm soil erosion

TERMS TO KNOW		
diversion ditch	water chute	mulching
berm	sediment basin	reclamation plan
waterway	sodding	

The previous chapter dealt with farm-related erosion and steps that farmers can take to help control it. This chapter discusses nonfarm erosion and some steps that nonfarm landowners can take to control their erosion.

Soil erosion can become a major problem whenever the soil is disturbed. Clearly, the soil is disturbed by farm cultivation; but it is also disturbed by home building, industrial expansion, urban construction, and many other nonfarm activities. Daily, as many as 8,000 acres of U.S. farmland are converted to nonfarm use. Much of this conversion involves construction and other changes in the shape of the land.

In addition, millions of acres of land annually are disturbed by strip mining for coal and other resources. Another source of potential erosion problems is along our massive highway system. As new highways are constructed and old ones are widened, moved, or repaired, millions of miles of road banks are subject to water-caused erosion.

After looking at all the nonfarm demands on our soil, it is little wonder that erosion is a problem on nonfarmland, too.

CONTROLLING NONFARM EROSION

Damage caused by erosion in nonfarm situations is much like that in farming. Construction, however, typically involves more extreme changes on the soil surface than farming does. (See Figure 7-1.) Natural drainageways are drastically changed overnight. Large surface areas are covered with concrete or asphalt, making the runoff problem severe. Soil structures are

FIGURE 7-1 Construction sites usually involve extensive grading and excavation.

completely destroyed. Soil horizons—the A, B, and C layers—may become irrelevant or nonexistent. Sedimentation or gullying quickly may become immense problems. (See Figure 7-2.)

Controlling nonfarm erosion involves runoff control, sedimentation control, and erosion control. Techniques, as before, involve both vegetative and mechanical practices. Because of the capital-intensive nature of most construction, more expensive erosion-control practices are not only possible, they are often economically essential.

MECHANICAL EROSION CONTROLS

Diversion Ditches or Berms

As water collects on a built-up site, it begins to run downhill. Without contour plowing or terraces to channel it to safe disposal areas, it flows directly down the slope. As it flows, it picks up more water and more speed.

A diversion ditch or berm of earth is placed across the slope much like a terrace. It collects the runoff and conducts it to a selected site for safe disposal. (See Figure 7-3.)

The **diversion ditch** or **berm** may be sodded or it may be covered by concrete or some other material. Regardless, it must be protected; otherwise, it too would be destroyed by erosion.

Waterways

In construction and in built-up sites, **waterways** serve the same purpose they serve in a plowed field. They collect excess soil water in the form of runoff and conduct it elsewhere for disposal.

Unlike farm waterways, it is not uncommon to find concrete-lined waterways in built-up areas. There are several reasons for this. First,

FIGURE 7-3 A diversion ditch or berm conducts runoff safely away from a construction site.

FIGURE 7-2 Erosion problems at construction sites can be severe.

the runoff rate on a parking lot is nearly 100 percent. Almost all the rainfall runs off. Second, the runoff must be conducted away more quickly. A water buildup is usually not acceptable, and the area must be used continuously.

A shopping center does not close down because the parking lot is wet. Third, because of the higher investment per square foot in business and industry construction, the costs of concrete waterways are justifiable. Of course, vegetated waterways are used off the farm too.

Where the slope is very steep, a concrete-lined waterway is known as a **water chute**.

Sediment Basin

On a field used for crop production, sediment can be a problem. But, if sediment is deposited one year, we can plant on top of it the next spring. In urban and commercial development, the same is generally not true. Sediment deposits often must be removed with costly machinery and labor. Damage to buildings and equipment from construction-related sediment can be very expensive indeed.

FIGURE 7-4 A sediment basin.

A **sediment basin** is like a small pond, with one major exception. It is designed to be filled up by sediment or mud, not to hold water or for recreation. (See Figure 7-4.) It catches excess runoff from diversion berms or ditches or from other waterways. The muddy water is held long enough for the sediment to settle out in the form of mud. The excess water can then be allowed to soak into the ground, evaporate, or be conducted away by another waterway or drainage pipe.

Sediment basins are normally temporary structures. As a construction development is in process, the erosion problem can be very severe. Grading exposes unprotected soil; no sod is present to slow the runoff or to speed the soaking-in of water. Parking areas are not yet paved, and drainage and sewage systems may not yet be in place. All these combine to make a severe erosion potential, and that erosion may lead to excessive sedimentation until the construction and landscaping are completed. Sediment basins are the obvious solution.

Bank Protection

Because construction sites often produce steep slopes, extra efforts to stabilize banks may be needed. (See Figure 7-5.)

FIGURE 7-5 Steep slopes must be stabilized.

The steepness of these banks, as well as the fact that no vegetation is present, represents an extreme erosion hazard. A number of mechanical solutions are available.

Where banks will be particularly steep, concrete or stone walls may be the answer. Where the banks are longer or less steep, other solutions are needed. Jute matting, wooden strips placed across the slope, or chemical binding materials can be used to hold the soil in place until grass or other ground cover can be established.

Terracing

On banks that are particularly long and steep, a special effort may be required. The length of the bank is the culprit. As runoff flows downhill, it picks up both more runoff and more speed. Runoff may start as a slow trickle at the top of a long, steep roadbank and become a racing stream before it reaches the bottom. The larger the quantity of water and the faster it flows, the more erosion danger is present.

Terraces on such banks present a potential solution. In effect, each terrace breaks the long slope into shorter slopes. Runoff is captured by the terrace and conducted to a safe location for disposal. The disposal area may be a grassed or lined waterway, a chute, a wooded area, or a storm sewer leading to a pond, lake, stream, wastewater treatment facility, or sediment pond.

Downstream Runoff

Very large construction projects such as shopping centers or large factories produce great amounts of runoff. Because of this, normal stream flow may be greatly increased downstream. If that happens, flooding may result. To prevent such damage, streambank clear-ing, stabilization, or other measures may be needed downstream.

VEGETATIVE EROSION CONTROLS

Vegetative erosion-control techniques on nonfarm sites are generally more extensive than in farming. Again, there are high investments in residential construction, urban development, and commercial or industrial construction. Thus, elaborate and costly erosion-control measures are economically more feasible than in farming.

Lawn

The most common vegetative control technique for erosion control on nonfarm sites is the lawn. Lawns are most often established by seeding a prepared seedbed with one or more varieties of grass. However, they also may be established by sprigging, plugging, or sodding.

The seedbed should be prepared by thorough tilling, smoothing, and removal of rocks and other debris. The seeds are then broadcast or the sprigs or plugs are planted. The seedbed is then firmed, mulched, fertilized, and watered. With good weather and good luck, the lawn grasses will sprout and begin to grow fairly soon.

Sodding

Under conditions in which quick development of a heavy growth of grass is needed, **sodding** is possible. (See Figure 7-6.) Sod is a living layer of mature grass and topsoil that usually is found in squares or long, rolled strips. It is produced on a sod farm and is harvested as live plants complete with stem, roots, and a thin layer of growing medium, or topsoil.

FIGURE 7-6 This truckload of sod can be used to establish a new turf quickly. *(Photo courtesy of Dr. Jack R. Hall, Virginia Tech)*

The area to be sodded is thoroughly prepared. The sod is applied in place, firmed, and watered. If the sod is kept well watered, healthy, and protected for a few weeks, the root system will grow downward into the soil. The result is an almost instant lawn. This is much more expensive than seeding, sprigging, or plugging a lawn to grass; but it is also more effective and much quicker.

It is very important that the soil be well prepared before sodding. Also, the sod must be allowed to dry out before the roots take hold in the receiving soil.

Mulching

When grass is to be established by seeding, heavy **mulching** may be necessary. Mulch, such as straw, may be applied as a lawn is being installed. The mulch holds the grass seeds in place, helps maintain surface moisture, and protects the soil from erosion until the lawn is established. On long slopes, stakes (perhaps with strings between them) may be used to hold the straw in place.

Mulch may also be applied by spraying a silagelike slurry of vegetable matter. Several chemicals are used as soil-binding agents, which serve the same function as mulch.

Where grass is not being seeded, other mulches may be used. Pine bark, pine needles, or even gravel may be used as a permanent mulch between shrubs or other plants, or such mulches may be used in place of plants to hold the soil in place and provide attractive landscaping features. (See Figure 7-7a.)

Ground Covers

Low-growing shrubs, vines, or other plants (see Figure 7-7b) can be used much like grass is

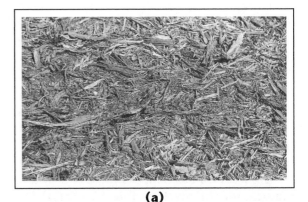

(a)

(b)

FIGURE 7-7 Many things can be used for ground cover: (a) shredded hardwood mulch; (b) low-growing shrubs.

used in a lawn. These are often combined with mulching to increase protection.

Temporary Cover Crops

Cover crops on construction sites are used to protect an area temporarily. Once the ground is graded, something must be done. If construction is delayed for any reason, a cover crop may be in order. Also, many plants cannot be successfully started except at certain times of the year. For instance, certain lawn grasses germinate only during warm weather; in this situation, a temporary cover crop is needed.

For example, most lawn grasses require a fairly warm soil for best germination. A house being completed during November needs a lawn right away. The builder may not want to try to sell the house with a "mud lawn." In that case, winter rye (annual ryegrass) might provide a temporary cover crop until warm weather when other, more permanent lawn grasses can be seeded.

HIGHWAY EROSION CONTROL

Because of the massive amounts of land involved in highway construction, expansion, and repair, special problems exist here. Highway surfaces are practically waterproof, and virtually all the rain falling on a highway either evaporates or runs off. Also, highways are constructed as nearly level as is practical. This almost always means grading high spots and filling low spots. For instance, when a highway crosses a hill, there are usually exposed banks. (See Figure 7-8.)

Both the exposed bank and the filled-in area are susceptible to erosion damage. On highways this presents extreme problems. Soil

FIGURE 7-8 Highway banks present a special problem.

washed from the banks may cover parts of the highway with sediment. Erosion in the fill underneath the highway could cause sags and cracks in the road. Both situations are quite dangerous to motorists.

Erosion-control techniques along highways are basically the same as those used elsewhere. In this case, however, the scale is much greater. Therefore, erosion-control, runoff management, and sediment control are especially important. Highway engineers and contractors must be aware of these problems and take continuous steps to keep them in hand.

An extreme example of a vegetative erosion control for roadbanks is Kudzu. Kudzu is a plant that grows very rapidly as a vine. It was imported from Asia to solve a very difficult erosion problem. As you can see from Figure 7-9, the vine also grows in places other than where it is intended.

STRIP-MINING OPERATIONS

The third and final erosion problem we discuss is the erosion caused by strip mining. During

FIGURE 7-9 Kudzu was introduced as an erosion control for roadbanks. It worked well but created its own problems.

the last quarter of the last century, a large increase occurred in the strip mining of coal, especially in Appalachia. The problems here are enormous because of the very steep slopes and, therefore, high-velocity water. (See Figure 7-10.)

Because of the high impact on the environment, numerous federal and state regulations control mining. The General Assembly of the Commonwealth of Virginia, for instance, has stated that unregulated strip mining of coal causes soil erosion and stream pollution, and, in general, creates conditions detrimental to life, property, and the public welfare. The Commonwealth of Virginia has created the Department of Conservation and

Economic Development, Division of Mined Land Reclamation to enforce its policies and laws. There are numerous do's and don'ts in the laws of each state, but strip mining is regulated in every state in which it occurs.

Each strip mine is different with different topography, vegetation, and habitats; each mining operation, therefore, must also be different. By law, each mining operation must have a **reclamation plan**. What are the alternatives? Rehabilitated lands can be utilized for recreation, forestry, range and pasture, cropland, and even urban development or expansion. Each plan will be influenced by the particular reclamation end

FIGURE 7-10 A strip-mining operation in North Dakota. *(Courtesy of the U.S. Department of Agriculture)*

result. Nonetheless, certain aspects are necessary for any plan.

First, the surface above the coal, the spoil, must be placed in a stabilized state. A complete drainage and erosion-control system must be engineered and installed during and after the mining operation. For example, sediment trap ponds must be built to catch sediment that does erode. Each operator in Virginia must post a cash bond before operation begins. After the coal is extracted, there is a 2-year period within which the land must be reclaimed. If the operator fails to comply, the bond is forfeited. Also, during the mining process, government officers make trips to the mine to ensure that all regulations are followed.

Again, in strip-mining operations as in highway construction, the erosion-control measures are the same as those discussed earlier in this chapter. Because of the scale of these operations, erosion-control must be well planned in advance if steps are to be effective.

SUMMARY

All control techniques for water-caused erosion rely on two basic factors: The impact of falling raindrops must be absorbed by something other than the soil surface, and the speed of the runoff must be kept under control.

If these two things can be accomplished, accelerated erosion can be kept at acceptable levels.

Urban, residential, commercial, and industrial construction present extreme erosion dangers. Because almost all such construction disturbs and reshapes the land, it exposes the soil to greatly accelerated erosion. The rate of erosion may be increased by as much as 40,000 times by construction activities in a given area.

Mining operations and highway construction present the same or even more extreme problems. Because of the scale of these operations, vast areas of land are disturbed. Huge amounts of runoff, erosion, and sedimentation may result.

But, in all these cases, reasonable and effective erosion-control techniques include both mechanical and vegetative practices. The best solution to the erosion problem is almost always a combination of both vegetative and mechanical measures.

DISCUSSION QUESTIONS

1. How does construction cause the danger of erosion to increase in a given location?

2. How much land is being converted from farm to nonfarm use in this country? What does this imply about the erosion problem?

3. What is a diversion ditch? A diversion berm?

4. A waterway may be either grassed or lined. When would a grassed waterway be good enough? When should it be lined?

5. What is a sediment basin? When is it important?

6. How can freshly graded banks be protected from erosion?

7. Where have you seen banks protected from erosion by terraces?

8. List and describe five vegetative erosion-control measures.

9. Why does highway construction present a special erosion problem?

10. Discuss the need for erosion-control and reclamation in strip-mining operations.

SUGGESTED ACTIVITIES

1. Visit a large-scale commercial construction site and examine the erosion and sedimentation problems. What conservation measures are in use?

2. Visit a local garden center to discuss recommended lawn grasses, straw, and seeding techniques. Does the garden center sell sod? What kinds of decorative mulches are carried? What plants are recommended for lawn ground covers?

3. Visit a well-landscaped shopping center or office building. What mulches, ground covers, and sods are being used? Are any mechanical erosion-control measures being used?

4. Find a road bank or hillside that is being damaged by erosion. Plan and implement an erosion-control plan for it.

8

RANGELAND MANAGEMENT

OBJECTIVES

After reading this chapter, you should be able to

- explain the importance of rangeland and its careful management

- discuss the history of the rangelands in this country

- define and discuss the three major types of grasslands in the continental United States

- describe the types of vegetation prevalent in the grasslands of the continental United States

- explain the relationship between grazing and grassland management

TERMS TO KNOW		
rangeland	Taylor Grazing Act of 1934	animal equivalent unit (AEU)
range	decreaser	grazing capacity
forb	increaser	
grassland	invader	
overgrazing	undergrazed	

For the vast majority of Americans who live in the eastern or midwestern states and in large cities, the word **rangeland** is almost a meaningless term. The vastness of the original rangelands in this country and the significance of their contribution to the development of our nation may very well be surprising. Indeed, the size and importance of the world's rangelands on a worldwide basis today can hardly be overstated.

In its broad sense, **range** refers to the land area that provides food, particularly in the form of forage, and browsing for animals. Using that definition, almost all land area must be considered a range for something. As we use it here, the term is more narrowly defined. For the purposes of this book, *rangeland,* also known as range, grassland, or prairie, refers to the land areas of the world that tend to be naturally covered by grasses, grasslike plants, **forbs**, and shrubs as the primary vegetation instead of trees. Naturally, that excludes the true deserts and forested areas of the world.

So we see that many parts of the world are considered rangeland. Natural rangelands occur in the form of tundra, meadows in openings within forests, marshes, wetlands, vegetated areas above the treelines on mountains, shrub lands, and grasslands. Of these, by far the most important to human development have been the grasslands.

Human-generated rangelands consist of pasture land and forage crop land produced on areas that were originally covered by forests or natural grasslands. Of the total land area on earth (about 32.2 trillion acres), approximately 26 percent (about 8.3 trillion acres) are managed as permanent pasture. In the North American continent, there is a total land area of about 5.3 trillion acres. Of that area, about 17 percent, or about 0.9 trillion acres, are managed as permanent pasture.

Before the advent of agricultural settlement in the area that is the continental United States today, the treeline of the eastern forests reached about as far westward as a line drawn roughly between Dallas, Texas, and Chicago, Illinois. West of that lay open grassland stretching as much as fifteen hundred miles to the Rocky Mountains. In your history books, the general area is frequently referred to as the Great Plains, a term that describes a geographical/physical region rather than a biome.

Today rangeland makes up about one-third of all the land in the United States, or about 770 million acres. Of that, about 200 million acres are in Alaska. Very little of Alaska's rangeland is used for pasture for livestock. Almost all of it is left for wildlife to graze. In contrast, Canada has only about 70 million acres of rangeland suitable for livestock grazing.

In the southwestern United States, as much as 85 percent of the land area is considered rangeland, and as much as 90 percent of that rangeland is managed as pasture for grazing livestock. Of the rangeland in the continental United States, about two-thirds is privately owned or owned by local or state governments. The most productive and accessible rangelands are generally privately owned. Land that

is less productive or that is less accessible has tended to be retained by the various levels of government. The remaining one-third is owned by the federal government. The federal rangelands are managed by various agencies. The Bureau of Land Management manages most of that federal land, leasing much of it to ranchers for grazing livestock. Other rangeland is managed by the U.S. Forest Service and the U.S. Park Service.

THE GRASSLANDS

In general, the **grasslands** in what is now the continental United States consisted of three broad types: tallgrass prairies, transition prairies, and shortgrass prairies. Over 25 percent of our land area was covered by rangeland when the first European settlers landed on the east coast. (See Figure 8-1.)

The most striking of the grasslands was the tallgrass prairie, which covered much of the area between the present states of Indiana to Nebraska and southward to Texas. The land that produced the tallgrass prairies was very rich. The topsoil was deep and abundant in plant nutrients. The climate produced a fairly regular and adequate supply of rainfall, with winters that varied from merely cold to extremely harsh.

The result of that combination of adequate rainfall and rich soil produces a lush growth of prairie grasses that might reach 6 feet in length at the peak of a good growing season. It also makes for very productive agricultural land, some of the richest farmland in the world. When the settlers came, they recognized the richness of the land and cleared it for farm production to feed a growing and hungry nation. Thus, the tallgrass prairies are almost gone today. (See Figure 8-2.)

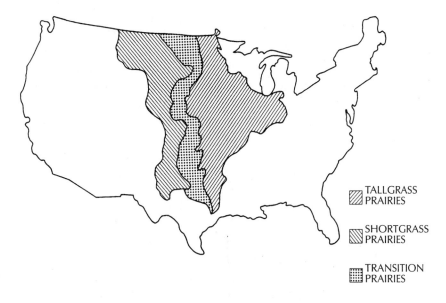

TALLGRASS
PRAIRIES

SHORTGRASS
PRAIRIES

TRANSITION
PRAIRIES

FIGURE 8-1 Major grasslands in what is now the continental United States, prior to European settlement.

FIGURE 8-2 This tallgrass prairie is located in the Flint Hills area near Riley, Kansas. (*Courtesy Dr. Jay McKendrick, University of Alaska, Fairbanks*)

The shortgrass prairies stretched from central Canada through parts of the Dakotas to central Texas. In general, the soils of that region were also very rich in plant nutrients with deep topsoils, as were the tallgrass prairies further east. However, this area has a much lower average annual rainfall and it is much less dependable than that of the tallgrass prairie region further east. At the peak of a good growing season, these grasses might reach a height of up to 3 feet. (See Figure 8-3.)

Between the shortgrass and tallgrass prairies were the transition prairies. This region experiences more average rainfall than the shortgrass prairie but it tends to be less dependable than that of the tallgrass areas. The maximum height of the grasses falls between 3 and 6 feet in good growing years, but good years are fewer and further between.

There still exist at least parts of all three original grassland types. However, at one time or another the vast majority of the land in all three types has been plowed under. There are still substantial areas of shortgrass prairie, primarily because it has less agricultural productive potential. Virtually all of the shortgrass region requires irrigation to make cultivated farming profitable. Almost all of the original tallgrass prairie land has been converted to farm production or other human developments, primarily because of its high productive potential.

In other parts of the world, the great grassland regions are known by different names. In the Asian continent and eastern Europe, the great steppes stretch for thousands of miles. The African veld is a vast and rich grassland. We have all seen television documentaries about

FIGURE 8-3 This is an example of a shortgrass prairie. It is located in the Custer State Park in South Dakota. *(Courtesy Dr. Jay McKendrick, University of Alaska, Fairbanks)*

the huge herds of herbivores in Africa. What is usually not pointed out by these documentaries is the absolute reliance of those wild animals on the grasses of the veld. In parts of South America, there are great regions of pampas. In Brazil, the grasslands are called campo. In other parts of the Americas, the term used to describe some grasslands is savannah. In Spain, it is the protero; in the Philippines, the kogonales.

HISTORY OF THE RANGES IN THIS COUNTRY

Early History

Before the arrival of the European explorers and settlers, the center of this continent was a land of vast, uninterrupted grasslands. (Refer to Figure 8-1.) There has been a long-standing controversy among ecologists as to whether grasses are the true climax vegetation of the region. Many believe that much of the region could easily support forests; in fact, in much of the grasslands the rainfall is quite adequate for that. It may be that lightning-caused fires brought about the destruction of the forests that could have flourished there in prehistoric times. Later, the combination of natural fires and fires set by the Native Americans as a hunting tool caused the eventual elimination of the trees that would have been needed to provide seeds for reforestation. (See Figure 8-4.)

Interestingly, spots of grasslands occurred all the way from the Pacific Ocean to the Atlantic Ocean. Wherever fire had destroyed the forest, rapidly growing grasses would quickly replace the trees until new trees returned to

FIGURE 8-4 This range is in the Flint Hills area near Riley, Kansas. *(Courtesy Dr. Jay McKendrick, University of Alaska, Fairbanks)*

dominate the smaller plants. Often the fires were caused by lightning or other natural causes; but, just as often, they were set by Native Americans for hunting (as will be discussed later) or specifically to establish grasslands to attract bison and other grazing animals. In these cases, the grasslands were temporary and did not truly belong to the great grassland biome.

These vast areas were virtually void of trees when the first European settlers arrived. At that time the only checks on the population of herbivores were predators, disease, and climate. Predators included the large cats, wild dogs, wolves, and Native Americans. The Native Americans hunted just enough animals for meat for food, hides for clothing and shelter, and bones for tools. Their meager needs would hardly have affected the population levels of wildlife on the prairie.

On the other hand, their hunting methods could be quite destructive. To harvest enough wild game—much of it larger and faster than the hunters themselves—they often resorted to the use of range fires. A "killing party" would be set up at a location selected to intercept the fleeing animals. A second party would set a fire on the opposite side of the herd of bison or other herbivores. The frightened animals would then flee into the ambush. Another technique involved forcing animals to run over a cliff or into a deep ravine. Falling either to their deaths or at least to severe injuries, the animals provided the food and hides needed.

When the fires did not spread very much, the technique was not too wasteful. With nothing to stop them, such fires sometimes burned out of control until they reached a river or other natural barrier wide enough to extinguish them, or until rain put them out.

That may seem a remarkably destructive way to gather food, but there were a number of reasons that it was done. In the first place,

the hunters were on foot. (The Spanish were the first to introduce horses to this continent.) A group of hunters on foot might not be very successful in harvesting deer, antelope, or bison in quantities large enough to provide a reliable food source for their people.

It is also important to remember that the grasses were not actually eliminated by the fire. The day after a range fire, the soil would be bare; but most of the nutrients in the plant matter remained after the fire and were returned to the soil. More important, many of the seeds and roots remained alive in spite of the heat from the fire. The new crop of grass would begin to grow with the next rain, and the range very soon would become green and rich again.

As a side effect, trees were virtually eliminated from most of the grasslands. Unlike grasses, trees take years to grow large enough to produce seeds. Range fires killed trees that would otherwise have become dominant in the most productive grasslands. Where the range fires were fairly frequent and regular, the shrubs and forbs were also eliminated for the same reason. A rangeland without shrubs and trees to compete with grasses for sunlight, nutrients, and water, produces lush grazing and has a much higher carrying capacity for herbivores.

Beyond the effect of the human-caused and natural fires on the carrying capacity of the grasslands were disease and other predators. Disease took weak animals and became important primarily when populations exceeded carrying capacity in conjunction with climate fluctuations. Also, whenever the population of grazers became very large, predators multiplied more rapidly. That meant more predation and thus downward pressure on the population of grazers.

Climate, in the form of periodic droughts coupled with years of unusually heavy rainfall produced reduced grazing during some years and abundant grazing during other years. The abundant years encouraged increased breeding, and the sparse years caused high death rates among the animal population. The result, though seemingly cruel, was an effective population control.

Even with the effects of predation, natural disasters, Native Americans, and disease, the grazing animals of the plains grew in numbers to the millions. Early pioneers spoke of herds of "buffalo" that stretched from horizon to horizon. The grasslands had a massive carrying capacity for herbivores. It was only with the coming of agriculturally based civilization that the plains came to be stretched beyond its ecological limit.

When the first Spanish explorers marched northward into this region in the sixteenth century, they found seemingly endless expanses of rich grazing land populated by millions of bison, antelope, deer, and other wild animals. When they came to the region, they brought donkeys for pack animals, cattle and other grazing animals for food, and horses to ride. Some of their animals escaped or were released into the wild. There the animals found a land of plentiful and nutritious grazing with few natural enemies. They reproduced in large numbers, and the result was **overgrazing** by cattle and horses in a number of limited areas. Interestingly, it was not until the Spanish introduced them to America that the Native Americans acquired horses to ride and as draft animals, making hunting fires less necessary.

The earliest Westerners came to the region as exploiters, not as settlers. They came to look for gold and silver or to harvest the fur pelts of beaver and other wild animals for the use of wealthy Europeans. These pioneers put little real pressure on the ecosystem of the plains.

After the Settlers Came

Later, the English, French, Spanish, and, later still, American settlers moved westward to make a living from those same great grasslands and herds of herbivores. In Chapter 2, we discussed the effect of trappers and market hunters on game populations. They took some game animals for food, but by far the major use by people for the wildlife was furs, hides, and feathers.

Soon, however, other settlers moved westward from the American colonies along the east coast and northward from Mexico. At that point in time, agriculture became necessary for human survival. Even if that had not been the case, most of the new settlers came with the specific intent of earning a living from farming.

The grasslands offered one particularly important source of potential wealth for enterprising newcomers. Tall, abundant grass on the open ranges offered free grazing for potentially millions of cattle. How else could one make a profit more surely than by putting cattle into an area where they could eat, drink, and multiply freely at no cost to their owner? The cattle could then be slaughtered for meat and hides, with only a minimum of financial investment by the cattle rancher.

The only problem was how to get the cattle to the eastern and midwestern markets. Near the middle of the nineteenth century, the coming of the railroad offered a solution. As a result, a number of great fortunes were made by ranchers who were strong enough and ruthless enough to survive in the open-range cattle industry.

A great part of American folklore comes from the time between about 1850 and 1890. That was the time, particularly during the two decades after the American Civil War, when the open-range cattle industry became a very important source of wealth in this country. By this time, much of the tallgrass regions had been cleared for cultivation. The somewhat less productive but still rich lands of the shortgrass prairies were still mostly in their original state.

It was also a time, coincidentally, when nature played a cruel trick on the people there. For over 20 years, there was a climatic cycle in which unusually dependable rainfall and mild winters prevailed. The cattle ranchers pushed their herds to sizes that overtaxed the carrying capacity of the grasslands in the shortgrass prairies. Great ranches sprang up, and the cattle barons grew rich and powerful.

Farmers who wanted to clear the land for plowing were unwelcome, and the cattle ranchers kept them away, often by force. Fencing material was largely unavailable in the region, and in any case would not have been allowed by the cattle ranchers. After all, the grasslands were a vast region that belonged to no one in particular. The grass was a gift of nature to anyone who was able to take advantage of it. This was a huge common area, and the cattle industry grew up explicitly to exploit the wealth of the commons. In Chapter 35, the concept of the commons will be explored in some detail.

By the mid-1880s, the effects of overgrazing were beginning to be obvious everywhere. The abuse of the grasslands occurred for a number of reasons. First, the cattle ranchers were not ecologists. They had little understanding of the impact of their activities on the ecosystem. Beyond that, the motive of the cattle ranchers was profit, not conservation.

Thus, cattle populations were pushed up to and even slightly beyond the ecological limits of the ranges, with the cattle ranchers hoping each year for good weather to make the grasses grow rapidly. The land began to dry out more than usual because of the loss of soil cover. The organic matter in the topsoil was depleted, further decreasing the soil's water holding capacity.

When the vast herds of cattle were beginning to be less healthy because of declining grazing, calamity struck. The winter of 1885-86 was a particularly cold and harsh one. Millions of cattle died or became sick as a result of the unusual cold combined with the lack of adequate grazing. The following summer the grazing was still inadequate because of the effects of the overgrazing of the previous decades. The winter of 1886-87 followed with even colder and harsher conditions.

The open-range cattle industry was shaken to its foundations by this triple catastrophe. True, the ranchers had made such an event almost inevitable because of their management practices. At the same time, one cannot blame any individual rancher for the disaster.

The whole thing occurred because the prairie was a rich common area with value to many persons. It belonged to no one but could be used by anyone who voiced a claim for it and was willing to stand behind that claim. Finally, it was an area in which no one was responsible for making and enforcing overall management decisions.

Barbed wire was invented in the 1870s but was violently resisted by the cattle ranchers. They were the ones, they reasoned, who had taken the land from the Native Americans and who had labored to develop huge herds of cattle. They had "tamed" the land. Settlers who wanted to clear the land for farming would have to build fences to keep cattle out of their fields. Thus, fences were seen by the cattle ranchers as invasions of "their" land, which was not legally their land at all.

Resistance to the settlement and fencing of the open ranges soon became so violent that laws against fence-cutting became extremely harsh. It was not until the disasters of 1885–87 that the fate of the open ranges was determined. After that, fences came to be largely accepted as necessary to the future of the cattle industry. Cattle ranchers came to understand that range restoration was not possible without fencing to control grazing. They also learned that without range restoration, the upgrading and return to profitability of the cattle industry was impossible.

Yet, overgrazing continued even after the range was settled and fenced. This was true for a number of reasons, not the least of which were a general lack of understanding of the long-term effect of overstocking, and the economic necessity to produce income for short-term survival of the farm and ranch families whose livelihoods were dependent on the land.

It was true also because much of the rangeland remained the property of the federal and state governments. That land remained essentially a common area under no one's direct management. Yet, the public land was still left available to ranchers for free grazing of their cattle.

The Taylor Grazing Act of 1934

The **Taylor Grazing Act of 1934** was designed to assert federal control to prevent continued overgrazing in the arid grasslands. When it was being considered by Congress, it was strongly opposed by much of the western livestock industry on economic grounds and by political conservatives in the name of states rights.

Responsibility for the administration of the Taylor Grazing Act was finally passed to the Bureau of Land Management (BLM) in 1946. Under that agency, the concept of the grazing district has become an effective tool for management of government rangelands. As of this date, the BLM controls the grazing on over 1.75 million acres of public lands.

Only ranchers who own land adjacent to grazing districts are eligible to lease land from BLM-managed ranges. Ranchers who hold leases to BLM-grazing rights can graze their livestock at very low costs on public domain grasslands. They must abide by strict management policies set by BLM officials based on the recommendations of government range conservationists. At the same time, the reality is that BLM decisions can often be swayed by political considerations under the pressure that can be applied by ranchers who hold substantial grazing rights. As many as 9 million head of sheep and cattle are allowed on public grazing lands each year.

TYPES OF VEGETATION

Several major types of vegetation predominate in the rangelands: grasses, grasslike plants with fibrous root systems, forbs, and shrubs (many of which have tap-root systems). Each of the types of vegetation must compete for sunlight, moisture, and nutrients in the arid plains region. Grasses are particularly valuable from two standpoints. First, the complex root systems are very adept at holding the soil in place and thereby preventing erosion while also preserving soil moisture. Second, grasses provide a higher-quality source of nutrition for grazing animals than do most other plants that can grow freely on rangelands. (See Figure 8-5.)

The proportion of grasses varies substantially from year to year, and even during a single growing season. Grazing rates particularly affect the relative amounts of the four major types of vegetation present. Light grazing actually benefits the grasses in comparison

FIGURE 8-5 This rangeland area near San Diego, California, has been invaded by chaparral and other tap-rooted shrubbery. The brushy conditions developed because of extreme undergrazing. *(Courtesy Dr. Jay McKendrick, University of Alaska, Fairbanks)*

to other types of plants. As long as over half of the length of a grass stalk is not eaten, grazing causes little damage to the grass plant. Unfortunately, heavier grazing is very detrimental to the grasses. When most of the leaf is eaten, it is more difficult for the plant to recover—particularly if the plant is already stressed by dry, hot weather.

Varieties of grass that are easily damaged by even moderate grazing are called **decreasers**. These are often the ones that grazing animals find most desirable. Given the

choice of which grasses to feed on, the grazers eat the palatable ones and leave the unpalatable ones. Thus, the more desirable grasses tend to be eaten until they are damaged. In other cases, the decreasers are simply more susceptible to damage from grazing because they are not very hardy plants.

Other rangeland plants that tend to thrive under heavy grazing are called **increasers**. Many increasers are successful merely because the grazers find them unpalatable and so avoid eating them. In some cases, the plants are simply better able to get at the limited soil, water, and nutrients.

Plants that move into an area after it has been badly overgrazed are called **invaders**. Oddly enough, badly **undergrazed** areas are also subject to being taken over by such less desirable plants. Prickly cactus and some shrubs are particularly potent invaders. (See Figure 8-6.)

The ideal situation involves a moderate level of grazing. A rangeland without grazing as well as with overgrazing will produce less total biomass than one with moderate grazing.

RANGE MANAGEMENT TECHNIQUES

Objectives of Range Management

As we will discuss in Chapter 35, management implies *doing* something. It implies decisions and actions. Those decisions and actions should be guided by objectives if they are to be consistent and productive. In the case of range management, there is generally a single major objective that is accepted by most range managers.

The primary objective of range management is the *long-term maximization of livestock productivity from managed rangeland.* That

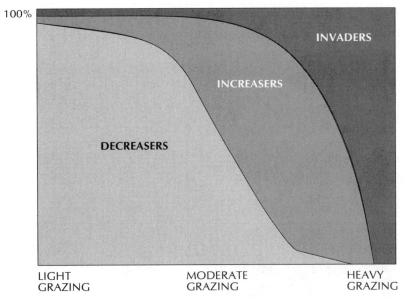

FIGURE 8-6 Relationship between levels of grazing and type of vegetation on rangeland.

sounds simple enough, but there are several additional objectives that are implied by this:

1. Current grazing must not damage the land's capacity for grass production.

2. Current grasslands not at their best must be given treatment so that they will improve. This can mean reseeding with natural grasses and fertilization.

3. Soil conservation techniques must be used to control soil erosion and depletion.

4. Water management must be used to ensure that the groundwater supply is not depleted.

5. Appropriate grazing pressure must be maintained to help reduce the population of invaders into the area.

6. Undesirable vegetation must be controlled in the grazing areas.

Grazing Capacity

The first step in any planned range management program must be the determination of the carrying capacity (or grazing capacity) of the area. This is not a simple task because it must take into account a number of factors: types of local vegetation, growth rate of desirable grasses, the effects of grazing rates on the specific local vegetation, climate, soil type, slope, and expected rainfall, to mention some of the most important ones.

The grazing capacity is then used to determine an acceptable stocking rate. An area's stocking rate is expressed in terms of **animal equivalent units (AEU)**. An AEU is the amount of forage that is required to feed a 1,000-pound animal for a given period of time. For instance, one AEU for a month is known as an animal unit month. (See Figure 8-7.)

In general, one steer is assumed to require one AEU. Five goats or five sheep are assumed

FIGURE 8-7 This grazing area for domestic cattle is located in the grasslands of the Monument Rock chalk flats of Grove County, Kansas. *(Courtesy Dr. Jay McKendrick, University of Alaska, Fairbanks)*

to require one AEU. A horse or a bull is assumed to require 1.25 AEUs; but wildlife must eat, too. One elk needs about 0.67 AEU, and four deer require about 1 AEU.

Grazing Management

Once the **grazing capacity** of an area is determined, grazing rates must be controlled to ensure that the rate is not exceeded. Even with controlled stocking rates, other management techniques also must be used.

Undergrazing or overgrazing discourages the growth of desirable grasses and promotes invaders in an area. Left to themselves, animals do not graze uniformly across an entire area.

They will tend to overgraze near water sources and near salt blocks. Separating the salt blocks from water sources helps to encourage better grazing patterns.

Even then, they will still overgraze parts of an area and undergraze others. A system that controls grazing distribution so that it is more even across an entire range area is necessary. There are a number of common management systems:

1. *Continuous grazing.* The livestock are kept in the same area year-round and are allowed to graze as they choose. The major disadvantage of this system is uneven grazing. The advantage is that

it is less work and less expensive than other systems.

2. *Deferred-rotation grazing.* This is a system of controlled grazing in which a range area is fenced into two or more separate grazing areas. The livestock are moved periodically to allow the grasses time to recover.

3. *Holistic management.* This is a newer method similar to deferred-rotation grazing. In this system, the principle is that healthy grass is not damaged by being bitten off once but may be damaged by being eaten down more than once. The livestock grazing area is divided into multiple paddocks. The animals are moved every few days to a new paddock according to a predetermined schedule or as the grass appears to reach the point where it needs to be allowed to recover. (See Figure 8-8.)

Range Restoration

There are a number of range restoration techniques. The best time to restore a range area is before it is damaged too severely. Once too much damage is done, major restoration efforts may be needed. (See Figure 8-9.)

Controlled grazing is the first step in range restoration. Overgrazing or undergrazing was what caused the problem in the beginning, so controlled grazing is essential in restoration. If the range must be reseeded, grazing will need to be withheld completely until the grasses are reestablished. (See Figure 8-10.)

If a range area has been stripped bare, it will require reseeding. Although many varieties of "improved" grasses have been tried over the years, in general, grasses native to the area have been found to be the only long-term solution for reseeding. Grasses the livestock will eat and are less susceptible to overgrazing should be selected.

If the area is overgrown by invaders, the undesirable vegetation must be eliminated first. Then reseeding can be done. Regardless, during the re-establishment of grasses in an area, application of fertilizers can make a substantial difference in the speed with which the recovery takes place.

In less severe cases, merely controlling grazing may be adequate to restore a productive range. Light overseeding and fertilization may be needed.

Paddock A, 5 Days	F, 5 Days	E, 5 Days
B, 5 Days	C, 5 Days	D, 5 Days

FIGURE 8-8 An example of a holistic management schedule designed for a 30-day rotation with 25-day recovery time per paddock. The number of days can be varied if the grass-growing conditions change.

FIGURE 8-9 This example of severe soil erosion developed as a result of the loss of grass cover from overgrazing. *(Courtesy Dr. Jay McKendrick, University of Alaska, Fairbanks)*

FIGURE 8-10 Extreme overgrazing like this can present major erosion and other hazards to the land. *(Courtesy Dr. Jay McKendrick, University of Alaska, Fairbanks)*

SUMMARY

The world's grasslands are known by many names. Whatever they are called, grasslands have been fundamental in the development of humans and civilization as we know it. They remain critical in feeding our growing human population. The productivity advantage grasslands have over forests is that the sun's energy is captured near the ground rather than in the crowns of trees. At ground level, grazing animals can feed in large numbers and move from one feeding site to another.

In what is now the United States, about one-fourth of our land area was covered by grasslands prior to European settlement. The most productive part of that area was the tallgrass prairies. Throughout the world, humans have tended to overuse grasslands, which resulted in damage to the soil and reduced carrying capacities. Today, we know how to manage grasslands to produce the maximum amount of grazing for animals useful to humans, while protecting the land from damage due to overuse.

DISCUSSION QUESTIONS

1. Why do some parts of the world tend to be naturally covered by trees and other parts by shrubs, grasses, and other small plants?

2. What happens to the vegetation in a range area if there is overgrazing for an extended period of time? Undergrazing?

3. What factors led to the development and eventual end of the open range cattle industry in this country?

4. What is the Taylor Grazing Act? What has its effect been in this country?

5. What factors affect the grazing capacity of a range area?

SUGGESTED ACTIVITIES

The learning activities you can do will be determined in large part by where you live. You will need to select those activities that are possible in your location.

1. Locate a grassed area that is not being grazed or mowed. Measure and mark off a typical 10′ × 10′ area. Estimate the percent of the 100-sq.-ft. area that is covered by grasses and by shrubs. Then locate a grassed area that is being grazed. Measure and mark off a typical 10′ × 10′ area. Determine the percent of the area that is covered by grasses and by shrubs. How do the two areas compare? What do you think made the difference?

2. Look through the last several years of *National Geographic* magazine. Find as many articles as you can about parts of the world with important grasslands. Bring the editions to class for an open discussion on the grassland ecosystems represented. What problems are they facing? How are people using them?

3. Find a natural range area. Dig up several grass plants and single examples of some of the other plants present there, particularly the woody ones. Try to identify the plants. Clean off the root systems. As you compare the different plants, which would be more effective in preventing erosion? Why? Which would be more useful to cattle? Why?

9 LANDFILLS AND SOLID WASTE MANAGEMENT

OBJECTIVES

After reading this chapter, you should be able to

- define and explain solid waste, municipal solid waste, and industrial solid waste

- explain why disposal of municipal solid waste is such an important and growing concern in our society

- explain how natural attenuation works and when it is an appropriate means of solid waste disposal

- explain how a containment landfill works

TERMS TO KNOW		
solid waste	radioactive waste	natural attenuation landfill
industrial solid waste	nonhazardous waste	attenuation
municipal solid waste (MSW)	percolate	containment landfill
	leachate	liner
hazardous waste	landfill	recycling

Many environmental situations that are cited as "problems" in our environment are debatable. Although climatologists generally agree that global temperatures are at their highest point in several thousand years, scientists disagree on whether greenhouse gases generated by human activity are actually causing significant global warming. Even if global warming is resulting from human activity, there is not universal agreement as to whether it will be as severe and cause as much devastation as many people contend. Many other scientific and political disagreements plague the environmental movement. Yet, one undeniable problem facing our society is that of solid waste disposal. In fact, with the progress we have made in water pollution and air pollution in the United States in the past 30 years, solid waste management may well be our most pressing remaining environmental problem.

SOLID WASTES

What is solid waste? **Solid waste** is nonliquid, nonsoluble materials ranging from municipal garbage to industrial wastes that contain complex, and sometimes hazardous, substances. Solid wastes also include sewage sludge, agricultural refuse, demolition wastes, and mining residues. Technically, solid waste also refers to liquids and gases in containers.

We have basically two major sources of solid waste: municipal waste and industrial waste. When you "take out the trash" and place it in a "trash can" you are participating in the solid waste disposal process. (See Figure 9-1.) Whether you live in a rural area and dispose of your own solid waste or live in an area served by a trash collection and disposal service, this is a traditional part of everyday life. What you are disposing of is what we refer to as municipal waste, even if you live in a rural area.

The first category of solid waste we will discuss is industrial. **Industrial solid waste** consists primarily of spoilage from mining, logging, and other industrial processes that is not disposed of in landfills. Mining is the larg-

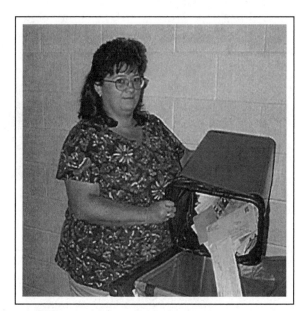

FIGURE 9-1 Most of our municipal solid waste starts with the homeowner, schools, and local businesses.

est producer of industrial solid waste, much in the form of rocks, soil, and other materials scooped from the earth and piled onto other locations. Mining waste may be as innocuous as topsoil scooped off a building site and spread onto a nearby field or it may be as extreme as toxic waste from a copper mining operation or radioactive waste from a uranium mine. Mineral mining, coal mining, gravel mining, and other forms of natural resource exploitation produce huge quantities of spoilage, but accurate data are simply not available to document how much is produced worldwide, or even in industrialized countries like the United States. According to the WorldWatch Institute, 571,000 hectares (approximately 1.4 million acres) of land were mined in one recent year worldwide. Of that about two-thirds was non-fuel mining (minerals, metals, gravel, etc.), and the remaining one-third was coal mining. Much coal is mined by tunneling underground, but the vast majority is strip mined. Strip mining is the process of excavating the coal seam by removing the soil and rocks above it and then digging out the coal, which is now on the surface.

Logging waste gives the physical appearance of being very destructive, but in reality it consists of leaves, branches, and bark that are all biodegradable. In the long run, solid waste from logging operations returns to the ecosystem. Indeed, one component of logging waste that has found a commercial value is ground or shredded bark. At one time bark was a solid waste, but now it is a by-product that many home and business owners prize as bark mulch. (See Figure 9-2.)

Municipal solid waste (MSW), though smaller in total quantity than industrial solid waste, actually presents a much more difficult problem in this country. In the United States, the total amount of municipal solid

FIGURE 9-2 This mulch is a by-product of logging and is very useful in landscaping around homes and businesses.

waste generated has more than doubled between 1960 and 2000, and per-capita waste generation has increased by half during the same time period. In 1960 each American produced an average of about 2.7 pounds of municipal solid waste per day. That rate increased steadily until about 1990 when it reached 4.3 pounds per person per day. In the decade of the 1990s the rate of waste generation in this country stabilized at about 4.4 pounds per person per day. With the increasing population, though, the total amount of municipal solid waste generated in the United States has continued to grow, increasing about 250 percent from 88 million tons in 1960 to just over 221 million tons in 2000. As we will see later, the increased generation of "garbage" over that time span has produced a problem in disposal much greater than 250 percent!

Much of this MSW is generated by households, but a major source of municipal solid waste is industry. This is an entirely different category from industrial solid waste that we discussed earlier. Think of the garbage generated by a large factory: paper, metal scraps,

wood scraps, cans, plastic, and more. Imagine the massive amounts of solid wastes that might be generated in such a setting. Imagine "taking out the trash" for a factory like this one. For another example consider a food service business such as McDonald's® Corporation. McDonald's has shown responsible environmental citizenship over the years by using recyclable products and minimizing waste, but regardless of how careful managers are, such a commercial operation cannot avoid generating huge quantities of solid waste. In fact, according to the Environmental Protection Agency, a typical McDonald's restaurant produces about 230 pounds of solid waste per day, with corrugated paper shipping boxes and food waste each contributing just over a third of the total. The remaining third consists of plastic wrappings, paper wrappers, napkins, polystyrene cups, and customer waste. (See Figure 9-3.)

The total waste stream includes both business/industrial waste and household waste. (See Figure 9-4.) If we look at the composition of the total waste stream, the biggest contributor to the MSW problem is still paper (39.2 percent) followed by yard waste (14.3 percent). Those two components make up over half of the MSW generated in the United States each year, with all other materials making up just over 46 percent of the total.

How do we dispose of all that municipal solid waste? The picture is getting a little brighter in this regard. In 1960, we recycled 6.4 percent of our MSW. By 2000 that recovery rate had gone up to 30 percent. That may not sound like much of an improvement, but if you think of the total amount of MSW generated, that means we are recycling about 66.3 million tons in 2000 compared to 5.6 million tons in 1960. Again, when we discuss the changed requirements for landfills between

WASTE GENERATION RATES — 1960 TO 2000

FIGURE 9-3 Waste generation rates, 1960–2000. (*Source:* U.S. Environmental Protection Agency, 1996, *Municipal Solid Waste Handbook,* Washington, DC)

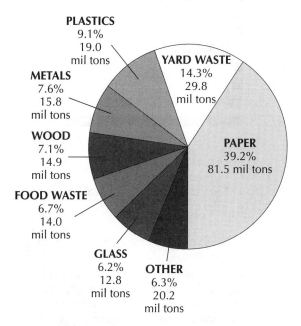

1995 TOTAL WASTE GENERATION — 208 MIL. TONS (before recycling)

- PLASTICS 9.1% 19.0 mil tons
- YARD WASTE 14.3% 29.8 mil tons
- METALS 7.6% 15.8 mil tons
- WOOD 7.1% 14.9 mil tons
- PAPER 39.2% 81.5 mil tons
- FOOD WASTE 6.7% 14.0 mil tons
- GLASS 6.2% 12.8 mil tons
- OTHER 6.3% 20.2 mil tons

FIGURE 9-4 1995 total waste generation, 208 million tons (before recycling). (*Source:* U.S. Environmental Protection Agency, 1996, *Municipal Solid Waste Handbook,* Washington, DC)

1960 and 2000, the importance of that change will become very clear. (See Figure 9-5.)

Where do we stand internationally? The United States is the biggest culprit in terms of trash generation. Figure 9-6 shows that the United States produced not only the largest total amount of MSW among the major industrialized nations but also produced the largest amount on a per capita basis.

The good news is that for the same year (1995), the United States also led the world in recycling. (See Figure 9-7.)

TYPES OF SOLID WASTE

There are three general types of solid waste: hazardous, nonhazardous, and radioactive. The state of Pennsylvania's definition of **hazardous waste** is typical:

> Hazardous wastes are wastes that, in sufficient quantities and concentrations, pose a threat to human life, human health, or the environment when improperly stored, transported, treated or disposed. In regulating hazardous waste, Pennsylvania uses a

WASTE MANAGEMENT PRACTICES, 1960-2000
(as a percentage of generation)

	1960	1970	1980	1990	1995	2000
GENERATION	100%	100%	100%	100%	100%	100%
RECOVERY FOR RECYCLING/COMPOSTING	6.4%	6.6%	9.6%	17.2%	27.0%	30.0%
DISCARDS AFTER RECOVERY	93.6%	93.4%	90.4%	82.8%	73.0%	70.0%
COMBUSTION	30.6%	20.7%	9.0%	16.2%	16.1%	16.2%
DISCARDS TO LANDFILL	63.0%	72.6%	81.4%	66.7%	56.9%	53.7%

FIGURE 9-5 Waste management practices, 1996–2000 (as a percentage of generation). (*Source:* U.S. Environmental Protection Agency, 1996, *Municipal Solid Waste Handbook,* Washington, DC)

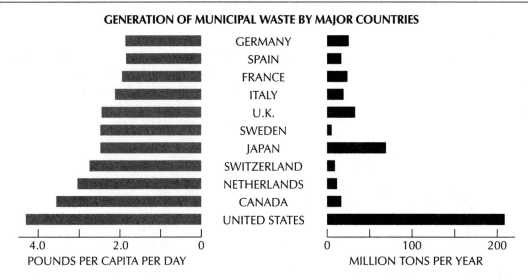

GENERATION OF MUNICIPAL WASTE BY MAJOR COUNTRIES

FIGURE 9-6 Generation of municipal waste by major countries. (*Source:* U.S. Environmental Protection Agency, 1996, *Municipal Solid Waste Handbook,* Washington, DC)

federal list of over 600 specific wastes. Other wastes are designated "hazardous" if they contain any of the following characteristics:

1. Ignitable—combustible under certain conditions
2. Corrosive—highly acidic, basic and/or capable of corroding metal
3. Reactive—unstable under normal conditions and capable of creating explosions and/or toxic fumes, gases, and vapors when mixed with water

4. Toxic—harmful or fatal when ingested or absorbed
5. Mixtures of hazardous and nonhazardous waste are also labeled hazardous. The hazardous waste designation does not include low-level radioactive waste, which is covered under separate state and federal rules.

(http://www.dep.state.pa.us/dep/
deputate/airwaste/wm/hw/
Facts/FS1961.htm)

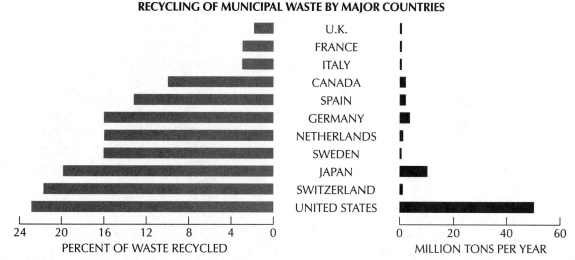

FIGURE 9-7 Recycling of municipal waste by major countries. (*Source:* U.S. Environmental Protection Agency, 1996, *Municipal Solid Waste Handbook,* Washington, DC)

According to the University of Maryland, **radioactive waste** is defined as follows:

Radioactive Waste—any waste that contains, or is contaminated with any radioactive material. This includes liquids, solids, animal carcasses and excreta, used scintillation vials/cocktails, etc. [*sic*] A radioactive waste must *not* be disposed of as regular waste. Non-Radioactive waste should not be disposed of as radioactive waste.

Mixed Waste—radioactive waste that also has the characteristics of a hazardous waste as defined by the State of Maryland or the EPA. There are several chemicals which are specifically regulated by the State and the EPA as hazardous waste, and many more which possess the characteristics of a hazardous waste because they are corrosive, reactive, toxic, or otherwise potentially harmful to the environment. There are currently no permitted disposal options for most mixed waste, therefore care must be taken to avoid the generation of these waste. Plans for the proper management of these materials should be reviewed with the Radiation Safety Office and the Hazardous Waste Division in the early stages of an experiment.

Mixed Wastes include contaminated lead pigs or other lead shielding and radioactively contaminated organic liquid waste that contains a regulated chemical. One common example is liquid scintillation media containing toluene or xylene.

Dry Solid Waste—Gloves, paper, plastic, glass, metal or other solids that contains [*sic*] radioactive material or is [*sic*] contaminated with radioactive material.

Liquid Waste—Organic and aqueous liquids containing radioactive material. Liquid waste must be segregated from solid waste. Organic and aqueous liquids must be stored separately in appropriate containers.

Liquid Scintillation Vial Waste—Capped scintillation vials containing either organic or biodegradable scintillation fluid contaminated with radioactive material.

Radioactive Sharps—Radioactively contaminated syringe, needle, surgical instrument or other article which has the potential to cut or puncture human skin.

(http://www.ehs.umaryland.edu/ HazardousWaste/whatisrw.htm)

In general, **nonhazardous waste** is municipal waste and industrial waste that does not meet the definitions of hazardous or radioactive waste.

DISPOSAL OF SOLID WASTE

For many centuries, human societies have managed their solid waste by burying it. In fact, archeologists, scientists who study human history, freely admit that archeology is largely the science of studying "trash" because most ancient artifacts are found either in old burial sites or in old trash disposal sites.

Until the 1950s, people generally believed that once trash was buried, it was safely disposed of. The assumption was that buried waste would eventually decompose under the soil into harmless materials and that any harmful chemicals or bacteria would be attenuated (purified and made harmless) by the soil (soil particles, bacteria, and fungi). As you will see later in this chapter, the soil does have a number of characteristics that led scientists to that belief. In addition, scientific instruments of that time were not capable of detecting very minute concentrations of contaminants in water. Thus, damage being done by waste decomposition and drainage to the groundwater was not easily detected.

As long as that belief prevailed, disposal of solid waste was a simple process. We simply dug a hole, placed the trash in it, and covered the hole with soil. For an individual home, the waste might be burned and the residue buried in a small ditch or hole. To dispose of the solid waste from a large city simply meant hauling it to a rural area, digging a huge hole with bulldozers, dumping the trash, pushing a layer of soil over it, and packing the whole mess down by driving the bulldozers over the top. For a city, the waste might be buried in a huge open-air pit and eventually covered with soil. As rain percolated through the open trash or through the soil covering, people assumed that the waste decayed and that the soil underneath took out any harmful chemicals. The term "dump" or "trash dump" is still very commonly used to describe solid waste disposal sites, although sanitary landfill is probably a more appropriate description.

As late as the mid-1960s, that was the way almost all solid waste disposal was done in the United States. In many parts of the world and in some parts of the United States, solid waste disposal is still accomplished using that method. The process was quick, cheap, and easy to manage. Many smaller towns and rural areas used even simpler and cheaper methods, like dumping the trash down the side of a hill or in a small natural valley and allowing it to decompose in the open.

Starting in the mid-1950s, several very important studies revealed that significant amounts of water **percolate** through landfills. That water picks up and transports minute particles of contaminants and biological agents such as bacteria and viruses and dissolves water-soluble chemicals. The water enters the landfill, mostly from rain and melting snow, and once it moves through the buried materials, it is known as **leachate**. From there, the leachate percolates downward until it either flows out of the landfill as seepage or enters the groundwater. Once contaminants are in the groundwater, they remain there indefinitely or until the water is removed for human use (see Chapter 13). Leachate from buried solid waste is one important reason that well water must be tested periodically for purity. (See Figure 9-8.)

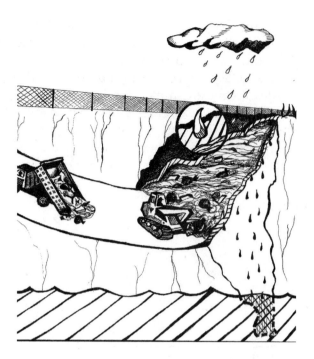

FIGURE 9-8 Rainfall percolates through the soil cover, collects in the buried waste, dissolves soluble materials, and seeps downward toward the groundwater as leachate. (*Courtesy of Sarah Michelle Williams*)

WHAT IS A LANDFILL?

In many parts of the world today, MSW is disposed of on any landform: ditches, open pits, hillsides, or even open areas. It may or may not be burned, and it may or may not be covered with a soil layer. A **landfill** is an open area into which garbage is placed, to be covered by a layer of some other material, typically soil. Until the 1950s in the United States, landfills were typically in open fields or along a slope. Earth-moving equipment such as bulldozers might scoop out a long trench or a large flat area. Waste would be deposited in the opening. Often the waste would be burned to decrease

the volume of materials. The soil that had been removed would then be pushed back over the partially burned garbage to bury it. Often heavy equipment would drive over the top of the covering to pack down the material as much as possible. With research that demonstrated the damage to the environment from leachate as well as other environmental concerns came much more restrictive regulatory requirements and much more sophisticated landfill designs.

LANDFILL DESIGN

There are two basic types of landfills today in industrialized nations such as the United States: natural attenuation and containment. Let us examine both types in some detail.

The **natural attenuation landfill** is designed to hold the waste material in a covered area, allow natural percolation of precipitation to pass through the waste and flow through the underlying soil and rocks. The expectation in a natural attenuation landfill is that the leachate will be attenuated (neutralized) by the microorganisms and soil particles. Although research has clearly shown that leachate can damage the groundwater and cause other environmental problems, certain kinds of MSW can be safely disposed of in natural attenuation landfills. As you will recall, municipal solid waste can be of three kinds: hazardous, radioactive, and nonhazardous. Such items as paper products and yard waste are generally nonhazardous and can be safely disposed of in a natural attenuation landfill. For managing such waste, the cost is substantially lower than for more dangerous materials.

Regardless of the type of landfill, the basic shapes will be similar. Landfills may be of three basic shapes: at grade, canyon fill, or trench fill.

An at-grade landfill is used when waste is placed at or nearly at the normal soil level. The waste is then covered with a layer of material, typically soil. A canyon fill landfill involves pushing the waste over the side of a hill or slope, then pushing a covering layer over it. In both the canyon fill and at-grade landfill, a berm of soil is constructed to contain the waste first; the waste is then placed in the holding area; and last, a final cover is pushed over the waste. The most common kind is a trench fill landfill, in which a trench is formed by removing the soil, the waste is placed in the trench, and the soil that had been removed is used to form the final cover. (See Figure 9-9.)

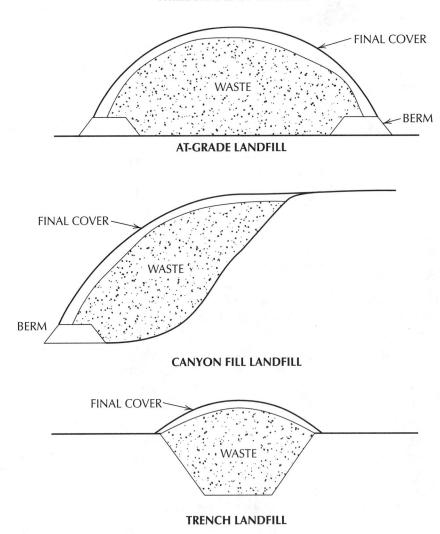

THREE SHAPES OF LANDFILL

FIGURE 9-9 Three shapes of landfill.

NATURAL ATTENUATION LANDFILLS

How does **attenuation** work? Six natural mechanisms occur in the soil as leachate percolates through it that attenuate the contaminants: adsorption, biological removal, ion exchange, dilution, filtration, and chemical precipitation.

Adsorption is the process by which the molecules of a chemical adhere to the surface of some other material. Clay is particularly important in this process. As Chapter 4 explains, clay is made up of extremely small particles. That means that a given volume or weight of clay has a huge surface area compared to the same volume or weight of sand or gravel. The increased surface area, as well as other chemical and physical properties of clay, mean that much of the contaminates in leachate simply "sticks to" clay particles so tightly that they are permanently removed from the percolating water.

Biological removal takes place when bacteria, fungi, or other soil microorganisms break down or absorb the leachate constituents. Soil microorganisms use for food some of the materials we would consider to be contaminants. Other chemicals are converted into materials the microorganisms need to survive. Still other materials are simply oxidized by the microorganisms into harmless products. In particular, carbonaceous wastes are broken down by microorganisms and released as carbon dioxide that then becomes carbonic acid, which then breaks down still more of the harmful constituents of the leachate.

As you learned in basic chemistry, ions are charged molecules or atoms. Also as was noted earlier, clay particles are very small. They are also chemically active and so are very important in ion exchange. This process neutralizes part of the constituents in leachate by changing the molecular structure of the ions involved in the exchange.

Dilution simply means that the concentration of the leachate is decreased by mixing it with large quantities of water. Too heavy a concentration of even the most harmless of chemicals may be dangerous. You like a little table salt ($NaCl$) in your food, but too much salt makes the food inedible. Too much salt in your diet can even be dangerous.

Filtration involves the physical removal of solid constituents from the leachate by trapping them in pores in the soil. This process can remove only small quantities of materials, because the pores eventually become clogged. To illustrate this process, place a kitchen sieve or paper filter over a container. Pour slightly muddy water through the filter and you will see that the soil particles are captured. Continue the process for a while, and the filter will become clogged so that the water will stop passing through it and will start to pour over the top. For that reason, filtration is of only limited importance in the attenuation process.

Precipitation is the process of a phase change in the leachate. To understand what a phase change is, let us consider a familiar example. A phase change occurs in the atmosphere when a gas, such as water vapor, changes to a liquid, such as rain, or a solid, such as snow. In the leachate, a phase change occures when a dissolved mineral crystallizes. The liquid contaminant in the leachate then becomes a solid material that is removed from the leachate by adsorption or to a lesser extent filtration.

CONTAMINANT LANDFILLS

A **containment landfill** is designed to minimize the seepage of leachate into the surrounding soil and groundwater. The term "minimize seepage" is key in this regard. Ordinary landfills will never provide containment to the point that seepage is prevented from entering the environment. This is the type of landfill that is required for any solid waste that may contain hazardous waste.

Total containment is necessary for certain very hazardous and radioactive waste. The discussion in this section DOES NOT APPLY to total containment waste management. Such very hazardous materials are disposed of in deep shafts, salt beds, and other extreme locations.

A natural attenuation landfill can be made into a containment landfill by the addition of some sort of **liner** to restrict water from percolating in a natural way. The liner can be natural, such as a very heavy, compacted clay or bentonite soil. It can also be constructed by using a synthetic material such as butyl rubber, chlorinated polyethelyne (CPE), chlorosulfonated polyethelyne (CSPE), ethylene-propolene rubber (EPDM), low density polyethylene (LDPE), high density polyethylene (HDPE), polyvinyl chloride (PVC), or several materials. In theory any liner will eventually allow some leachate to seep through, but the amount of seepage can be cut to near zero in several ways.

The first way to minimize seepage is by using multiple layers. A layer of very fine, compacted clay or bentonite at the bottom will cut the seepage down dramatically. If a PVC or other synthetic layer is applied over the clay, the synthetic material will stop most of the seepage. The clay will have to restrict only the seepage that gets through the synthetic layer. Naturally, the synthetic material will have to be covered by a layer of clay, bentonite, or even sand to prevent damage from the large equipment used to move the waste and to apply the final cover. In general, the more layers, the less seepage will occur over time.

In addition to multiple layering, leachate collection pipes can be used to capture the seepage collecting above the liners and drain it away. Typically that leachate is treated in a wastewater treatment plant similar to a sewerage treatment plant, until it is harmless and can be disposed of in some environmentally safe way. In some cases, not only multiple liners but also multiple leachate collection systems are used. (See Figure 9-10.)

Why should we care how landfills are constructed? It should be very obvious by now that the least expensive way to dispose of municipal solid waste is simply to dump it on the ground and leave it there. Although that solution was acceptable for thousands of years, it is no longer a solution we can live with in modern society. The next cheapest way is to push it over a canyon wall and cover it with a layer of soil or bury it in a trench. Though that may be acceptable for limited kinds of solid waste, the mixed waste that comes from households, factories, and other businesses (in other words, almost all the municipal solid waste produced in this country) simply cannot be safely disposed of in natural attenuated landfills any longer. Containment landfills are the only acceptable method of disposing of the majority of municipal solid waste today and in the future. Containment landfills are very expensive to build, and the leachate must be captured in the drainage pipes and treated for 30 to 50 years before the landfill is safe to simply abandon.

WHY RECYCLING?

Given the obvious costs of solid waste disposal, the argument for **recycling** should be clear. We do not really recycle paper to save trees. As you can read in the chapters that deal with forestry, there is no shortage of trees for paper. As you will see from the chapter that deals with energy sources and use, there is no real shortage

A MULTIPLE LINED LANDFILL WITH A SINGLE LEACHATE COLLECTION SYSTEM

FINAL COVER,
COMPACTED CLAY

WASTE

SYNTHETIC
LINER

SAND DRAINAGE
BLANKET

BENTONITE SOIL
AMENDED LINER

COMPACTED
CLAY LINER

LEACHATE
COLLECTION PIPES

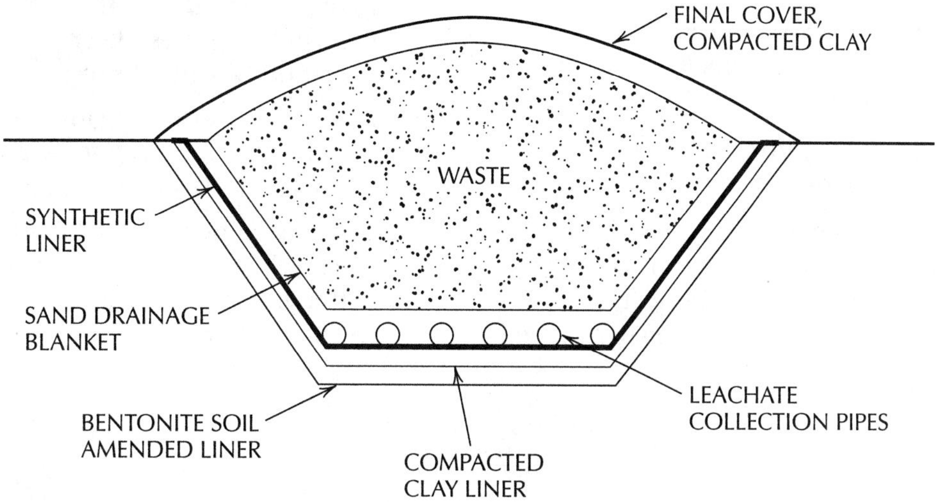

FIGURE 9-10 A multiple lined landfill with a single leachate collection system.

of energy sources. Given the undeniable fact that our natural resources are not really benefited in any substantial way by recycling, then why is recycling so important? The cost of solid waste disposal is the primary reason.

SUMMARY

There is no shortage of land on which landfills could be constructed. There is a severe shortage of communities in which the people are willing to allow for the construction of a landfill. Would you want a landfill in your backyard? Most people would not. Even when a politically acceptable location for a landfill can be found, the cost of constructing a multiple lined containment landfill is tremendous. Even after it is constructed, when the landfill is closed, the costs of capturing the leachate and treating it can go on for up to half a century.

DISCUSSION QUESTIONS

1. If we did not have landfills, what would you do with your household garbage?

2. Why do natural attenuation landfills work for some kinds of waste but not for others?

3. Why do so many local governments try to avoid constructing their own landfills and instead ship their municipal solid waste elsewhere for disposal?

4. Most people want to have adequate landfills available but "not in my backyard!" Why do people not want a landfill near their homes?

5. How do you construct a landfill if nobody wants one near their home?

SUGGESTED ACTIVITIES

1. Find and visit as a class a location in your community where people are dumping trash improperly.

2. Organize a cleanup campaign for a community park or other recreation facility.

3. Collect all the trash discarded at your home for a week. Lay it out on a large plastic sheet and inventory it to determine the kinds of waste materials included such as paper, food waste, containers, and so on. Weigh the various components and figure percentages of the total. Be sure to use rubber gloves when handling the trash and be careful not to get cut.

4. As a class project, organize an adopt-a-park program to keep some community area clean on an ongoing basis.

5. Invite an official from the local landfill or recycling center to speak to the class. Better yet, arrange a class field trip to visit the landfill or recycling center.

10 WETLAND PRESERVATION AND MANAGEMENT

OBJECTIVES

After reading this chapter, you should be able to

- explain what a wetland is

- describe the difference between on-site and off-site wetland identification procedures

- explain the three common types of wetlands

- list and explain the major causes of wetland areas

- explain the major wetland management practices

- list the major governmental programs and agencies regulating wetlands

TERMS TO KNOW		
hydrology	ecological wetland	bog
hydrophyte	marsh	prairie pothole
hydric soil	floodplain	vernal pool
jurisdictional wetland	swamp	natural wetland

Wetlands are a premier, underrated, and overlooked natural resource. They are home to birds, amphibians, waterfowl, small mammals, and various plants and mushrooms. The variety of life surrounding wetlands should be reason enough to realize their importance. However, wetlands have a major ecological role in controlling floods, acting as a filter for pollutants, adding to underground water sources, providing habitat for many species (most notably waterfowl and amphibians), and providing recreational use. According to the Environmental Protection Agency (EPA), more than one third of all threatened and endangered species live only in wetlands while one half of threatened and endangered species use wetlands at some point in their lives. (See Figure 10-1.)

WHAT ARE WETLANDS?

There are many types of wetlands that can be found in a variety of habitats across the United States. Depending on what group or individual you ask, you can get many different definitions of wetland. Even the "official" definition of a wetland varies because different governmental

FIGURE 10-1 Bicyclists on the Tidelands Trail in the San Francisco Bay National Wildlife Refuge, California. This saltwater wetland area serves as habitat for migratory birds and endangered species, and offers superb recreational activities. *(Photo courtesy of the United States Fish and Wildlife Service)*

FIGURE 10-2 These green pitcher plants in Little River Canyon, Alabama, are an example of hydrophytes. *(Photo courtesy of the United States Fish and Wildlife Service)*

agencies use several differing definitions, as you will read later. Farmers have many different ways of defining wetlands, depending on where they live and on what kinds of crops and livestock they produce. Environmentalists use still another set of definitions. Individuals' and groups' perspectives and agenda drive what they perceive to be a wetland.

Characteristics

Most experts agree that wetlands exhibit three characteristics. First, there will be a prolonged presence of water. An area whose **hydrology** (water characteristics) includes frequent saturation with free water is probably a wetland. Second, this prolonged presence of water affects how the soil develops and what plants will grow on the water-saturated soil. Plant types that are attracted to these growing conditions are generally called **hydrophytes**

(see Figure 10-2). A soil that is covered by hydrophytes is probably a wetland. Third, because the soil is subjected to frequent saturation and may have a permanent water table close to the surface, it develops into hydric soils. **Hydric soils** are soils that tend to be saturated with water most of the time. They are usually low in air content and are generally colored differently than other soils in the same region and may have a completely different soil structure. Hydric soils may be mottled with white or gray coloring or may be very yellow where you would expect a darker coloration. The structure may be sticky and wet rather than granular. An area with hydric soils is probably a wetland.

Two Definitions

The first definition is the jurisdictional, or legal, definition. The U.S. Army Corps of Engineers defines wetland as *an area that has frequent flooding or saturation, is covered by hydrophytes, and includes hydric soils.* According to this definition, wetlands exhibit all three characteristics previously noted: a wet hydrology, hydrophytes, and hydric soils. (See Figure 10-3.) The Corps is responsible for **jurisdictional wetlands**.

For an ecological standpoint, some lands may be wetlands, but they are not jurisdictional wetlands. As noted before, there are many different ecological definitions of wetlands. For our purposes, we will use the United States Fish and Wildlife Service (USFWS) definition because it is broad and this branch of government maintains a number of wetland refuges across the country. Wetlands as defined by the United States Fish and Wildlife Service are *lands transitional between terrestrial and aquatic systems where the water table is usually at or near the surface or the land is covered by shallow water.* For purposes of this classifica-

FIGURE 10-3 This photograph of the Okefenokee National Wildlife Refuge in Georgia (known locally as the Okefenokee Swamp) shows hydrophytic plants as well as permanent shallow water cover. In wet seasons the area covered by water expands. If you were to dig into the soil, you would find hydric soils. Because it has all three characteristics of a wetland, this is a jurisdictional wetland. *(Photo courtesy of the United States Fish and Wildlife Service)*

tion, wetlands must have one or more of the following three attributes:

1. At least periodically, the land supports predominantly hydrophytes

2. The substrate is predominantly undrained hydric soil

3. The substrate is nonsoil and is saturated with water or covered by shallow water at some time during the growing season each year.

As you can see, the ecological definition used by the United States Fish and Wildlife Service is much broader that the legal definition used by the Corps of Engineers. According to the United States Fish and Wildlife Service definition, **ecological wetlands** may or may not have all three characteristics of the jurisdictional wetland definition. As an example, the Corp of Engineers would not consider a mudflat or coral reef as a wetland, but the Fish and Wildlife Service would.

HISTORY OF WETLANDS IN THE UNITED STATES

Since the beginning of government in the United States, wetland areas have been perceived as harsh, disgusting places everyone should avoid. Wetlands were considered a problem rather than an asset until the last half of the 1900s.

A perfect example of the government's position on wetlands occurred in the 1770s. The District of Columbia was partially covered by swampland, which the government decided to fill in to build the nation's capital.

In another example, the whole southern part of Louisiana is designated as wetlands, but the citizens of Louisiana felt that building a city on a water crossroads site was important. New Orleans is that city, and it is still surrounded by various types of wetlands (swamps, marshes, rivers, and lakes.) The fact that New Orleans was built on wetlands and is still surrounded by wetlands leads to constant problems with flooding.

The very names of some wetlands show the early Americans' attitude: The Great Dismal Swamp of southeastern Virginia and northeastern North Carolina is an example. We would never name a wetland area "The Great Dismal Swamp" today. (See Figure 10-4.) On the other hand, Americans of 1700s would hardly consider the Everglades a national treasure worthy of safeguarding. It is important to understand that our social values change over time; sometimes they change drastically. Until recent years, governmental programs were aimed at "reclaiming" swamps rather than at protecting wetlands.

Prior to the 1970s the federal government encouraged the idea that wetlands are unhealthy, unproductive areas that should be avoided. The governmentally authorized and subsidized draining of wetlands began in 1849 with the passage of the Swamp Lands Act for the state of Louisiana, which was later broadened to cover the whole of the United States. This act gave the states permission to fill in and change areas that "were unfit for cultivation." For over a century, the process of draining wetlands was thought of as the recovery of swampland for productive use.

A recent article in the *Washington Post*'s Parade section (May 21, 2000) recommended that during "bug season," the majority of the summer months, people should stay away from all areas of standing water, such as swamps and small pools. These areas are to be avoided due to the fact that mosquitoes and other nuisance bugs breed in stagnant water. Each year more foreign insect-borne diseases seem to appear here in the United States, due to increased foreign travel and trade. Traditional insect-borne diseases include malaria, encephalitis, and many others. The newest fear in this country revolves around a disease that killed a handful of people in New York the summer of 1999 and is called West Nile Fever.

In 1972 the federal government changed its policy about wetlands and their importance and took steps to protect and restore wetlands by requiring permits with Section 404 of the Clean Water Act. This not only helped to protect current wetlands from being filled but also allowed for the restoration of degraded wetlands.

Under the Wetland Conservation provision (commonly known as the Swampbuster) of both the 1985 and 1990 farm bills, farmers are required to protect the wetlands on their farms and ranches in order to be eligible for USDA farm program benefits. According to the Swampbuster guidelines, producers will not be eligible for benefits if they plant an agricultural crop on a wetland that was converted by drainage, leveling, or any other means after December 23, 1985, or if they convert a wetland to make agricultural production possible after November 28, 1990.

WETLAND IDENTIFICATION

Because of the wide variety of governmental legislation that governs wetlands, it is important that each site be correctly identified. How are wetland sites identified? The two most common techniques for wetland identification are off-site identification and on-site identification methods.

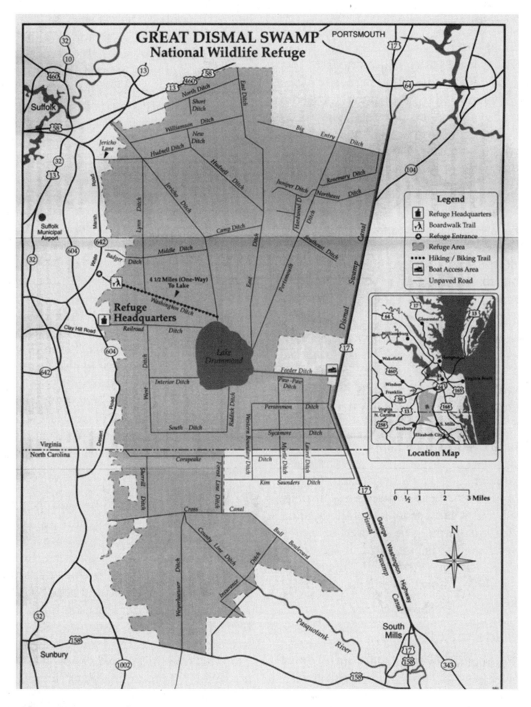

FIGURE 10-4 The Great Dismal Swamp on the border of Virginia and North Carolina. *(Map courtesy of the United States Fish and Wildlife Service)*

Off-Site Identification

Off-site inspection involves checking maps and wetland inventories maintained by the relevant federal agencies. Though off-site identification will not determine the exact size and location of a wetland, it can be used as a screening device to find possible wetland locations. Three principle resources are available to the landowner when researching for wetland sites.

1. The U.S. Fish and Wildlife Service produces the National Wetlands Inventory (NWI). Following the enactment of the Emergency Wetland Resources Act of 1986, the service was required to map wetland areas every 10 years. The NWI is concerned with wetlands and open water in the United States.

2. National Resource Conservation Service Soil Survey List. The NRCS maintains a list of hydric soils in the county-by-county survey maps. Individual landowners can visit the local office to view the maps for their land.

3. U.S. Geological Topography Maps. These maps look at vegetative covers, surface characteristics, bogs, and marshes. The map uses specific terminologies. The map spells out what is growing on the land or if water is standing in the area. By using several map years, a historical record of the area can be determined. This will be helpful if a restoration program is undertaken.

On-Site Identification

The location of the wetland will control what agency is responsible for the site analysis. (See Figure 10-5.) If the site in question concerns dredging, filling, or discharge into a suspected wetland, the U.S. Army Corps of Engineers will be called upon. If the site is close to a lake or inland water, the U.S. Fish and Wildlife Service will be called. The Natural Resources Conservation Service (NRCS) from the United States Department of Agriculture (USDA) will be needed to identify wetlands on agricultural lands or nonagricultural lands that border on agricultural lands. The NRCS determines the use of the wetland under the Swampbuster guidelines. They manage the wetland use for agricultural crops as pastureland, orchards, cranberries, and rice as well as for aquaculture purposes such as crawfish fields.

An individual who knows and understands plant identification and soil science should complete an on-site inspection. That person should determine if hydrophytic vegetation is present. He or she will determine what plant species are present and identify the plants that tolerate wetland conditions. The condition of the plants will also be an indicator of wetland status. Swollen tree bases, trees with multiple trunks having adventitious roots (roots that branch above the ground and help provide support to the plant), and shallow roots, or trees with leaves that are various shaped are all examples of wetland adaptations for plants.

The soil will need to be examined. Hydric soils are often highly organic soils. A large amount of black humus indicates a less decomposed plant life, which shows that the soil has been subjected to long periods of saturation by water. The soils also could be poorly drained with the water table less than 6 inches from the surface. It is not uncommon for the soil to emit a sulfur odor and smell like rotten eggs. Closer examination would show iron pockets (rustlike) with gray to black looking color spots, called mottling.

The last part of the on-site inspection would deal with the hydrology or water sup-

FIGURE 10-5 This photograph of Rock Creek Park in Maryland would be very easy to classify as a wetland because it has shallow standing water and numerous hydrophytes directly visible, and an accumulation of organic matter in the soil surface. The soil is almost certainly hydric. This area also appears in the National Wetlands Inventory, so it could be classified either on-site or off-site. *(Photo courtesy of Susan Aksamit)*

ply of the site. Standing water is definitely an indicator of a possible wetland location. A closer examination may need to be conducted. The inspection might be conducted during the dry part of the year. Further field investigation could show water-stained leaves, water-borne sediments or plants, watermarks on trees and other fixed objects, or observation of soil saturated areas.

By looking at the plants present, the condition and development of the soil, and the water makeup of the area, a technician can determine if the site should be classified as a wetland area.

TYPES OF WETLANDS

As mentioned earlier, there are many types of wetlands. We will discuss marshes, floodplains, lakes and ponds, rivers and streams, swamps, bogs, prairie potholes, and vernal pools. Keep in mind that there are many more depending on how you categorize a wetland.

Marshes

Marshes are found throughout the United States. There are freshwater, saltwater, and tidal marshes. The largest area of freshwater marshes is found in the Dakotas and Minnesota. The majority of saltwater and tidal marshes are found along the East Coast and in the Gulf of Mexico. Marshes do not rely on rainfall for their water supply, and soft-stemmed plants are the dominant plant type. Examples of marsh plants are reeds, sedges, rushes, grasses, and cat-tails. Other plants that may be found in marshes are arrowhead, lizard's tail, water parsnip, common pipewort, spider and swamp lily, sweetflag, marsh marigold, tall meadow rue, Virginia meadow beauty, thread-leaved sundew, Turk's-cap lily, marsh Saint-John's-wort, pickerelweed, marsh fern, marsh cinquefoil, arrow arum, water pennywort, and the carnivorous Venus flytrap. The Florida Everglades is a well-known marsh. (See Figure 10-6.) Urbanization, conversion of wetland to agricultural production, and use of underground water supplies for the cities in South Florida have all combined to cause serious degradation of the Everglades Wetland ecosystem.

FIGURE 10-6 This photograph of the Everglades National Park shows dense marsh grasses in a freshwater marsh ecosystem. You can see from the cranes standing in the foreground that the water is only inches deep. *(Photo courtesy of University of Florida, Institute of Food and Agricultural Sciences / © 2000 IFAS. Photo by Milt Putnam.)*

Floodplains

Floodplains are those areas that border rivers, lakes, and streams and that are flooded periodically. The common trees found in floodplains are American sycamore, paw paw, black willow, American elm, cottonwood, bald cypress, water tupelo, several oak species, and various birch species. One of the more common plants found in floodplains is Virginia bluebells. Another less common plant is "Phlox" (*Phlox divaricata*). The most well-known floodplain is the Mississippi River floodplain. The Mississippi floods sporadically, therefore leaving some warning for those residents who choose to live in a well-known floodplain. The last major Mississippi flood occurred in the summer of 1993. That flood was judged to be a 500-year flood—meaning that we would expect such an extreme flood only once every 500 years. Many towns were evacuated permanently, while in other places, the former residents chose to build their houses on higher ground. Those who live in a floodplain take a great risk because floodplains exist for a reason, and no matter how many dams and levees the Army Corps of Engineers builds, the floods will inevitably reoccur from time to time. (See Figure 10-7.)

FIGURE 10-7 This farm near Winfield Kansas lies in a floodplain that is protected by a levee. In 1998, the levee was breached by floodwater and this damage resulted. *(Photo courtesy of the Federal Emergency Management Agency [commonly known as FEMA])*

Ponds

Not all lakes and ponds should be considered wetlands although some do have wetlands present around their edges. Some ponds may even turn into marshes if they become shallow enough. In that process, cattails will grow around the edges as water lilies and duckweed may take over the water's surface. Lakes and ponds are home to an abundance of life that will usually draw people seeking outdoor recreation. Examples include, but are not limited to, a variety of fish species, frogs, salamanders, aquatic insects, birds, waterfowl, and small mammals, such as beaver and muskrat. A variety of plant species may also be found along the shores, including various species of alder and birch, blueberry, and winterberry. Lakes and ponds are permanent bodies of water whereas a prairie pothole is not. Potholes will be discussed later.

Rivers and Streams

Once again, not all rivers and streams are considered wetlands. However, some do move slowly enough that certain types of vegetation start to take over. Sometimes this process leads to a larger area that turns into a marsh, bog, or forested wetland. A beaver may build a dam and cause a small pond, which will also change the characteristics of the river or stream. Rivers and streams also harbor a great variety of plants and animals. To the list of animals found in lakes and ponds may be added various snake species, as well as shorebirds. As for plants, you may find watercress, turtlehead, cardinal flower, globe mallow, checker mallow, water hyacinth, monkey flower, wood nettle, and red osier dogwood. The slower and warmer a river or stream becomes, the more prolific the plant life becomes as well.

Swamps

Now we come to the wetland that is most synonymous with dark, eerie, and scary. There is even a Dr. Pepper™ television commercial that takes place in a swamp where a man jumps out of the boat to save the Dr. Pepper and is eaten by an alligator. (See Figure 10-8.) Unfortunately, the media has also added to the population's generally adverse reaction towards swamps. Movies and television shows such as *Swamp Thing* did not help the swamp's image either.

There are various types of **swamps**: cypress, southern bottomland hardwood, shrub, and northern. Each is different in the type of vegetation found there. However, they all have the same general characteristic

FIGURE 10-8 *Many people associate alligators with swamps, such as this one in Mississippi. Alligators are found in wetlands in many parts of the southeastern United States. They can live 30 or more years. They can grow up to 14 feet and weigh as much as 1,000 pounds. At one time, people had killed so many alligators for their skins to make purses, boots, and shoes, that they were placed on the U.S. Endangered and Threatened Wildlife and Plants List. Conservation measures have resulted in the recovery of the alligator to the point that alligators are becoming a problem in many areas. (Photo courtesy of the United States Fish and Wildlife Service)*

of standing water with trees or shrubs growing in the water. This is very stagnant water that tends to be dark and nontranslucent. Along with the shadow of the trees, the swamp leans toward being dark and eerie. Great Dismal and Big Cypress Swamps are the well-known swamps in the United States. These swamps, however, are highly productive areas. They represent small remnants of earlier times when the areas surrounding river systems were swamps. Some areas have giant trees that are untouched by foresters. There are also animals found in swamps that are rarely found in other areas; these animals remain unchanged due to the lack of human interaction. The cougar, alligator, red wolf, alligator gar, ivory-billed woodpecker, and sirens (similar to salamanders), are examples. There are many plant species that thrive in swamps: spicebush, cowslip, swamp buttercup, skunk cabbage, elderberry, swamp dewberry, showy lady's slipper, butterfly orchid, swamp rose, hibiscus, red iris, spotted touch-me-not, great hedge nettle, spotted joe-pye weed, large purple fringed orchid, blue flag, cinnamon fern, and jack-in-the-pulpit.

Bogs

Bogs are areas that are very damp, usually with evergreens present, with a floor covered with moss and peat. This makes for a spongy walk, if you are lucky enough to not get stuck in the mud. Most bogs are found in the Northeast and in Michigan, Minnesota, and Wisconsin. If you were to take a hike on the Appalachian Trail through the eastern states, many boardwalks or logs have been placed by trail clubs to traverse through these areas. In other areas you may be forced to pick your way through bogs. That is often quite difficult when you step in the wrong place and almost lose your shoe to the suction of the muck.

Bogs are known for a few plants that are found nowhere else, such as cranberries and carnivorous pitcher plants. Typically, you will find only acid-loving plants in bogs, due to the high acidic level produced by the leaching of nutrients and slow fermentation process of organic matter. Small mammals such as shrews and voles are typically found in and near bogs. Trumpets, cotton grass, round-leaved sundew, leatherleaf, bog rein orchid, swamp pink, and sheep laurel are among the few plant species you will find in a bog.

Prairie Potholes

Prairie potholes rely on periodic rainfall for their water supply. They are usually full in the spring and early summer before water levels start to drop off and the potholes start to disappear for the rest of the year. Prairie potholes are found mainly in North Dakota, South Dakota, Minnesota, and Nebraska. They are best known for providing habitat for migrating waterfowl. Unfortunately, many waterfowl species rely on this critical habitat that had declined steadily until recent years. Prairie potholes have always been in the middle of debates between landowners and naturalists. Recently, many people have come to understand the importance of potholes as critical waterfowl habitat, but not before many potholes were reclaimed for agricultural production and lost forever.

Vernal Pools

Vernal pools are a special type of wetland that may last for only a few months each year. Like prairie potholes, vernal pools rely on periodic rainfall to form in the spring. They disappear in early summer. These pools become a haven for amphibian species because they lack predators. Many species of amphibians rely on these pools for breeding grounds, including the tiger, spotted and marbled salamanders, spring peepers (a frog), and wood frogs to name just a few. A very interesting animal, which also uses these pools for its entire life cycle of about two weeks, is the fairy shrimp. According to Jim Petranka, a biologist at the University of North Carolina at Asheville, "About one-third of all amphibians in the eastern U.S. are strongly dependent on vernal ponds as breeding sites."

STATUS OF WETLANDS IN THE UNITED STATES

It is estimated there were over 200 million acres of wetlands in the United States in the 1600s. According to the Environmental Protection Agency (EPA) about one half of that area remains in wetland today. Many states have lost far more than half of their original wetlands. Many inland freshwater wetlands in the southern states were drained for agricultural or urban use.

What are the major causes of the loss of wetlands? The main causes of wetland loss can be attributed to the following:

- urbanization
- industry
- agriculture
- timber harvest
- mining operations

Urbanization

As our cities grow out toward rural areas, the construction of roads, parking lots, and buildings move and shift rainfall flows. The runoff from these areas moves fertilizers, wastes, organic matter, and road salts. These compounds cause a change in the water quality even if structures are not on a wetland area. Because wetlands were considered low-value land for so long, roads and bridges were constructed through wetlands rather than around them. Road runoff becomes a problem to the maintenance of the wetland ecosystem. Many times non-native plants and animals invade the area and disrupt the normal succession of the wetland. (See Figure 10-9.) The construction may not even be in the wetland itself, but in the watershed area draining into the wetland.

Industry

Industry can affect the wetland habitat. As industry calls for increased use of water, the hydrology of the area can be changed. Industrial manufacturing plants may discharge warmed water (thermal pollution) or change the acidity levels of the water. Pollutants from abandoned industry sites may significantly damage wetland areas.

Agriculture

Agriculture has been a major factor in the loss of wetland areas. It was common for drainage devices to be installed to "drain" wetland areas, making them available for tillage operations. As we discussed earlier, the U.S. Congress enacted the Swampbuster provision with the 1985 Farm Bill. This program has prevented the conversion of wetland to agricultural production. The Act stated that agriculture production could not alter the water hydrology, water quality, and the plant and animal life of an area. Areas that provide habitat for waterfowl were also protected. Farmers were encouraged to allow prior wetland areas to revert to their original state. Since the Swampbuster program

FIGURE 10-9 Many wetlands lie near urban areas. This is the John Heinz National Wildlife Refuge at Tinicum, near Philadelphia. The refuge offers trails, fishing, and environmental education. *(Photo courtesy of the United States Fish and Wildlife Service)*

started, duck and migratory bird populations have increased by over 25 percent.

Farmers were also encouraged not to graze wetland areas. Wastes from cattle as well as their physical destruction of wildlife homes damage wetland areas. Programs were established that help farmers manage the grazing. It was found that the grazing could coexist with wetlands, therefore allowing a benefit to farmers as well as protecting the wetland area.

Timber Harvest

Most wetland loss from timber harvesting is only temporary. The destruction usually lasts

for three to five years. During the harvesting, new timber roads are constructed and rutting of existing roadways may impair wetland quality. The use of heavy equipment in the wetland area will affect the water quality, soil structure, and development. Through careful management practices, timber harvest can be accomplished with a limited damage to the wetland.

Mining

The most common mining practice that affects wetlands is the mining of peat moss. Peat moss is used in the horticultural area as a planting medium for greenhouse production.

The mining operation involves the removal of vegetation, the drainage of the area, and the extraction of the peat. The wetland is totally transformed from a wetland to an open water area. The area surrounding the wetland is also changed, thus affecting the habitat of many species of wildlife.

WETLAND PRESERVATION

The lower 48 states have lost over half their original wetlands. According to the EPA, the annual loss of wetlands is estimated at 70,000 to 90,000 acres on nonfederal lands. In the last 20 years wetlands have been a focus for environmentalists. Wetlands give us those amphibians that are indicator species of degradation in the environment. Sometimes environmental problems are finally noticed when amphibians are born with extra limbs or missing limbs or die at a quicker than normal rate.

In recent years, the preservation of our nation's wetlands has drawn increased national interest. With the federal government's goal of a "no-net-loss" policy, a closer examination of wetlands has taken place. Wetland preservation efforts can be classified into three major areas:

- the protection of our natural wetland areas
- the construction of new wetland areas to enhance water quality
- the restoration of previously altered wetlands

Natural Wetland Protection

Natural wetlands can be defined as those that have not been constructed by human activities. Most government regulations are in-tended to prevent the destruction of these wetlands. These regulations work to keep construction projects such as roads, bridges, and buildings from invading the area. Regulations also examine the watershed around the natural areas and try to keep them as intact and natural as possible. The constant challenge to preserving the natural wetland is urban growth and all the construction that accompanies that growth.

Construction of New Wetland Areas to Improve Water Quality

A relatively new approach to wetland management is the construction of natural structures to treat all forms of water pollution. This might include everything from animal wastewater to municipal effluent. Most modern municipal programs use the wetland as a secondary treatment after the primary removal of solid waste. The wetland is a way nature can clean the nutrient-rich effluent from wastewater by using plants. The cost of wetland construction is considerably less than that of the construction of water treatment facilities.

Restoration of Previously Altered Wetlands

As we lose wetlands to the pressures that were previously discussed, researchers look to areas that may have been overlooked as altered wetland areas. Some restoration may be demanded due to requirements that an industry must replace or restore a wetland at a location other than the areas they altered. These "new" wetland areas will be a welcome addition in helping to implement the no-net-loss goal.

The major concern for the new wetland area is for the development of a large variety of wildlife. This will include mammals, fish, migratory game birds, and waterfowl. A constructed marsh wetland will achieve these habitat goals for the wetland. (See Figure 10-10.)

The United States Forest service has also promoted the redevelopment of cleared forested areas. These areas help in maintaining water quality in shallow ground water. These reforested areas are allowed to develop naturally, ensuring that livestock are excluded from the area.

Urban areas also are included in the restoration efforts. Urban restoration deals mainly with water quality in the landscape. Other concerns include sedimentation, buffer zone construction, and the development of smaller land areas.

WETLAND MANAGEMENT

Wetland management is a concern of many government and private agencies—Environmental Protection Agency, United States Fish and Wildlife Service, Army Corps of Engineers, Ducks Unlimited, and The Nature Conservancy, to name a few. Most difficulty in management occurs when wetlands are found on private land. Such agencies, whether private or part of the local, state, or federal government, cannot force a private landowner to manage their wetland in any particular way. This is why so many conservationists are deeply concerned about the future of wetlands. However, the Army Corps of Engineers is trying to tighten policies and permits regarding wetland development and to help in the conservation of such vital ecosystems.

Since June 2000 the Army Corps of Engineers has been enforcing a tighter permit program regarding wetlands. This new program allows destruction (grading, draining, filling, etc.) of one half acre of wetland without an individual permit—that is down from three acres, which was previously allowed. Another change requires landowners to notify the Corps of Engineers of activities that impact any wetland area one tenth of an acre or larger in size. These changes allow the government to more closely monitor development in wetland areas. The intention is to limit developers from building on wetland sites.

The causes of wetland loss and degradation include drainage, dredging, stream channelization, tilling for crop production, logging, mining, construction, air pollution, water pollution, introduction of non-native species, and grazing by domestic animals. Natural threats that play a part in wetland loss and degradation are erosion, droughts, hurricanes, tornadoes, and other climate and weather conditions.

For personal management of wetlands, the best thing to do is to leave the wetland alone. The best way to protect a wetland area is to provide a buffer zone around the wetland, in which destruction or degradation is not allowed. Do not pollute the waterways or wetlands. If you know of pollution that is taking place, report it to your local Fisheries and Wildlife or Natural Resources Conservation Service agency. Take the initiative to start an Adopt-a-Wetland program in your area. Have the school and community become involved by participating in a wetland awareness program.

Restoration plays a big part in the management of degraded wetlands. Restoration may involve the removal of non-native species, planting native trees, shrubs or other aquatic plants, and creating walkways to minimize or eliminate further degradation. Along with these activities, one should also place the proper signs to alert others that this area is being restored, hopefully to the original state.

FIGURE 10-10 Wetlands on private land are very important. This shows two scenes of a farm, before and after a restoration project. The dark area in the first photo shows a wetland area that had been drained for farming. We used to call that "reclamation." The second photo shows the same farm after the wetland area was restored; it is now a farm pond. *(Photos courtesy of the United States Fish and Wildlife Service)*

Private landowners working with local, state, and federal government agencies, as well as private companies and agencies, to manage and preserve wetlands will ensure the future of these pristine and critical habitats. For more information on your wetlands and what you can do, call the EPA's Wetland Hotline, 1-800-832-7828.

OTHER GOVERNMENT PROGRAMS

In 1899, the Rivers and Harbor Act required approval from the Secretary of War before any construction activities were started that disposed wastes into a navigable river. In 1967 the Fish and Wildlife Coordination Act required the Army Corps of Engineers to study the ecological effects of all water related areas. Then in 1969, the Environmental Protection Agency began conducting environmental impact statements that affected wetlands. This legislation led to the Federal Water Pollution Control Act of 1972. This bill is commonly referred to as the Clean Water Act. Several sections in the bill address the regulations of wetland areas. This bill instructs that both the U.S. Army Corps of Engineers and the U.S. Environmental Protection Agency will regulate the wetlands of the United States. The bill was amended in 1977 to exempt some farming and mining activities. The agencies administer the permits to both companies and individuals when a wetland will be affected by some human-directed activity.

Other legislation affecting wetland management is given here:

- 1973 Endangered Species Act
- Flood Disaster Protection Act
- Floodplain Management Order

- 1985 Farm Bill
- 1986 Emergency Wetland Restoration Act
- 1989 North American Wetland Conservation Act
- 1990 Coastal Wetlands Planning, Protection, and Restoration Act
- 1990 Farm Bill
- 1990 Water Resources Development Act

State governmental agencies have also taken the lead in wetland protection. They are also concerned with similar areas as the federal regulations, but are more aware of their individual state needs. Most states direct educational programs that help the public understand the value and function of the wetlands. The states can help fill the gaps that federal programs might not address. Examples could be the establishment of sanctuaries, refuges, or wilderness areas that would coincide with state owned wetlands.

Local governments can also develop local wetland areas. They can help the local citizenry to develop workable wetland management plans that will be of value to all. Local governments can lend support to private wetland areas on an individual basis.

SUMMARY

Historically, wetlands have been regarded as swamps and bogs that were breeding grounds for unwanted insects and reptiles. The main management procedure was to install drainage devices so that agricultural crops could be grown, or structures could be built on the land. Through careful examination and research, it has been determined that wetland areas provide valuable assets to the ecosystem. The added population of wildlife, the natural water filtration qualities and the buffering possibilities make the wetland areas worth the regulation and concern.

DISCUSSION QUESTIONS

1. Define wetland, both from legal and ecological perspectives.

2. What are the three characteristics of a wetland area?

3. What would an on-site evaluation consist of in determining a wetland?

4. What resources are available for off-site wetland identification?

5. Explain how marshes, floodplains, lakes and ponds, rivers and streams, swamps, bogs, prairie potholes, and vernal pools are alike and how they differ.

6. List and explain the major causes of loss of wetlands.

7. What governmental agencies regulate wetland management?

8. What is the primary federal law that regulates wetlands?

SUGGESTED ACTIVITIES

1. Contact your local NRCS office for the location of wetlands in your area. Visit one of the sites and observe the components of the wetland area.

2. Design a wetland plan on a previously altered area.

3. Obtain a set of maps and make an off-site evaluation of a possible wetland area in your school district.

4. Start an Adopt-a-Wetland project for your class or school.

11

LAND-USE PLANNING

OBJECTIVES

After reading this chapter, you should be able to

- explain why land-use planning is important to our ecosystems and to our economy

- differentiate between farmland-use planning and political land-use planning

- compare farming for immediate income and farming for long-term income as they relate to soil conservation

- explain why economic development for short-term profit can be damaging to the economy as a whole in terms of long-run soil erosion

TERMS TO KNOW		
land-use planning	farming for today	regional land-use planning
farmland-use planning	conservation farm planning	irreversible change

Land-use planning can have two distinct meanings. First, it can be used to refer to the planning done for an individual farm or ranch, that is, **farmland-use planning**. In this sense, it refers to the development of a cropping and livestock production system. In the context of this unit it would, of course, include soil conservation in the planning.

The second general meaning of land-use planning is more political in nature. It deals with questions of political priorities. Who will control the land? How will it be used? Who makes those decisions? These and other questions dealing with the land-use planning process are also discussed.

NON-FARMLAND-USE PLANNING

Most of the time, when we speak of land-use planning, we are talking about the process of making political land-use decisions. In most counties, townships, cities, and towns, some sort of land-use planning is conducted. Typically, the governing body in each locality establishes its own set of rules and regulations concerning how the land within its jurisdiction can be used. The most common result of such a decision-making process is a set of "zoning regulations."

When a community has zoning regulations, the use to which each part of the community can be put is determined in advance. This is done to force like-activities to be grouped together. For instance, in one part (zone) of a city, zoning regulations may allow only single-family dwellings. That means that only individual homes can be built in that area. Apartment buildings, condominiums, stores, factories, and other structures are not allowed in that "zone." In another zone of the city, single- or multifamily dwellings may be allowed. In that zone, a person building a single-family home would have to understand that apartment buildings could also be built in the same neighborhood. Zones may be established for commercial building, such as groceries and gasoline service stations; light industry, such as small shops; heavy industry for large factories; and so on.

Land-use planning of this kind should consider not only economics but also the environmental impacts of the zoned uses of the community. The capacity of existing water purification systems to supply water must be considered. Requirements for additional sewage facilities must be considered in making zoning decisions. Traffic patterns and congestion are important considerations. Increased rainfall runoff from parking lots can overtax storm drainage systems and cause excessive erosion and sedimentation, unless planners take that factor into account.

You can easily see that political land-use planning is very complicated. Unfortunately, because land-use planning is so difficult, many localities have no land-use planning regulations. In those communities, there is no organized way of grouping like-activities together. As a result, service stations, factories, homes, and apartment complexes can all be side-by-side. That situation creates difficulties for everyone concerned.

FARMLAND USE

The Many Hats of the Farmer

Today's farmer "wears many hats." He or she is breadwinner, parent, business manager, financial director, technician, carpenter, driver, scientist, equipment operator, and many other things. One of the most important of the farmer's responsibilities is that of caretaker of the land.

In a very real sense, the farmer is custodian of much of our nation's most important natural resource—our land with its thin covering of soil. Thus, farmers are ultimately responsible not only to themselves but to society as a whole for the care of that soil.

Clearly, the farmer has many responsibilities:

- to provide an immediate income for the farm family
- to plan for a long-term income for the farm operation
- to consider the effects of the farming operation on his or her neighbors and community as well as on the rest of society
- to use the land wisely so that it will still be available to feed and clothe future generations

Farming for an Income Today

Obviously, the first requirement of any business is to make a profit. The money earned from the business provides an income for the operator. For the operator, that means being able to pay the bills, buy food and clothing, and in general do those things that make life possible and pleasant. Farming, in that respect at least, is no different from any other business. The farmer must earn enough income from the farming enterprise to pay the costs of operation and to meet the immediate needs of the farm family.

This means that today's business decisions cannot be based on the good of humanity. They cannot always be based on the long-range good of the farm itself—**farming for today**, at times, takes preference. The reality of the situation is that the farming operation must produce enough income each year to

- pay operating costs
- pay land lease costs
- pay for machinery and equipment purchases
- meet debt payments
- cover insurance costs
- maintain farm buildings and equipment
- support the farm family and provide for the farm home
- update the operation to meet environmental requirements

Farming for the Future

Clearly, those costs and many others must be met each year if the farmer is to stay in business. On the other hand, the farmer must also consider his or her other responsibilities in making day-to-day business decisions. Obviously, the bills must be paid today and this year, but they will still need to be paid in 20 years or in 100 years. The farmer's children will need to be fed and clothed—and their children, and their children's children. The farmer's job is im-

portant to the rest of society and to the future of humankind. It is too important to be used *only* as a source of income for *just* today.

The experiences of farmers in this country have shown that business decisions cannot be based solely on maximizing profits from the current year's production. Where such a short-sighted approach has been taken, the soil was soon ruined and the farmer had to move on to other farmland. As a result, vast areas of marginal-to-productive farmland have been lost in this country to soil erosion.

Conservation Farm Planning

The individual farm's cropping program is the single most critical part of the soil management effort. It is through the cropping program that the farmer translates into action a commitment to **conservation farm planning**.

The first step in developing the farm plan is to know the farm's physical features. The best solution for that is a well-trained, observant, conservation-minded farmer. To assist the farmer, the soil maps provided by the Natural Resources Conservation Service (NRCS) can be quite valuable. With the help of the local soil conservationist, the cropping system can be matched to the soil's capability to produce as well as to other soil characteristics. The basic principle is to match the crops to be grown to the soil's capability. The better the land, the more row crops can be safely produced. (See Table 11-1.) If the farm includes a livestock operation, the plan should include housing, feeding, and waste management locations and methods.

The details of the soil maps can be interpreted by using the information contained in Figure 11-1.

TABLE 11-1 Land Capability Classes and the Maximum Percentage of Time That Cultivated Row Crops Should Be Grown

Land Class	Comments	Maximum Cultivation (%)
Class I	No limitations	50–100
Classes II, III, IV (level)	Drainage problems can be corrected, drought-prone soil can be irrigated	50–100
Class II (gently sloping)		33–50
Class III (sloping)		25
Class IV (sloping)		17–20
Classes V, VI, VII, VIII	Should not be cultivated for row crops	

An aerial photo of a farm near Blacksburg, Virginia, is shown in Figure 11-2. The classification of the soil type is given by the first numbers of the code. The other symbols can also be interpreted by using Figure 11-1. Using the information presented in Chapters 4, 5, and 6, as well as Table 11-1, we can estimate the maximum amount of row cropping that can be done safely in each area.

Clearly, the farm has several logical fields. Parts of the farm are suitable for row cropping, whereas other parts should never be used for row cropping. The use of these maps allows the farmer to plan the layout of the fields in such a way that each field is fairly uniform. (See Figure 11-3.) Once the field layout is mapped, the land-use and treatment plan can be developed.

REGIONAL LAND-USE PLANNING

Where We Stand Now

The United States has a total of 2.264 billion acres of land. On that land area we all must live, produce food and fiber, move about, work, and play.

Cities, towns, and other housing areas take up just 35 million acres. That is less than 2 percent of our total land area, so housing alone does not take much space. To that figure, we must add the land in use for highways, streets, factories, parking lots, shopping centers, and schools. In addition, the nation's parks, wildlife areas, public lands, and military reservations take up much land.

Even with all those uses, only about 8 percent of our land is taken up for nonfarm use by people. That is about 178 million acres. So why all the fuss about the loss of our farmland? Let us look further.

Another 32 percent of our land area is not usable for farming. Mountainous areas, deserts, swamps, and most land under water cannot be farmed profitably today.

Of our total of nearly 2.3 billion acres, only 1.3 billion is suitable for farming—60 percent of the total. Of that, about 300 million acres are federal grazing lands. One billion acres are available for other types of farming.

That one billion acres of farmland is being used in the following manner: permanent grassland, 6 percent; nongrazed forest, 5 percent; miscellaneous, 2 percent. Only 44 percent is used for crop land. Of that last figure, almost one fifth is being used for forage production.

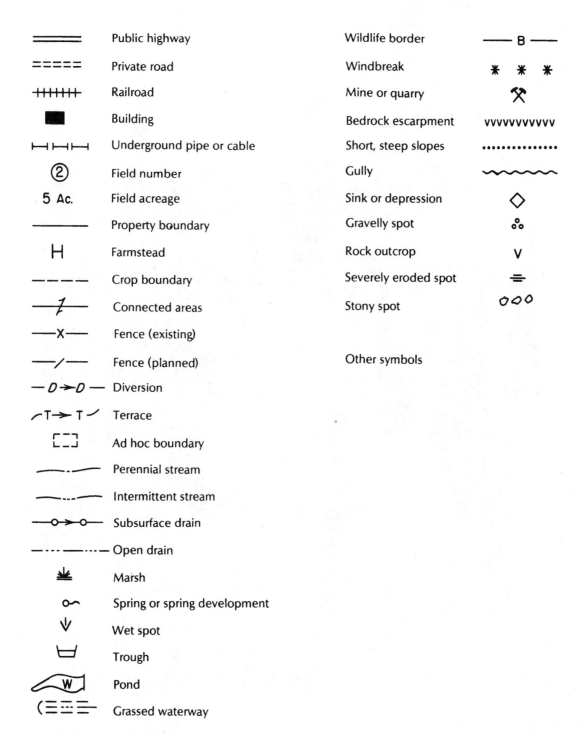

≡≡	Public highway	Wildlife border	—— B ——
=====	Private road	Windbreak	✳ ✳ ✳
++++++	Railroad	Mine or quarry	⚒
■	Building	Bedrock escarpment	vvvvvvvvvv
⊢⊣⊢⊣⊢⊣	Underground pipe or cable	Short, steep slopes	••••••••••••
②	Field number	Gully	∿∿∿
5 Ac.	Field acreage	Sink or depression	◇
——	Property boundary	Gravelly spot	⚬
H	Farmstead	Rock outcrop	V
— — — —	Crop boundary	Severely eroded spot	⇛
—∕—	Connected areas	Stony spot	○○○
—X—	Fence (existing)		
—∕—	Fence (planned)	Other symbols	
— D ➤ D —	Diversion		
⌒T ➤ T⌒	Terrace		
⌐⌐⌐	Ad hoc boundary		
——•—	Perennial stream		
——•••—	Intermittent stream		
—○➤○—	Subsurface drain		
—•••——•••—	Open drain		
☀	Marsh		
⌀⌐	Spring or spring development		
⇓	Wet spot		
⊔	Trough		
⌒W⌒	Pond		
(≡≡≡	Grassed waterway		

FIGURE 11-1 A legend for soil and conservation plan maps.

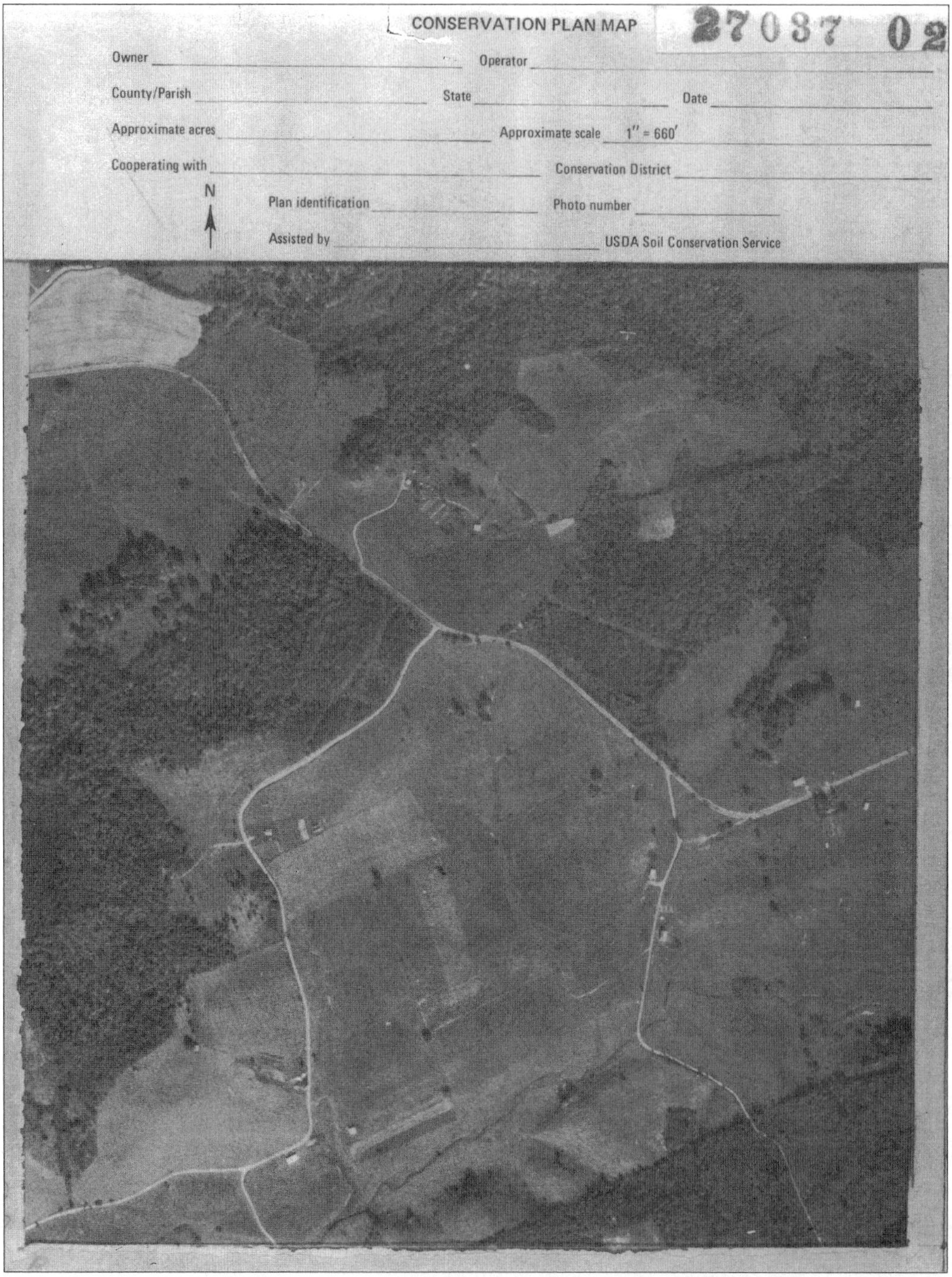

FIGURE 11-2 Aerial photograph of a farm. (*Courtesy of the United States Department of Agriculture, Natural Resources Conservation Service*)

FIGURE 11-3 Field layout map. (*Courtesy of the United States Department of Agriculture, Natural Resources Conservation Service*)

In all, our nation's crops are being produced on about 385 million acres, and not all of that is prime farmland.

The streets, industries, highways, and so on tend to be in the best farming areas. Why? Because people usually prefer to build houses, factories, and roads on level or gently rolling land. The best farmland is like that. They also generally prefer moderate climates with adequate water supplies. The best farmland is also like that. So nonfarmland users compete for the best, not the worst, farmland.

Our nonfarm use of these good farmlands has not seriously damaged our food supply—yet. But, at present rates, farmland losses could reach that point in the not-too-distant future.

Each year, 1 million acres of prime farmland and 2 million acres of lesser-quality farmland are taken for nonfarm use. That works out to about 4 square miles of prime farmland and 8 square miles of marginal farmland lost *each day*—or about 960 total acres per hour. Imagine that. A large farm being taken out of production every hour of every day and night, on the average.

Good farmland is the basis of our nation's strength and wealth. Once it is covered with concrete, it is lost to farming, at least for our lifetime. According to USDA projections, we have more than enough good crop land left for today. In fact, we appear to have enough to provide our food and fiber needs through this century. If our industrial and urban growth continues after that, students of the 2000s may face serious problems as adults.

Planning for Land Use

Earlier in this chapter we discussed land-use planning on the individual farm. In that case, the farmer makes the decisions. Who will make the decisions about **regional land-use planning** for our counties, states, and nation? Certainly, individual farmers will not.

There would be a simple solution. A federal law reserving prime farmland for farm use would solve the land-use planning problem. Unfortunately, that would not work. As we have seen, prime farmland is also prime urban, suburban, and industrial land.

One method would be to provide property tax breaks to farmers. Unfortunately, that technique encourages corporate and absentee land holders. Several states have tried this approach, and their efforts have had little effect on land-use conservation.

Federal programs designed to discourage urban sprawl could be designed. The same could be done by individual states, but that approach would have sporadic effects at best.

Local zoning boards, encouraged by state and federal policies, can decide which farmlands should be preserved. The problem with this approach is clear. The national trend toward loss of farmland is a problem for all of our citizens. For a given local community, the overall problem is not acute. If a shopping center is better for the local economy than a 100-acre field, the zoning board will allow it to be built. After all, loss of 100 acres of farmland is not much. It won't affect our total national food production; but as many thousands of such decisions continue each year, the total effect builds. (See Figure 11-4.)

The question thus becomes threefold: (1) What lands should be reserved for farm production? (2) Who will make that decision? (3) How will the decision be implemented?

Some True Stories

In one midwestern state, a retired couple decided to move from town into the country to live. They took their life savings and built their

FIGURE 11-4 Zoning boards can determine the land use for areas under their control. *(Courtesy of Raines Real Estate, Inc.)*

Land-Use Issues

Land Prices

In general, commercial and industrial land use is more profitable per acre than is farm use. Thus, a developer of a shopping center can almost always afford to pay more than a farmer can for even the very most productive farmland. Since that is the case, unrestricted land use usually favors the conversion of farmland to non-farm uses. Is it always best to let such decisions be made in the market place?

In this situation, zoning regulations can provide a solution. If a zoning commission determines only farming or forest production will be allowed in a certain zone, other land uses are prohibited in that zone. For instance, a 20-acre minimum lot size would mean housing developers would be unlikely to purchase farms and subdivide them into small lots for the construction of homes in the "farming only" zone.

Most Profitable Use

Land naturally tends to be put to its most profitable use. For instance, farmland near a city or along a major road system may be more profitable if converted to housing, commercial, or industrial use. As the amount of prime farmland decreases because of that conversion, less productive land will be converted to farm use. The result must be a decrease in the quality and quantity of food produced. This might lead to higher food prices. More and more forest land will eventually be converted to farm production. This may lead to higher costs in wood and wood products. Recreation and natural wildlife areas will have to be converted to food and fiber production, too.

dream home beside a country road. Over the next few years, other houses were built along the road. A service station and a junk yard were added. A small apartment building was constructed nearby. Over the space of less than 10 years, their "dream" became a "nightmare." They had retired from the city to the country and the city had followed them. Not only that but, because of the undesirable developments, their property value fell, so their savings were lost, too.

In a southern state, a farmer was operating a farm in the same place his family had farmed for several generations. He was making a good living with a hog operation. As the suburban expansion of a nearby city got closer to his farm, the new neighbors began to complain about the smell. Eventually, he was forced out of the hog business because of the neighbors' pressure.

Land Taxation

Should a developer with one hundred 1-acre building lots pay the same level of property taxes as a farmer with 100 acres of farmland? Should a land speculator be allowed to buy prime commercial property, declare it to be a farm, and pay farm-rate taxes on the land? Should a shopping-center owner pay property taxes at the same rate per acre as the neighboring cattle farmer?

Land values are largely affected by location and use. Tax boards usually set the value of (appraise) land based on its highest profit potential use.

It may be impossible to farm profitably on a farm that is being taxed at developed-land rates. In such a case, a solution may be for the zoning commission to declare that land being used for farm production be taxed at a special, low "farm rate." That would allow farmers to continue to farm land even though the land might be more valuable if it were sold for building lots or for commercial development.

Similar land not being farmed could be taxed at the so-called "fair-market" rate. The fair-market rate is established in one of two ways. One is to compare a parcel of land to similar land that has been recently sold in the same general area. The second way of establishing fair-market value is to have a real estate appraiser estimate how much the land would be worth if it were sold. To discourage land speculators from purchasing land and having it taxed at lower farm rates, the zoning commission would need to set a minimum value of farm product that must be sold from the land.

Irreversibility

If a wildlife area is converted to farming, that decision can be reversed: A field left fallow for a few years becomes a veritable "wildlife heaven." But once a field is paved with asphalt or concrete, it is lost for the rest of our lifetime as far as food production is concerned. This is an **irreversible change**. Does the public in general have a right to have input on decisions of this nature? This book will not try to give answers to such questions as these, but the questions themselves are too important to leave out of any discussion on land-use planning.

SUMMARY

Land use is a concern not only of the individual landowner but also of our society in general—at local, state, and national levels.

The individual landowner, particularly the farmer, should use the land carefully with conservation in mind. Certainly the farmer must operate the farm for a short-term profit, but the long-term effects of day-to-day decisions must be considered too. As caretaker of our nation's precious soil, the farmer must farm not only for today but with an eye toward the future—his or her own future and the future of our nation as a whole. Wise land use through conservation planning allows both profitable and safe farmland use.

Other nonfarm landowners should consider conservation of the soil and water just as the farmer does. Chapter 7 discussed this important topic in detail.

On a larger scale, local, state, and national land-use issues are becoming more and more important each year. As we look to an uncertain future, we must safeguard that most basic natural resource—our soil. Important policy questions must be asked and answered. In searching for the best answers to those questions, certainly we will look to the present needs of our society and of individuals within that society. However, we must also look to the future of our people and our nation. The day must surely come when the choice will have to be made between more conversion of prime farmland to nonfarm uses or a reliable, inexpensive food supply produced by Americans.

DISCUSSION QUESTIONS

1. Explain some of the many different "hats" of the farmer.

2. Why is the farmer's role in soil conservation so important?

3. Farming for a profit today and planning for the more distant future may sometimes produce conflicting business decisions. Why? How should those conflicts be resolved?

4. What is the most important part of the farmer's conservation effort? Why?

5. In the United States, how many total acres of land are there? How much is being used for non-farm purposes? How much for farm use?

6. How fast is farmland being converted to nonfarm use in America per year? per hour?

7. How do zoning regulations affect land use?

8. Are there zoning regulations in your local area? Try to get a zoning map of your local community.

SUGGESTED ACTIVITIES

1. Invite a member of your city, county, or regional commission to visit your class to discuss land-use planning.

2. Invite a Natural Resources Conservation Service agent to your class to discuss farmland-use planning.

12

CAREERS IN SOIL MANAGEMENT

OBJECTIVES

After reading this chapter, you should be able to

- outline several major career areas in soil and range conservation

- describe some of the things people in those careers do

- list the kinds of training needed for entry into those careers

TERMS TO KNOW

soil conservationist	soil conservation technician	soils engineer
range manager		percolation rate
permeability	soil scientist	

Soil conservation workers are the primary group employed in occupations involving soil management. A smaller group is made up of soils engineers. Another smaller group would be a highly specialized group of soil scientists, teachers, and researchers. Most of these are employed by the United States Department of Agriculture Natural Resources Conservation Service; the Bureau of Indian Affairs of the Department of the Interior; by various state agencies concerned with soil conservation, land-use planning, or farm assistance; or by colleges of agriculture in the various universities around the country. Let us examine a few of these specific jobs.

SOIL CONSERVATIONIST OR RANGE MANAGER

Nature of Work

Soil conservationists provide advice and assistance to farmers, ranchers, and others concerned with the conservation of soil and water. They help to develop farm plans that take the greatest advantage of the ranch or farmland's capability without damaging it.

Range managers have many of the same kinds of responsibilities as soil conservationists, except they specialize in conservation in the grasslands. Range managers, or range conservationists as they may also be known, help ranchers, farmers, other landowners, and public officials in the range areas plan for the wise use of soil and water resources. They also work with wildlife management and provide technical assistance and professional advice in governmental land-use planning. Because rangelands are both a very important part of our natural resources and a very fragile ecosystem, careful range-management planning is important.

Soil conservationists and range managers do most of their work outside. If a farmer or rancher needs help in planning, the conservationist visits the farm or ranch. He or she studies any available soil type and land capability information for the particular farm or ranch. The farmer's needs and business goals are taken into account. The conservationist then prepares a farm plan to allow the wise use of land to help meet those business goals.

If a farm or ranch has an existing erosion problem, the conservationist can help. He or she studies the nature of the problem. The source of the erosion must be found, then a plan is developed to correct the erosion problem and restore the land to productive use.

On land that needs terracing (see Chapter 6), the conservationist can plan a terrace system. First, the size of the field and its slope are considered. This allows the conservationist to plan the size, slope, and location of each terrace.

As the terraces are constructed, the conservationist, or more likely a technician, can still help. If the farmer or rancher constructs the terraces, the conservationist offers assistance and advice on "how to." If the farmer hires a contractor to construct the terrace, the conservationist may check for the correctness of the job. Simple surveying work is needed to lay out the terraces at the proper locations. The conservationist can do this.

Planting of windbreaks and shelterbreaks should be planned. The types of trees and shrubs to plant are important. The locations, directions, and distance between these breaks are also critical. The conservationist provides assistance in developing the plan and can also help supervise the planting. (See Figure 12-1.)

In many parts of the country, the construction of farm ponds is very important. Wherever rainfall is not regular or plentiful enough, irrigation may be needed. Farm ponds provide one source of irrigation water. Also, if livestock are being produced, farm ponds may provide water for those enterprises.

The conservationist can help the farmer or rancher plan and build farm ponds. First, the rainfall and drainage patterns of the farm or ranch are considered. The soil type and permeability are important factors. **Permeability** refers to the ability of the soil to allow water to move through it. If a soil is not very permeable, it will probably not have good drainage. The uses to which the water will be put are also considered. The conservationist

FIGURE 12-1 A soil conservationist is busy developing a conservation plan for a local farm. *(Photos courtesy U.S. Natural Resources Conservation Service)*

then develops a plan specifying the size, shape, and location of the dam, spillways, banks, and other parts of the pond. He or she then checks during the construction to ensure that the pond is being built correctly.

In areas where snowfall accumulations are large, the conservationist may be responsible for predicting spring runoff. Where this is the case, he or she must make snowmobile or ski trips into backcountry areas to measure the accumulations. This allows the amount of snowmelt to be estimated. When the snowfall is very light, the conservationist alerts farmers and other water users about possible summer water shortages. When the snowfall is particularly heavy, the conservationist must estimate the danger of flooding. People in possible danger areas must be warned in advance.

The conservationist also works with other agencies. Local governments frequently consult with the conservationist in developing land-use plans and zoning plans. The soil and water conservation district depends heavily on the professional conservationist. He or she provides advice and assistance in watershed management planning, land-use planning, mapping, and soil conservation planning.

Employment Locations

There are nearly 9,000 professional soil and range conservationists in the United States. Many of these work for the Natural Resources Conservation Service of the United States Department of Agriculture. Others also work for the Bureau of Indian Affairs or the Bureau of Land Management, both of the Department of the Interior. These federal employees are located in almost every county in the United States. Local natural resource agents are listed in telephone directories under

United States Government, Department of Agriculture, under the NRCS or Natural Resources Conservation Service office listings. Range managers are listed under the United States Department of Interior, Bureau of Land Management. If you live near or on a reservation, look under United States Government, Department of the Interior, Bureau of Indian Affairs.

In addition to federal jobs, many natural resourceists work for state, county, or city governments around the country. Some are also employed by banks with large farm programs, insurance firms, lumber and paper companies, and public utilities.

Training and Qualification

A bachelor's degree is a minimum for entry into soil conservation work. Some land grant and other colleges or universities offer degrees in soil conservation, but most schools do not offer this specific degree. A degree in agronomy, agricultural education, or natural resources management will probably qualify you for this type of work. A degree, or at least some courses in agricultural engineering, would be helpful as well. On the other hand, degrees in range management are available from as many as 35 colleges and universities.

A farm background is helpful but not necessary. The range manager or soil conservationist, however, must be able to communicate well with the public. Part of the job is public relations—convincing the people in the area that soil and water conservation and range management are important. He or she may be called upon to talk at club meetings, schools, civic functions, and social gatherings. Radio interviews and newspaper interviews are common requests.

Outlook

The annual replacement requirement need for soil conservationists averages about 400 for the USDA Natural Resources Conservation Service. It appears that the number of jobs in this field may be holding steady over the foreseeable future, but this is a very attractive occupation for many people. The pay is adequate, the job security is high, and the working conditions are usually pleasant, outdoors, and flexible. Starting salaries for new college graduates ranged from $19,500 to $24,200 in 1997. The average salary for all soil conservationists was about $39,000 in 1997. Thus, there are usually plenty of applicants for these jobs.

If you are interested, call, write, or visit your local natural resources soil conservation agent. Or write

Soil and Water Conservation Society
7515 Northeast Ankeny Road
RR #1
Ankeny, IA 50021-9764

SOIL CONSERVATION TECHNICIAN

The **soil conservation technician** is a technical assistant to a soil conservationist. The technician helps in routine office and field matters. He or she must be able to operate simple surveying equipment, transit levels, and other measuring devices.

Supervision of soil conservation construction, such as terraces, farm ponds, diversion ditches, and waterways, are performed. The technician checks on the accuracy and quality of the work of contractors on those jobs. He or she assists farmers who desire to perform these or similar jobs on their own.

The technician is directed and supervised by the soil conservation agent. High school graduation, but no other specific training, is required for entry into the job of soil conservation technician. This is a civil service job with the United States Department of Agriculture, Natural Resources Conservation Service, or a state civil service job with appropriate state conservation agencies.

SOIL SCIENTIST

Remember that land is classified by its capability as well as by soil type. The **soil scientist** is responsible for this activity.

Soil scientists deal with the physical, chemical, and biological nature of the soil. They use this knowledge to analyze the soils in specific locations. After studying the soils in an area, they examine soil samples in the laboratory. They then use aerial photographs, topographic maps, and the information previously derived to develop soil maps. A collection of soil type maps and other related information is put together into a "soil survey." This is essentially a book that can then be used by the soil conservationist in developing farm plans as well as larger land-use plans.

Thus, the soil scientist develops the technical and scientific information that makes possible much of the field work of soil conservation and land-use planning. For instance, the farmer needs to know the land capability classes of his or her fields. This information is useful in planning the crop and livestock enterprises.

The soil conservationist needs to know the soil types in planning farm ponds and terracing systems. Builders need to know the firmness of the soil and its permeability in planning construction projects.

All these types of information are provided by soil scientists. Part of the information is available from existing NRCS soil surveys. Other parts must be examined on a case-by-case basis by on-site study.

In addition to these applied soil scientists, there are many research soil scientists. These latter professionals conduct studies on fertilizers, soil chemistry, soil physics, and soil biology. They may design experiments to find answers to new and changing questions about the soil on which we all live.

Soil scientists at universities and colleges may teach as well as write articles and books. They also work on research projects and write research reports. They provide technical assistance to agricultural extension, agricultural education, and soil conservation workers.

Employment Locations

In 1997, there were over 29,000 agricultural scientists, including about 3,500 soil scientists in this country. About one half of these worked for the federal government in the USDA Natural Resources Conservation Service.

Some worked for state and local governments. Others worked as professors at agricultural colleges and universities, and still others worked for construction firms, especially large building contractors and highway construction companies. A few worked as private consultants, researchers, or foreign advisors.

Training and Qualifications

At least a bachelor's degree is required for federal employment but, because of the highly specialized nature of this work, advanced degrees may be needed. Work as a research soil scientist will probably require both master's and doctor's degrees. In addition, a strong background in chemistry is important.

The soil scientist will probably deal less with other people than does the soil conservationist. The soil scientist must be meticulous and thorough. He or she must have an inquiring mind and enjoy laboratory work and writing. The soil scientist must be able to examine a problem, apply theoretical reasoning and experimentation, and arrive at a logical solution.

Outlook

The Natural Resources Conservation Service needs about 100 new soil scientists each year. In addition, a similar number is required for other governmental and industry jobs annually, while a far smaller number of research scientists is needed. Starting salaries in the 2000s for new Ph.D. holders will range from $40,000 to $45,000.

This is a fairly small field, and the educational requirements are very specific. One doesn't normally become a soil scientist by accident. Thus, the competition for these jobs may not be as severe as that for the more generalized job of conservationist.

If you are interested in this profession, contact the following:

Soil Science Society of America
677 S. Segal Road
Madison, WI 53711

or write or visit the agricultural engineering department or the agronomy department at any of your state's colleges of agriculture.

SOILS ENGINEER

A **soils engineer** is a professionally trained engineer who analyzes and evaluates soils for construction sites, pond and lake sites, erosion-control construction, and other soil and water conservation activities.

The soils engineer must be able to evaluate the percolation rates, compaction problems, drainage, structural weights, and the ability of the soil to support structures. **Percolation rate** refers to the speed at which surface water can soak into the ground and move downward; that is, it depends on the soil's permeability. He or she must then take that sort of information as well as many other considerations into account in planning and evaluating plans for specific land uses.

To become a soils engineer, a degree in mechanical or civil engineering or in agricultural engineering is required.

SUMMARY

There are opportunities for employment in the field of soil conservation. They range from the conservation technician to the soil scientist. The former requires little formal training beyond high school graduation; the latter usually requires a bachelor's, master's, and finally a doctor's degree.

Opportunities for advancement are available, just as in any government work. These typically require successful experience in one of the jobs discussed here. You should hold in mind that most, but not all, soil conservation jobs are with the federal or state governments.

DISCUSSION QUESTIONS

1. What are some of the responsibilities of the soil conservationist?

2. What are some of the responsibilities of the soil conservation technician?

3. What are some of the responsibilities of the soil scientist?

4. What training is required for each of these three jobs?

SUGGESTED ACTIVITIES

1. Find the current edition of the *Occupations Outlook Handbook*, published by the United States Department of Labor, and look up the occupations listed in this chapter. (Your school library or counselor should have a copy. If not, ask them to try to get one.) What are average starting salaries?

2. Write to the USDA, NRCS office nearest you for current information about employment possibilities.

3. Visit an NRCS office or ask the agent to visit your class. Find out what she or he does on the job and what he or she thinks about the profession.

Section II — CASE STUDY

LET'S GO TO THE BEACH

Every day our coastlines are bombarded by nature. As the tides move in and out, our coastlines shift (growing and shrinking) by small degrees. (See Figure II-A.) While nature continuously shifts our coastal soils, people continue to build on them. In the United States alone, millions of people call the coast home or build vacation homes there. What does this mean for the land and our tax dollars?

Americans spend billions each year for damage done to property along the coast and in the flood zones surrounding this property. In some areas, people expect major repairs or full home replacements every couple of years. All Americans share some of the costs for cleanup and reconstruction. Many are tired of paying for these repairs. Opponents to national spending on cleanup and repairs feel that we are fighting a losing battle. The tides will always rise and fall. Thus, the coastline is continuously in danger of severe erosion, and the buildings built on these shifting sands are constantly in danger. (See Figure II-B.)

The National Flood Insurance Program was enacted in 1968 to provide flood insurance programs to coastal communities in danger of severe flooding and damage. Two major problems plague the program:

- Participation in the program is not mandatory, and many property owners either do not understand it or have chosen not to purchase the flood insurance. This widespread participation problem has left much at-risk property uninsured, as

FIGURE II-A Spending time at the beach is not a uniquely American pastime. Children of all ages in all parts of the world love to be near the surf and the sand. "Let's go to the beach" is a common refrain heard by parents everywhere. *(Photo courtesy of KC Photo, Emerald Isle, North Carolina)*

FIGURE II-B The beach is not always pleasant. This beachfront home on the Great Lakes was destroyed by beach erosion. *(Photo courtesy of the United States Environmental Protection Agency)*

the floods from Hurricane Floyd in September 1999 showed.

■ Many contend that the program has encouraged coastal development with the result of increased residential development in coastal areas most prone to damage from storm surge and beach erosion.

In 1994 this program was reformed and mortgage-lending institutions were held more accountable for keeping insurance policies in place. Another important change increased waiting periods from 5 to 30 days for coverage to become effective after purchase. Why? Some people would buy coverage as a flood threat grew. Then, once the danger had passed, they would let the insurance policies die. Flood insurance purchased when real estate changes ownership does not require a waiting period.

Opponents of coastal development fear that insurance programs offer incentives for people to build on the dangerously unstable coast. They feel that the government should find ways to discourage people from building in these areas rather than promoting such development. Other people love the ocean and want to have homes and businesses near the beach. They insist that increased taxes generated by the economic development of coastal areas more than justify the cost of such programs.

What do you believe? What are the advantages and disadvantages of living on the coast? Describe the kind of person who would be in favor of the government's plan. Describe the kinds of people who would oppose it. What factors would contribute to these opinions? Should government promote or discourage coastal development? You decide.

Section III
Water Resources

13 WATER SUPPLY AND WATER USERS

OBJECTIVES

After reading this chapter, you should be able to

- explain the components of the hydrologic cycle

- explain the main water users

- identify and discuss the common types of irrigation systems

TERMS TO KNOW		
hydrologic cycle	runoff water	semiportable sprinkler
evaporation	groundwater	stationary irrigation sprinkler
meteoric water	zone of aeration	
transpiration	surface irrigation	drip irrigation
surface water	sprinkler irrigation	hydroelectric power

Water is one of the most fascinating compounds found on earth. Water is a necessary ingredient for all living organisms. It covers about 70 percent of the earth's surface, and there are about 340 million cubic miles of water on this planet. Of that amount, about 97 percent is located in the oceans and seas, 2 percent is freshwater, and about 1 percent is frozen in glaciers and icecaps.

THE WATER CYCLE

The amount of water on and around this planet is fairly constant, but the availability of water is not nearly as constant. Water is continually moving from place to place by means of the water cycle. This water cycle is more correctly referred to as the **hydrologic cycle**. (See Figure 13-1.)

The hydrologic cycle is powered by solar energy, which heats the water causing it to rise into the atmosphere. This movement of water into the atmosphere is called **evaporation**. The water will eventually return to the ocean in the form of rain, sleet, or snow. This returned water is called **meteoric water**. If the oceans were not recharged, their water level would decrease over 40 inches per year.

The oceans are the largest reservoir of water we know. As stated, the oceans comprise about 97 percent of the water on earth. This water is unsuitable for drinking, agricultural use, or industrial use because of the high salt

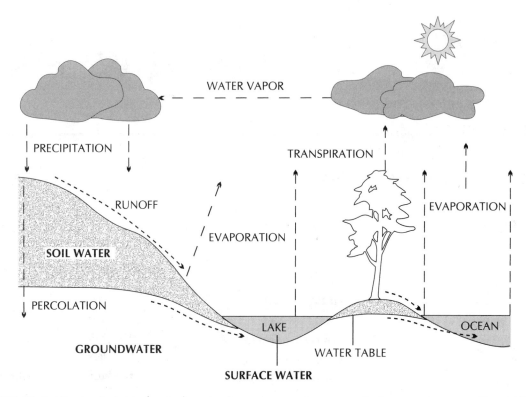

FIGURE 13-1 The hydrologic (water) cycle shows how water moves between oceans and land. Water evaporates from the ocean surface and falls on the land as some type of precipitation.

content. This salt is left behind when the ocean water evaporates.

Once the water has evaporated, it forms clouds; but it doesn't remain there. It returns to earth as meteoric water (rain, snow, sleet, etc.). The clouds may drop their water back into the ocean, or they may move inland and drop fresh rainwater. This rainwater is very important to crop land, even though it makes up less than 1 percent of the total water supply. This air-to-land interchange is vital in keeping a water balance on earth.

Plants and animals are also an important part of the hydrologic cycle. Plants absorb water through their roots and release it through tiny openings in their leaves called

stomata. This evaporation of water from plants is called **transpiration**. Animals also need water for their growth. Water promotes chemical activity, regulates temperature, and transports nutrients through the bloodstreams of animals. About 65 percent of an animal's weight is water.

The next component of the water cycle is that of **surface water**. About 30 percent of the rainfall in the United States falls into our lakes, ponds, and streams. This surface water is the most important element to the conservationist. It is used by people over and over as it makes its way toward the ocean, fulfilling agricultural, industrial, and domestic needs. It is here that pollution becomes a problem, rendering much of the water unsuitable for our needs. If the surface **runoff water** is too great, it can be detrimental, as seen with floods and soil erosion.

The final part of the water cycle is that of **groundwater**. If the water does not either evaporate or run off, it soaks into the soil to add to the groundwater supply. When rainfall arrives at the soil surface, the soil accepts water into three areas. The first area, located at the surface, contains the relatively dry soil particles. These particles accept the water until they become saturated. The next zone encountered by the water is the **zone of aeration**. This zone is made up of empty spaces, filled with air, between soil mineral particles. This zone extends from the deepest plant roots to just above the water table. Below the zone of aeration is the groundwater zone. In this zone, water is trapped in water-saturated rocks called aquifers. It is here we drill our wells. This supply of water can equal about 100 years of the surface runoff volume. The top of the aquifer area is called the water table.

Thus, the main components of the hydrologic cycle are

- oceans
- evaporation-transpiration
- plants and animals
- surface water
- groundwater

Worldwide, the total amount of meteoric water during an average year is so vast that it is usually expressed in cubic miles rather than in gallons. To give you an idea of how much water is in a cubic mile:

- 7.48 gallons of water = 1 cubic foot
- 5,280 linear feet = 1 mile

- 5,280 feet × 5280 feet = 27,878,400 square feet = 1 square mile

- 5,280 feet × 27,878,400 square feet = 147,197,952,000 cubic feet = 1 cubic mile

- 7.48 gallons × 147,197,952,000 cubic feet = 1,101,040,680,960 gallons = 1 cubic mile

Worldwide, about 9,804 cubic miles of rain, sleet, snow, and other forms of precipitation move to the surface of the land each year. That does not count the water that falls on the oceans, because we have no way of capturing that water, but it does count rain falling in lakes and rivers, for instance. Of that amount, about 588 cubic miles of meteoric water falls in the United States each year. Somewhat more water (about 681 cubic miles) falls in Canada. The United States withdraws from the water cycle about 112 cubic miles of water. What that means is that, in the United States, we use about 19 percent of our potential water supply and almost 81 percent continues in the hydrologic cycle. Domestic uses account for about 13 percent of U.S. water use. Industrial uses account for another 45 percent, with the remainder (about 42 percent) being withdrawn for agricultural use.

It is important to realize that "used" does not mean used up. Almost all water withdrawn from the hydrologic cycle eventually returns to the cycle. In reality, almost any use is temporary, so "borrowed" might be a more accurate description of what happens to water. The next section discusses water users, and after reading that, you should get a clearer picture of why we say that water that is used is really borrowed.

WATER USERS

Water has many users and has caused people to move from one area to another. Water has caused variation in population densities inhabiting an area. The U.S. population drinks about 100 million gallons of water per day, but this is only a tiny fraction of the water used in other ways in the country. The main users of water include

- agriculture
- industry
- hydroelectric plants
- fish and wildlife
- recreational users
- domestic users

Agriculture

The main agricultural use of water is irrigation. The idea of artificially applying water to land that does not receive enough rainfall is not new. The Hopi Indians of northeastern Arizona practiced watering the dry washes. The most common irrigation methods include surface and sprinkler techniques. **Surface irrigation** involves building a series of large and small ditches to transport water. The largest ditch, called a canal, brings the water to the fields from a stream or reservoir. From the canals, small ditches, or laterals, carry the water to the growing crops.

The surface irrigation technique can use either a flood or a furrow method. The flood method involves flooding with a continuous sheet of water on a strip down the field. (See Figure 13-2.) The furrow method uses land furrows to move the water to the field and is often used in orchards and truck farming. (See Figure 13-3.)

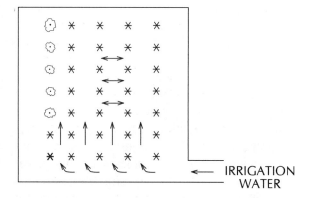

FIGURE 13-2 In flood irrigation, fields are irrigated by flooding with a continuous sheet of water.

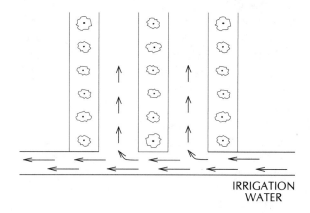

FIGURE 13-3 Furrow irrigation uses land furrows to channel irrigation water to a field.

The **sprinkler irrigation** system involves applying water over the top of the crops. The system can range from fully portable to fully stationary. In a fully portable system, sprinklers, laterals, and even the pumps are movable. This may involve moving the pumping mechanism along an open ditch from which the lateral sprinklers receive their water. Over one-half the irrigation systems in the United States are of this type. (See Figure 13-4.)

FIGURE 13-4 Portable irrigation system.

Semiportable sprinkler systems consist of several types. The most common method is to supply water to a lateral that pivots around the supply. This system is then moved to a new supply pivot when a change is necessary. About one fourth of the sprinkler systems use this method. (See Figure 13-5.)

The **stationary irrigation sprinkler** involves burying underground waterlines in

FIGURE 13-5 Semiportable irrigation system. *(Courtesy of Deborah M. Goetsch)*

FIGURE 13-6 Stationary irrigation.

the area to be irrigated. Both the mainlines and laterals are buried with only the sprinkler heads exposed. This system is most commonly found in berry patches, orchards, and nurseries. The stationary system is the most costly to install and represents only about 10 percent of the sprinkler irrigation systems in use. (See Figure 13-6.)

Drip irrigation consists of water supply pipes with lateral tubes going to individual plants. (See Figure 13-7.) A "dripper" is attached to the end of the tubes, which supplies water at a prescribed rate. Drippers are rated by the amount of water (gallons per hour) they supply to the plants. The larger the plant, the more water that needs to be supplied. By controlling the water to the individual plants, we can do a better job in getting the maximum benefit for the least amount of resources. Drip irrigation was originally developed for greenhouse use; but as water conservation became more important, outdoor applications have been used.

At first sight, irrigation appears to be the solution to all our problems connected with food supply and crop land. It would seem that

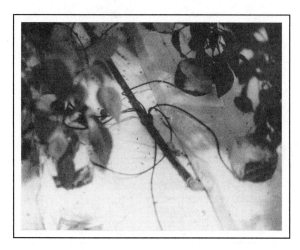

FIGURE 13-7 Outdoor drip irrigation application.

all we have to do is bring water to the desert, but it is not as simple as it seems; there are problems connected with irrigating crop land. The most common problem is the use of our groundwater supply. Earlier it was stated that our groundwater supply amounted to about 100 years of surface water runoff. This estimate, however, was under normal conditions and use. The continual sinking of vast wells to tap this water supply is creating a drain on that supply. Lowered water tables, some by as much as 400 feet, have been reported in Texas, Arizona, and even in California's San Joaquin Valley. In areas close to the ocean, this depletion of groundwater has resulted in salt water entering the water table, rendering the supply unsuitable for domestic use.

If the irrigation water is not being drawn from groundwater, it must be drawn from surface water sources. This, too, can be a problem. Most surface water is not close to the lands in need of irrigation, and long transport of water can be very wasteful. Irrigation canals may have to carry irrigation water hundreds of miles to reach the fields, and much of the water may be lost to evaporation before it reaches

its destination. The USDA estimates that only 25 percent, or one gallon in four, ever reaches the crop it is intended to irrigate.

Most of the water is lost to the soil around the canals, weeds, and evaporation. To cut this loss, closed pipes or lined canals have been used.

Once the water has reached the irrigation fields, another problem arises. Most fresh water contains traces of minerals and salts. When water is lost to evaporation, this mineral and salt content intensifies in the remaining water because it does not evaporate as does water. Add to this the additional minerals found naturally in the transport drainage ditches, and the problem readily becomes evident. The high concentration of irrigation salts in the water is termed salinization. Unless the canals are periodically flushed, the water they carry cannot be used for crop production.

Industry

Water is a very important part of the industrial world. Industry uses more water than any other raw material in its production of goods. Only a small part of this water is actually consumed; instead, it is used as a medium for other processes. For example, in a canning factory the water is used to clean both the product and the containers, cool the ovens, boil the product, and wash away the wastes. It even may become part of the mixture. Most of the water is either recycled for another use or returned to the natural water cycle in a drainage ditch or stream.

A large portion of the industrial use of water is in the manufacturing of goods. Paper is actually made in water. According to the USDA, industry draws about 40 billion gallons of water per day from our water supply. It is used as many as two-and-a-half times

FIGURE 13-8 Factories use water in many ways. *(Courtesy of Bethlehem Steel Corporation)*

FIGURE 13-9 A hydroelectric plant. *(Courtesy of Dominion Engineering Works, Montreal, Canada)*

before being returned as waste water. It is also estimated that only about 2.5 billion gallons per day are actually consumed. The supply of water is usually a determining factor in the selection of the site for a factory. (See Figure 13-8.) Strong pollution-control programs have caused industry to develop methods to recycle water rather than draw from an outside source.

Hydroelectric Plants

A major use of water is in the production of electricity from water-driven turbine generators. (See Figure 13-9.) The United States leads the world in **hydroelectric power**. Total worldwide generation of electricity in 1995 was 13.1 trillion kilowatt hours (KWH). Of that total, about 2.5 trillion KWH, or about 19 percent of the total, were produced by hydroelectric generator plants. In the United States, of the 3.3 trillion KWH of electric power generated, only about 0.3 trillion KWH, or about 13 percent of the total was generated by hydroelectric power.

The principle behind the production of electricity from water power involves tapping the energy from water moving from a high place, such as a dam or reservoir, to a lower place, such as a stream or river; therefore, large dams must be constructed across major rivers to obtain this power. Although the water is not heated, polluted, or destroyed, much concern has developed in recent years. Once the rivers are dammed, the natural wild rivers are tamed, fish runs are interrupted, and wilderness areas are altered. Which is the highest priority item?

Fish and Wildlife

Water plays a very important role in fish and wildlife populations. The way that water is handled not only affects fish in the water but also all the wildlife around the water. (See Figure 13-10.) This can include land wildlife and inland birds as well as waterfowl. A more important concern is that different species of wildlife require different types of water. Some wildlife and fish prosper in quiet, cold mountain streams, whereas other species prefer fast, running rapids. When an area is dammed or drained, the habitat is changed; if wildlife cannot adapt to the changes we impose, it will soon perish.

FIGURE 13-10 Water is home for many forms of wildlife, such as these female mallards.

In recent years, legislation has been passed to protect the valuable wetland areas. Before a wetland can be altered, an intensive study must be carried out to examine the impact of the changes. In many cases, no changes are allowed.

Recreation

Water provides many recreational activities: boating, water skiing, fishing, sailing, and swimming are only a few. Many times these activities have been an additional benefit from reservoir construction for flood control or from dam construction for hydroelectric plants. Chapter 3 deals with this concept under the term *multiple use*.

Recreation also has placed a burden on the water supply. The outdoors is not the only place where water is used. Thousands of swimming pools have been constructed in this country, and it takes a great deal of water to keep these in compliance with health regula-

tions. The water must always be changed, or chemicals added, to ensure that a safe place to swim is maintained. In addition, consider the thousands of gallons removed through splashing and evaporation.

Water is not only used in its liquid form. Consider all the frozen water activities we enjoy. We skate on frozen indoor rinks, watch ice hockey, and even see hit shows on ice. All these activities use our valuable water resources.

Domestic Uses

People use more water than we need just to stay alive. We cook, bathe, clean, and carry away wastes with water. Most people in the United States are privileged to have running water in their homes. Most have both hot and cold running water. This is not the case in many other countries of the world. Each American, on average, uses more water than any other person in the world. Think about your own water use. How many times have you let a leaky faucet run, or left one on part way? In the United States, each person uses approximately 173 gallons of water per day. Each bath you take uses 30 to 40 gallons of water. If a shower is your preference, you use 5 gallons per minute. Each time you flush a toilet, you use 1.5 to 3 gallons of water. When you wash your clothes, you use about 30 gallons of water. On top of all the water usage, we wash our cars, sprinkle our lawns, and run water fountains in landscape designs. (See Figure 13-11.) Some pump groundwater to run their air conditioners in the summer and to heat their buildings in the winter time. We use water in fire protection and in washing our city streets. How wasteful can we be and still have enough water to survive?

FIGURE 13-11 Is this a wise use of water? The answer to this question depends on many factors: location of the house, water supply available, and source of power to pump the water, to name only a few.

SUMMARY

Water is one of the most important compounds found on the face of this earth. Its supply and use should be as closely balanced as possible. The water cycle, or hydrologic cycle, is responsible for supplying water needs throughout the earth. Its main components are the oceans, evaporation/transpiration, plants and animals, surface water, and groundwater. Often too much water falls in some parts of the earth, and floods occur. Sometimes there is not enough precipitation and droughts occur. We have seen fit to move water to better suit our needs.

The main uses of water are for agriculture (through the use of irrigation), industry, hydroelectric plants, fish and wildlife, recreation, and domestic needs. Each of these areas is responsible for using great quantities of water. This vast use calls for good conservation and water management programs, which are discussed in other chapters in this book.

DISCUSSION QUESTIONS

1. List and explain the main components of the hydrologic cycle.
2. What are the three zones of groundwater supply?
3. What is an aquifer?
4. List and explain the main types of irrigation systems in agriculture.
5. How does industry use water?
6. List three ways hydroelectric plants affect our environment.
7. What type of wildlife is affected by water use?
8. What are the common ways water is used in recreation?
9. List five domestic uses of water.

SUGGESTED ACTIVITIES

1. Keep an account of all the water you use in one day. Look at the list and evaluate the areas where you might have saved water.
2. Make a report of the major water-using industries in your area. Compare this to the water resources available. Answer this question: "Is there a water shortage in my area?"

14

WATER POLLUTION

OBJECTIVES

After reading this chapter, you should be able to

- identify the three major water pollution groups

- explain the four major categories of industrial pollution

- explain the function of a cooling tower and cooling lagoon

- list and explain the major agricultural pollutants

- explain the common water pollution control measures

TERMS TO KNOW		
point source pollution	industrial pollution	radioactive material
diffuse source pollution	thermal pollution	organic waste
background pollution	cooling tower	BOD test
urban pollution	cooling lagoon	eutrophication

For many years, we have regarded water as a means of disposing of wastes we no longer want around. It seems very simple. If the wastes are put into water, they cannot be seen; therefore, they must be gone. After all, think of the billions and billions of gallons of water on this planet. There is no way we could ruin it. This type of thinking has caused a terrible water problem. After seeing lakes die and rivers turned into running cesspools, we have finally come to realize that water pollution is a problem that must be dealt with.

Water pollution results from many different sources. In this chapter we look at water pollution two different ways. First we discuss the broad classification of the sources of pollution. Then we examine in more detail the primary kinds of human-generated water pollution.

CLASSIFICATION OF POLLUTION SOURCES

There are three basic sources of water pollution: point, diffuse, and background. We tend to think of the first two of these as being associated with human activity and the third one as being natural pollution. In reality, all three kinds of pollution occur in nature. (See Figure 14-1.)

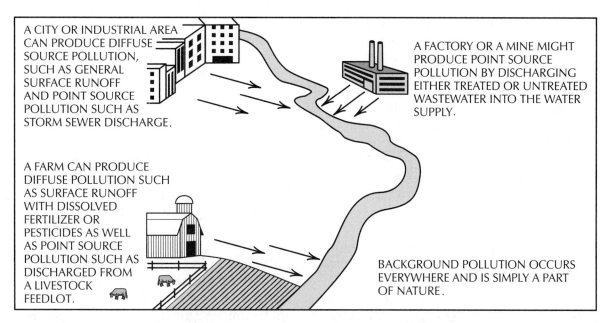

A CITY OR INDUSTRIAL AREA CAN PRODUCE DIFFUSE SOURCE POLLUTION, SUCH AS GENERAL SURFACE RUNOFF AND POINT SOURCE POLLUTION SUCH AS STORM SEWER DISCHARGE.

A FACTORY OR A MINE MIGHT PRODUCE POINT SOURCE POLLUTION BY DISCHARGING EITHER TREATED OR UNTREATED WASTEWATER INTO THE WATER SUPPLY.

A FARM CAN PRODUCE DIFFUSE POLLUTION SUCH AS SURFACE RUNOFF WITH DISSOLVED FERTILIZER OR PESTICIDES AS WELL AS POINT SOURCE POLLUTION SUCH AS DISCHARGED FROM A LIVESTOCK FEEDLOT.

BACKGROUND POLLUTION OCCURS EVERYWHERE AND IS SIMPLY A PART OF NATURE.

FIGURE 14-1 Sources of pollution may be point, diffuse, or background.

Point source pollution results from the direct introduction of contaminants into the water supply at an identifiable location or multiple locations. An example of point source pollution would be wastewater from a factory being dumped into a stream through a discharge pipe. For thousands of years, towns and villages have discharged their wastewater directly into streams or rivers or into the ocean. In many parts of the world, that practice is still followed. In the United States, city sewer systems may sometimes overflow, particularly after unusually severe rains. When that happens, untreated or partially treated wastes may be discharged into the water supply. In Chapter 9, we discussed landfills and solid waste management. A natural attenuation landfill would be a point source of contamination.

Most often we see point source pollution in terms of human activity, but an underground stream containing heavy concentrations of dissolved metals that surfaces and pours into a surface river could accurately be considered a point source of pollution for the river. In any case when the source of contamination can be precisely located, we refer to it point source pollution.

Diffuse source pollution is also referred to as non-point source pollution. Diffuse pollution results from the introduction of contaminants across a wide area. This typically results from human behavior, but again, can occur in nature. An example of diffuse source pollution would be surface contamination from a farm or city that seeps into the groundwater supply or into a stream, river, or lake in the form of runoff. Diffuse pollution can generally be at least somewhat localized but cannot be tied to a specific input location.

For an example of diffuse source pollution in nature, consider this. A wildfire resulting from lightning might destroy the vegetative cover of a large land area. Subsequent rains or snowmelt would then result in erosion. The soil particles as well as the organic matter suspended in the water as well as the dissolved salts and hydrocarbons (see Chapter 15 for more on this) would be diffuse source pollution.

Background pollution is not associated with human activity. It results from ongoing, natural processes. Background pollution occurs because of the properties of water. Water is one of nature's most universal solvents. In addition, it flows freely. As water moves, it generates forces that pick up particles and carry them along.

Almost the entire planet contains living organisms. When those organisms die, they decompose. One by-product of organic decomposition is a whole array of water-soluble hydrocarbons and other compounds. When those compounds contact water, they dissolve and are carried away into the water cycle. Many minerals and rocks are water-soluble. As water runs along the surface, circulates in a body of standing water (there are currents in even the smallest lakes), or percolates through the ground, those materials dissolve and move into the water cycle. General, ongoing contamination of the water supply from natural sources is termed background pollution.

URBAN POLLUTION

During the early history of this country, people were relatively self-sufficient. Families were responsible for supplying their own needs. As populations grew, however, the work force became more and more specialized. With the onset of the Industrial Revolution in the 1850s, cities started springing up over the countryside. Multitudes of people moved to these cities to seek employment, and this great shift of population brought problems to the cities. One

problem was supplying clean, fresh water in the city.

In the beginning, each family had its own well and its own waste disposal system, namely the outside toilet. Outdoor toilets, or "outhouses" as they are also known, might need explanation. All homes need toilets. For homes without plumbing for indoor toilets, outdoor toilets are the most common solution. These are normally small structures set away from the house far enough that odor is not a problem. Outhouses were common in the United States until the middle of the 1900s. Some still exist in remote locations in this country today and are still quite common in most developing countries. Eventually, the outside toilet gave way to indoor plumbing, with the aid of an underground septic tank. As populations grew and grew, cities had to install sewage disposal systems to take care of the large amounts of wastes. Over the years, research has improved the sewage discharge water quality. The techniques of waste disposal are covered in Chapter 9 and wastewater treatment is discussed in Chapter 15.

Urban pollution can also include chemicals applied to the roadways. This could be anything from detergents used to clean the roads to salts used to melt the snow and ice in the winter. The runoff of these chemicals enters the storm sewers and some eventually ends up in the rivers and streams.

Pollution from cities affects not only the surface water runoff but also the groundwater supply. Municipal dumps and landfills are potential sources of groundwater pollution. (See Figure 14-2.) If the landfill comes into contact with the water table, solids can infiltrate the supply. More common is the leaching of compounds such as methane and ammonia. Today, pollution of groundwater supplies from landfills is being monitored by test wells close to

FIGURE 14-2 Landfills are a potential cause of groundwater pollution.

the landfill sites. Most problems are removed when care is taken in selecting both the location and cover materials for the landfill.

INDUSTRIAL POLLUTION

The most common question a person might ask is "Who causes the most water pollution?" That question will probably be answered "industry." But because industry has more to lose from water pollution than any other segment of our economy, industry has also done the most to control water pollution. Some of the control measures were accomplished because of new federal water control standards.

Different industries use water in different ways; the pollution they create is, therefore, different. **Industrial pollution** of water can be divided into four major categories:

- thermal pollution
- radioactive materials
- organic chemicals
- inorganic materials

Thermal pollution involves returning heated water to a stream or river. Many industries use water to cool compounds or machinery. Although some thermal pollution occurs naturally in the summer, industrial processes create many more problems. The Federal Water Pollution Control Administration has attributed the largest industrial thermal pollution to chemical companies, steel manufacturing plants, and electric power plants. Of these three industries, electric power plants have the largest total. Whenever energy is changed from one form to another, heat is lost. In the electric power plant, the change of chemical energy (oil, coal, nuclear) to heat energy (burning) to mechanical energy (steam turbine) and then to electric energy involves a heat loss four times before electricity is produced. The power plant uses water to cool and carry the heat by-product away. The water is removed from a lake or river, heated, then returned warm, back to its source.

The dumping of heated water into a stream can create problems for both the stream or lake and the wildlife in and around the water. The most common problem involves disrupting the normal functions of the fish in the water. Some fish use water temperature to trigger spawning and migration. As they travel between the artificially warmed water and unaffected cooler water, their internal instincts are disoriented. Along with problems in fish reproduction, the warmed water will contain less oxygen, thus limiting the fish population the water can support. If the heated water is discharged for a long period of time, increased growth of algae will start to choke the stream.

If the growth becomes substantial, the algae will emit toxins that will eventually kill the fish population in the stream.

Even though the effects of thermal pollution are many, there are limited advantages to the warmed water. Research into this area has shown that the water could be placed in lagoons to produce tropical fish. Other possible uses include an ice-free water source for waterfowl, a heat source for greenhouses, and water for sewage plants.

The common techniques used to control thermal pollution are cooling towers and cooling lagoons. (See Figure 14-3.) The **cooling tower** employs the same principle an automobile radiator uses to cool your engine. The water passes through a series of coils around which air is forced by a fan, and the moving air cools the water before it is returned to the stream. A **cooling lagoon** is a pond beside the plant into which heated water flows. The water stays in the pond until it cools to the desired temperature. The goal of both methods is to return the heated water to the same temperature it was before entering the plant.

The second type of pollutant from industry is **radioactive material**, elements that emit radiation as a result of the disintegration of their atomic nuclei. The production of radioactive wastes has greatly increased over recent years. Medical workers use radioactive materials in the treatment of cancer. Many electrical power plants use nuclear power, and the wastes from these plants must be disposed of. If the wastes get into the water, fish and wildlife accumulate radioactive material in their tissues. If humans eat the contaminated fish, the radioactive material will pass on to them. The only way radioactive materials are destroyed is by the passage of time. Some radioactive materials take hundreds or even thousands of years to become safe. Radiation leaks are very rare in this country, but they can happen when radioactive waste from hospitals, factories, or research laboratories are disposed of improperly. Occasionally, a leak of radioactive gases occurs at nuclear power facilities.

The third industrial pollutant is **organic waste**, and both industry and agriculture are responsible for adding organic wastes to water. When a tree dies or a deer is killed, it starts to decay and decompose. Fungus and bacteria attack the dead organism and start the decaying process. In the process, the fungi and bacteria need oxygen to grow. On land, this is no concern; but in the water it can be a problem. The bacteria are much better users of water oxygen than are fish. If the waters are polluted with organic wastes either from industry or from a feedlot, bacteria will grow rapidly, placing great demands on the oxygen in the water.

If oxygen is not available to fish because of a high bacteria content, the fish will actually suffocate and die. Scientists have developed tests to determine the oxygen demand on water that directly indicates the amount of wastes in the water. This test is called the biological oxygen demand test, commonly referred to as the **BOD test**.

FIGURE 14-3 A cooling tower is used to control thermal pollution.

Inorganic industrial wastes are common in the manufacture of soaps and detergents, drugs and pharmaceuticals, paints, and fertilizers. Most control in this area is by complete stopping of dumping any chemical into water. These wastes must be destroyed by some other method.

AGRICULTURAL POLLUTION

Over the years, agriculture has changed greatly. The need to produce more and more food on less and less land has led to the development of intensified farming practices. If farmers are not careful, these practices can lead to water pollution. The most common agricultural pollutants include animal wastes, pesticides and fertilizers, and silt sediments.

Animal agriculture has seen many changes in the past years. Huge buildings and giant feedlots spot the nation. (See Figure 14-4.) These intense livestock areas create problems in the disposal of the waste materials. If the wastes are not properly disposed of, both surface water and groundwater will be affected. The most common disposal is to spread the manure on the fields. If great amounts are spread on small areas, leaching of the wastes may reach the groundwater supply. The most effective methods are to keep as much of the runoff water as possible away from livestock areas and to stockpile manures for later spreading only after bacteria have had an opportunity to decompose the wastes.

With increased food production came increased use of pesticides and fertilizers. The farmer knows how damaging insects, diseases, and weeds can be to a crop yield and has adopted the use of chemicals to help solve this problem. However, another problem arises with the overuse or incorrect application of

(a)

(b)

FIGURE 14-4 (a) Careful handling of animal wastes must be maintained to prevent either surface or groundwater contamination. This swine operation uses slotted floors to allow for collection of solid and liquid wastes. (b) The wastes are conducted to a nearby lagoon where the action of microorganisms breaks down the wastes to prevent them from becoming a source of pollution. *(Courtesy of Dr. Cindy Wood, Department of Animal and Poultry Sciences, Virginia Tech)*

chemicals. If chemicals and fertilizers are correctly used, they attach to the soil particles and stay there. (See Figure 14-5.) Some people, however, believe if the labeled directions say to use two pounds of chemical, then four pounds will do twice the job. When a rain comes, that extra, unattached chemical may either run off into a stream or leach into the groundwater supply.

FIGURE 14-5 In this photo, only minute amounts of pesticide are being applied to plant surfaces. This helps minimize leaching.

All pesticide users are required by federal law to comply with the printed directions on the package. This applies to homeowners as well as to exterminators.

When great amounts of fertilizers, especially nitrates and phosphates, reach the water in ponds or streams, the water becomes nutrient rich. This excessive enrichment of water is called **eutrophication**, and a nutrient-rich lake is called a eutrophic lake. What was once useful for livestock watering or swimming turns into an ugly, algae-filled lake. The longer the algae grow because of the nutrient-rich conditions, the more clogged the lake becomes. Eventually, the algae starts to die and decompose. This releases even more nutrients, thus causing the lake to become more eutrophic. Fish can no longer live in the water because of the toxins produced by the algae, and the lake starts to fill with vegetation. Within a short time, a once-beautiful lake becomes nothing more than a swamp area. (See Figure 14-6.)

FIGURE 14-6 A eutrophic lake.

The last agricultural pollutant we discuss is sediment. As was discussed in Chapter 5, erosion is very costly to the farmer due to lost farmland topsoil. Where does the soil go? Most of it moves into nearby lakes and streams. Once the water slows down, the soil will drop to the bottom of the lake or stream. This silt will soon fill the reservoir or stream and damage public water supplies. (See Figure 14-7.) Erosion-control techniques become very important when water pollution is added to the problems of soil loss.

WATER POLLUTION CONTROL MEASURES

The principal problem of water pollution control is to determine how much pollution is

FIGURE 14-7 Silt can fill in ditches, streams, and lakes.

acceptable. Do we want to have pollution-free water, or can we tolerate some pollution? If the latter is the case, how much pollution can we tolerate? Although governmental agencies continue to explore these issues, some comments on control can be mentioned here. Most industrial processes produce wastewater that must be disposed of. Research is needed to find better, less-polluting processes or even alternative uses of wastewater.

Research also needs to be conducted as to what can be done when accidents do happen. Even when we try our best, accidents, such as oil spills, will occur, and we must be ready with the best technology to control the damage.

DETECTING WATER POLLUTION

Much research has been done in trying to find a simple but effective way of detecting water pollution. One method developed involves examining streams for populations of invertebrates.

Invertebrates are animals having no spinal columns (vertebrae). The invertebrates we are concerned with are macroinvertebrates, which are large enough to have complex bodies and are visible to the naked eye. The descriptions of the classification system and specific organisms shown below are courtesy of the Virginia Museum of Natural History, Blacksburg, Virginia.

Group 1 organisms are generally pollution intolerant. That means they must have clean water to survive. (See Figure 14-8.) Examples of pollution intolerant macroinvertebrates include:

1. Stonefly: Order Placoptera
2. Caddisfly: Order Trichoptera
3. Water penny: Order Coleoptera
4. Riffle beetle: Order Coleoptera
5. Mayfly: Order Ephemeroptera
6. Gilled snails: Class Gastropoda

Group 2 organisms are somewhat pollution tolerant. That means you might find group 2 macroinvertebrates in water ranging from moderately polluted to very clean. (See Figures 14-9 and 14-10.)

1. Dobsonfly (Hellgrammite): Family Corydalidae
2. Crayfish: Order Decapoda
3. Sowbug: Order Isopoda
4. Scud: Order Amphipoda
5. Alderfly larva: Family Sialidae
6. Fishfly larva: Family Corydalidae
7. Damselfly: Suborder Zygoptera
8. Watersnipe fly larva: Family Athericidae (Atherix)

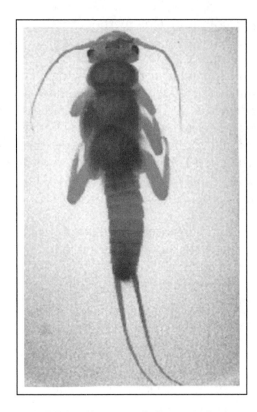

FIGURE 14-9 This dragonfly larva is a Group 2 organism and an indicator of a wide range of water-quality environments. They can survive in clean or moderately polluted streams and lakes. *(Photo courtesy Illinois State Department of Natural Resources)*

FIGURE 14-8 This stonefly larvae indicates a very clean water environment. Stonefly larvae are the least tolerant of organic stream pollution compared to all other stream-quality indicators. They have six legs, two tails, two sets of wing pads, and two claws on each foot. *(Photo courtesy Illinois State Department of Natural Resources)*

   9. Crane fly: Suborder Nematocera
  10. Beetle larva: Order Coleoptera
  11. Dragonfly: Suborder Anisoptera
  12. Clam: Class Bivalvia

Group 3 organisms are pollution tolerant. That means you may find these macroinvertebrates in water ranging from very polluted to very clean. (See Figure 14-11.)

   1. Aquatic worm: Class Oligochaeta
   2. Midge fly larva: Suborder Nematocera

   3. Blackfly larva: Family Simulidae
   4. Leech: Order Hirudinea
   5. Pouch snail and pond snails: Class Gastropoda
   6. Other snails: Class Gastropoda

By identifying the macroinvertebrates present in a stream or lake, you can roughly determine the water quality in the stream. In general, if only Group 3 organisms are present, the water quality is poor. If many Group 2 organisms are also present, but very few from Group 1 are present, the water quality is probably fair. If a larger number of Group 1 organisms are present, the water quality is probably good. If large numbers of all kinds of organ-

FIGURE 14-11 This pond snail (Family Lymnaeidae) is an example of a Group 3 organism. This is referred to as a right handed snail. When you hold the snail upright with the point up, the aperture (opening) is on the right hand side when it is facing you. Pond snails can survive in water ranging from clean to very polluted. *(Photo courtesy Illinois State Department of Natural Resources)*

FIGURE 14-10 This scud is an example of a Group 1 macroinvertebrate. It is very intolerant of pollution and indicates a clean water environment. The scud is sometimes called a side swimmer because it lies on its side when moving about. It has a segmented body and many legs. The body is flattened from side to side. Scud are white to tan colored and are an important food source for fish. *(Photo courtesy Illinois State Department of Natural Resources)*

isms are present, the water quality is probably excellent.

With a fine mesh net, a person can quickly examine a stream or other shallow body of water, to capture and identify the macroinvertebrates present in the water. Based on the macroinvertebrates identified, you can estimate the quality of the water environment. If the organisms present indicate very poor qual-ity water, then more elaborate laboratory tests may be called for. Your local water board or state university should be able to test the water at little or no cost to make scientific determinations of the water quality.

SUMMARY

Water pollution stems from many sources and can be grouped into pollution from cities, industry, and agriculture. Other urban pollutants include chemicals from roadways, dumps, and landfills.

Industry is working on water pollution but has a long way to go. Industrial pollutants include thermal or heat pollution, radioactive waste materials, and organic and inorganic wastes.

Intense agriculture has also led to pollution on the farm. Agricultural pollutants include animal wastes, pesticides, fertilizers, and silt sediments. Uncontrolled use of these materials can create eutrophic lakes and streams, destroying valuable water supplies.

Control is difficult. Some pollutants can be readily seen, but others can be found only through laboratory analysis.

DISCUSSION QUESTIONS

1. What are the main ways the urban environment pollutes our water supply?
2. What is thermal pollution?
3. How does thermal pollution affect fish populations?
4. How is thermal pollution controlled?
5. How do radioactive wastes affect our water?
6. What is the BOD test? What does it measure?
7. What are the three major agricultural pollutants?
8. What is meant by eutrophication? What causes it?
9. Why are water pollution control measures difficult to implement?
10. How are macroinvertebrates classified?
11. How does determining macroinvertebrate populations help in detecting water pollution?

SUGGESTED ACTIVITIES

1. Survey your area. Where is water pollution taking place?
2. Prepare a report on possible ways of correcting water pollution in your area.
3. Travel to a stream that has water runoff from some field or drainage file. Conduct a macroinvertebrate field study.

15 WATER PURIFICATION AND WASTEWATER TREATMENT

OBJECTIVES

After reading this chapter, you should be able to

- outline the kinds of impurities that must be removed from water for use by humans

- explain how water is purified in municipal water treatment systems

- list and describe the kinds of wastewater generated

- describe how septic systems are designed and how they work

- discuss how wastewater is treated before it is returned to the water cycle

TERMS TO KNOW		
human water cycle	turbidity	absorption field
solution	potable water	primary treatment
suspension	coagulation	secondary treatment
chemical impurities	flocculation	tertiary treatment
hard water	septic tank	
biological impurities	field line	

Water is one of the most basic natural resources. We use it to drink, to bathe, to water our crops, to manufacture goods and to move them to market. We use water to dispose of our wastes. In this chapter we examine the human water cycle as a subset of the water (hydrologic) cycle we discussed in Chapter 13. We then consider the processes by which water is treated for human use. Finally we discuss how wastewater is collected and treated before it is returned to the hydrologic cycle.

THE HUMAN WATER CYCLE

The hydrologic cycle is a natural process. Yet, the natural water cycle is only part of the story. To survive, humans must use small amounts of water from the hydrologic cycle. To thrive as an organized society, humans must remove significant amounts of water, use it in many different ways, and return almost all of it to the hydrologic cycle.

If you refer to Figure 13-1, you will see a drawing that illustrates the hydrologic cycle.

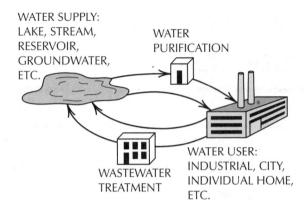

FIGURE 15-1 The human water cycle is really just a small part of the larger hydrologic cycle.

There is another part of the hydrologic cycle not shown in that drawing: the **human water cycle**. (See Figure 15-1.)

WATER AS A RESOURCE

In the entire world, there are about 332,000,000 cubic miles of water. That includes water in the oceans, rivers, lakes, and inland seas. It also includes water in the atmosphere in the form of clouds and water vapor, soil moisture, groundwater, water frozen in polar ice and glaciers, and water contained in plants and animals.

Each year about 119,000 cubic miles of water enters the water cycle through evaporation and falls to the earth in the form of precipitation. Of that amount, humans "use" about 700 cubic miles of water annually.

When we say that humans use that much water, we really should say we "divert" the water. It does not mean that we actually "use it up." It only means that we remove it from the natural water cycle, use it, and then return almost all of it to the natural water cycle. In fact, somewhere between 85 percent and 98

percent, depending on whose estimate you believe, of the water is returned immediately and unchanged to the natural water cycle in the form of surface water. Almost all of the remaining 2 percent to 15 percent returns soon thereafter, either by evaporation into the atmosphere, seepage into the groundwater, or as discharge from wastewater treatment plants. This 700 cubic miles of water represents the annual human water cycle.

As we discussed in Chapter 13, domestic use makes up only a small portion of water use by humans. In fact, of the 700 cubic miles of human-used water each year, only about 36 cubic miles went to domestic use in the latest year from which global estimates are available. It is that 36 cubic miles of domestically-used water that we discuss in this chapter. It amounts to about 5 percent of the annual human water cycle, 0.0003 percent of the natural water cycle, or 0.0000001 percent of the total world supply of water.

Almost all human activity relies on a reliable, abundant supply of clean water. Throughout history, human societies have prospered only beside fresh water. For most of our history, rivers or lakes have been precursors of all human development. Later, as humans developed technology, man-made canals or water transport systems such as the Roman aqueducts were constructed to make possible cities further away from natural water supplies. Still later, our developing technology made possible the use of groundwater in development of organized societies.

Along with the development of human society comes the creation of waste products. Water has long been the medium that transports much of our waste. Human waste products can come in any form: solid, liquid, or gas. The introduction of those waste products in the human water cycle that means that they

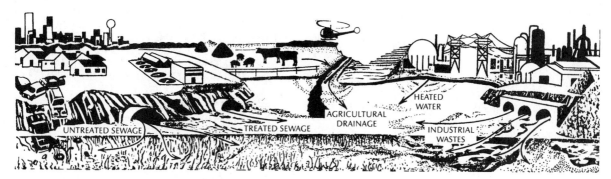

FIGURE 15-2 Various forms of water pollution.

will eventually enter the hydrologic cycle in the form of pollution. (See Figure 15-2.)

PROCESSING WATER

How water is treated before human use is determined in large measure by where it comes from. Water from clean groundwater needs little if any pretreatment before it is suitable for use. Water from municipal reservoirs, upland streams, and lakes with limited inflow may need only minimal treatment. Water from rivers into which industrial, agricultural, and municipal wastes have been introduced upstream may require very extensive treatment. (See Figure 15-3.)

Once the water has been used in an industrial, household, or municipal setting, it must be treated before it can be safely returned to the hydrologic cycle. There are two basic wastewater treatment systems in use in this country: individual household systems and large-scale wastewater treatment systems. Individual systems consist almost exclusively of septic treatment systems. Large-scale treatment systems can vary greatly based on the nature and size of the municipal or industrial system in which the water was used.

"PURE" WATER

We often talk about "clean" water or "pure" water. You may find those terms on bottled water for sale in your local grocery. In nature water is never pure and clean. If you want to test that statement, get a clean, clear glass and fill it with "pure" water from the drinking fountain nearest your classroom. Look at the water closely in front of a window or a bright light. The floating materials in the water should be obvious to you. If not, get a magnifying glass to examine the water more closely.

Water is never pure in nature because water is one of the most effective of all natural solvents. Solvents are substances that dissolve other substances. When one substance is dissolved in another, it is said to be in **solution**. In a solution, the molecules of one substance are dissipated among the molecules of another substance. To illustrate this concept, pour some table salt or sugar into a glass of water and stir the liquid. The salt or sugar "dissolves," and if you taste the "solution," you can taste the salt or sugar solution.

Water is also never pure in nature because it holds many other substances in **suspension**. When water moves (as in a stream or river), it can generate great amounts of force. Falling

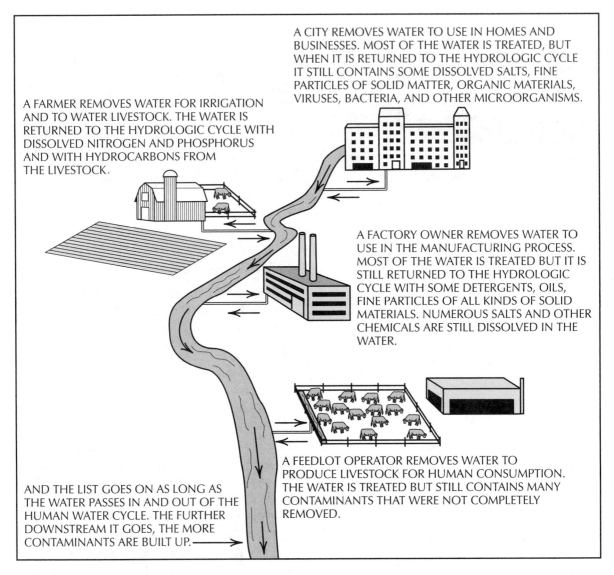

A CITY REMOVES WATER TO USE IN HOMES AND BUSINESSES. MOST OF THE WATER IS TREATED, BUT WHEN IT IS RETURNED TO THE HYDROLOGIC CYCLE IT STILL CONTAINS SOME DISSOLVED SALTS, FINE PARTICLES OF SOLID MATTER, ORGANIC MATERIALS, VIRUSES, BACTERIA, AND OTHER MICROORGANISMS.

A FARMER REMOVES WATER FOR IRRIGATION AND TO WATER LIVESTOCK. THE WATER IS RETURNED TO THE HYDROLOGIC CYCLE WITH DISSOLVED NITROGEN AND PHOSPHORUS AND WITH HYDROCARBONS FROM THE LIVESTOCK.

A FACTORY OWNER REMOVES WATER TO USE IN THE MANUFACTURING PROCESS. MOST OF THE WATER IS TREATED BUT IT IS STILL RETURNED TO THE HYDROLOGIC CYCLE WITH SOME DETERGENTS, OILS, FINE PARTICLES OF ALL KINDS OF SOLID MATERIALS. NUMEROUS SALTS AND OTHER CHEMICALS ARE STILL DISSOLVED IN THE WATER.

A FEEDLOT OPERATOR REMOVES WATER TO PRODUCE LIVESTOCK FOR HUMAN CONSUMPTION. THE WATER IS TREATED BUT STILL CONTAINS MANY CONTAMINANTS THAT WERE NOT COMPLETELY REMOVED.

AND THE LIST GOES ON AS LONG AS THE WATER PASSES IN AND OUT OF THE HUMAN WATER CYCLE. THE FURTHER DOWNSTREAM IT GOES, THE MORE CONTAMINANTS ARE BUILT UP. ⟶

FIGURE 15-3 The degree of contamination in a water supply is dependent on many things, including how many human uses the water has passed through before arriving at a given point.

raindrops striking the surface of unprotected soil dislodge soil particles, as we discussed in Chapter 5. Running water tears at the surfaces it moves across causing erosion and picking up pieces of all kinds of solid materials. (See Figure 15-4.) Large particles may be held in sus-

pension only while the water is moving very rapidly. When the water moves more slowly, these particles quickly settle out. Smaller particles may remain suspended for longer times. Very fine articles may remain in suspension for extended periods of time.

FIGURE 15-4 Major flooding occurred in January 1996 in the Upper Ohio River Basin. The sediment suspended in this water because of the flooding resulted in increased turbidity for the entire length of the river system. *(Photo courtesy of Pat Murphy, Meteorologist—Pittsburgh Office of the National Weather Service, National Oceanographic and Atmospheric Administration)*

WATER IMPURITIES

Before we can use water for human consumption, many impurities must be removed and many naturally occurring conditions must be changed in the water. Some of the impurities are dangerous, and we remove them for health reasons. Others are merely undesirable, and we remove them for aesthetic reasons. Three broad categories of impurities that are removed from water during the treatment process are

- chemical Impurities
- biological Impurities
- turbidity

We have come to think of "pure" water as water that has no color, no taste, and no odor. Anything that causes a strange taste, color, or odor in water is aesthetically undesirable. People will go to great lengths to remove those impurities before drinking the water.

Chemical Impurities

Chemical impurities in water can result in one of three conditions: water that is dangerous to drink, undesirable color, or undesirable smells. Water may also be too acidic or alkaline.

To be usable for human consumption, water should be neither very acid nor very basic. In fact, the pH of water should be near 7.0 (neutral). Water with a pH much lower than 7.0 is acid and can produce corrosive effects on metal equipment and pipes used in transporting the water to our homes. Water with a pH much higher than 7.0 is alkaline. Very alkaline water is not only corrosive but also tends to produce scale on pumping machinery and insides of pipes in the form of carbonate deposits.

In nature, there is no such thing as "pure" water. Water is one of nature's most versatile solvents, so a wide range of minerals, metals, and salts are almost always present in even the cleanest water. Distilled water comes as close to being pure water as most people will ever encounter, but even distilled water almost always contains a few "impurities." In fact, relatively pure water has no taste whatsoever, so most people find drinking distilled water to be unpleasant. We like a little "taste" in our water. In addition, low levels of dissolved minerals and metals in their drinking water are usually healthful for people. It is only when the concentrations of dissolved materials get too great that water no longer tastes good or becomes detrimental to human health.

Water from ground supplies often contains dissolved iron or manganese. Iron and manganese dissolve in water that has very little oxygen, such as in groundwater, in their ionic forms (Fe^{++} and Mn^{++}). When the water is exposed to oxygen once it reaches the surface, the ions combine with the O_2 to form oxides.

You may have heard of ferrous oxide—if not by that name, you are familiar with it as "rust." When water with iron or manganese in solution is exposed to air, the metals "oxidize" and can form deposits on toilet bowls, sinks, and other exposed plumbing fixtures causing red or brown stains. In addition, both oxides have very unpleasant tastes and people do not like drinking highly-concentrated "iron" water. If the concentration of metals is very high, the water will have a reddish color and even smell like "rust."

Another very common chemical problem results in what we call **hard water**. This condition is usually a result of excess calcium or magnesium. In water, excess calcium (Ca) combines with oxygen to form calcium carbonate ($CaCO_2$). Excess magnesium combines with the water itself to form magnesium hydroxide [$Mg(OH)_2$]. Both of these compounds precipitate out to form scale deposits that clog up water pipes and plumbing fixtures. They also react with soap making it less effective. They also taste bad. Hard water is treated with "softening" agents.

A problem in many areas is sulfur content in water. When sulfur is exposed to oxygen in water, it forms sulfur dioxide (SO_2) gas. You may be familiar with this, if not by its chemical name, then more likely by its smell. Sulfur dioxide is the odor that results when eggs begin to rot. It is not very likely that you would want to drink water if it smells like "rotten eggs."

Dissolved nitrogen and phosphorus are both common contaminants in surface water supplies. Both nitrogen and phosphorus are included in common fertilizers and are water-soluble. They are applied to the land to help plants grow, then they dissolve in rain water and are carried by runoff into streams, rivers, and lakes. Excessive nitrogen and phosphorus must both be removed from water before it is usable for human consumption. In addition, both promote the growth of algae in water, and excessive algae result in the formation of the hydrocarbons. Excessive nitrogen dissolved in drinking water can present serious human health problems.

Dissolved hydrocarbons are another common problem in water. Dissolved methane, methanol, and various alcohols may occur in water that contains or flows across organic matter. A troublesome hydrocarbon in some regions is tannin, which results from decaying organic matter. Other problem chemicals in water include excess chlorine and fluorine. Excess chlorine causes bad taste and smell, and excess fluorine can cause discolorations on teeth.

Biological Impurities

Biological impurities in water range from mammals to viruses and from aquatic trees to algae. In fresh water that will be used for human consumption, those organisms must be removed or neutralized.

In water treatment, fish, crustaceans, worms, and macroinvertebrates are removed at the very beginning by screening. The waste products from those animals are considered chemical contaminants and are included in the previous section. (Where else would a fish "go to the bathroom"?) Larger plants and plant parts are also removed by screening. Small plants, including algae are not completely removed by screening and require treatment.

Algae are tiny green plants that grow in sunlight and in the presence of air. Algae are generally harmless but in excessive concentrations can extract almost all oxygen from water so that other organisms such as fish are killed. This is not a problem in terms of the human water supply, but algae do cause the formation of hydrocarbons that enter water and must be removed, as we discussed previously.

Bacteria and fungi are a critical component of all healthy water supplies. They serve as decomposers and help break down organic matter and chemicals into harmless compounds. But they can also be dangerous to humans as pathogens (disease-causing organisms.) As far as we know, viruses do not serve similar valuable uses, but they may become pathogens when humans or animals drink the water, or even when they are exposed to the water. Treatment of water for virus contamination is more difficult and, in fact, effective treatment for viruses was not even possible until recent years. Common bacterial diseases carried in water are typhoid fever and cholera.

Protozoa are single-celled animals that occur naturally in all healthy water supplies in nature. In humans, certain protozoa can be pathogenic. Common human diseases caused by these organisms are diarrhea and dysentery. Similar to these organisms are helmiths, microscopic wormlike organisms that can infest the intestinal tract in infected humans causing illness.

It is easy to see why treatment of water for human consumption must include removing or neutralizing biological contaminants.

Turbidity

Turbidity refers to solid matter suspended in a liquid. In nature, water always contains some solid matter held in suspension. Thus all naturally occurring water has at least some degree of turbidity. When the concentration of solids in suspension become great enough to be visible, the water is aesthetically unpleasing. It may have color, odor, or taste. If the turbidity is more extreme, the water may appear to be muddy. Removal of turbidity is an important step in processing water for humans to use.

WATER TREATMENT

The objective of water treatment is to produce a potable water supply. **Potable water** is water that is chemically and microbiologically safe and that is otherwise suitable for human consumption.

In some rural areas, water is captured by people directly from precipitation in farm ponds, cisterns, and other kinds of catch basins for use in watering plants and livestock. People use wells and pumps to remove still more water from the ground for drinking water or other domestic uses. Much of that captured water is relatively clean and safe when left untreated, and can be used for human consumption without purification.

Potable water taken directly from nature is the exception and not the rule, however, in developed parts of the world and particularly in industrialized nations. The vast majority of water must be purified before it is safe for human consumption.

Well Water

Much well water is safe to drink as it comes from the ground. Well water should be tested periodically by local authorities for biological or chemical contaminants to ensure that it is safe to drink. Even wells that have been used by families for many years should be tested. In the United States, your local government or your state environmental agency provides such testing at little or no cost.

Well water may contain undesirable levels of chemicals such as iron, manganese, or sulfur. It may contain excess calcium or magnesium. Treatment systems are available from commercial sources to remove metals and minerals from water. More common are so-called water softeners that are also available commercially. You can find commercial water softeners, iron and sulfur removal systems, and activated charcoal filter systems for household use in your yellow pages and on the Internet.

Municipal Treatment Plants

Municipal treatment plants vary greatly, depending on the size of the community being served and on the nature of the water source. Water that is moving into a treatment system is referred to as an influent. Water that comes out the other end of the system and is ready for use is called effluent. That can be confusing because the same terms will be used later for sewage. Just remember that influent is flowing "in" and will be treated and effluent is flowing "out" because it has been treated. Following is a generic description of a typical municipal water treatment plant.

Influent is first filtered to remove the largest particles of solid matter and large organisms. The influent is mixed with a coagulator in a rapid mixing tank (much like a huge kitchen blender). A coagulator is a chemical that promotes the coagulation of solid materials in suspension. **Coagulation** is the physical process of smaller particles clumping together to form larger particles that will later be allowed to settle out of the water.

If excess metals or salts are present or if the water's pH is too far from neutral, auxiliary chemicals are added at the coagulation stage also. The water/coagulator mix passes into a flocculator. The flocculator is a large tank in

which paddles gently stir the water to encourage the clumping together of large particles of solid matter that will settle out later.

The coagulated, chemically-treated water flows out of the flocculator into the bottom of a vertical settling tank. In the settling tank, as more water is added to the bottom, the older water gradually and gently moves upward. As it moves upward, the coagulated debris settles out to the bottom of the tank. A very gentle scraping system mechanically continually removes the accumulated settlings from the bottom of the tank. The water is removed from the top of the settling tank and passed through a final filter made up of graduated layers of sand and gravel, and still more activated carbon. From there it will be treated with chlorine and fluorine and moved into the distribution system for disposal or use. (See Figure 15-5.)

Once the water leaves the settling tank, it is passed through an activated carbon filter that starts to neutralize odors and tastes and then it passes through a final filter. If the water source was particularly contaminated, such as a river, the coagulation/**flocculation**/activated carbon filtration process may be repeated one more time before final filtration.

WASTERWATER TREATMENT

Septic Systems

In rural areas and in many small communities, the basic form of human sewage disposal is through the use of a septic system. A typical septic system consists of four parts: the septic tank, distribution box, field lines, and absorption field. (See Figure 15-6.)

The raw sewage flows downhill from the home into an underground **septic tank**. The solids in the sewage settle to the bottom of

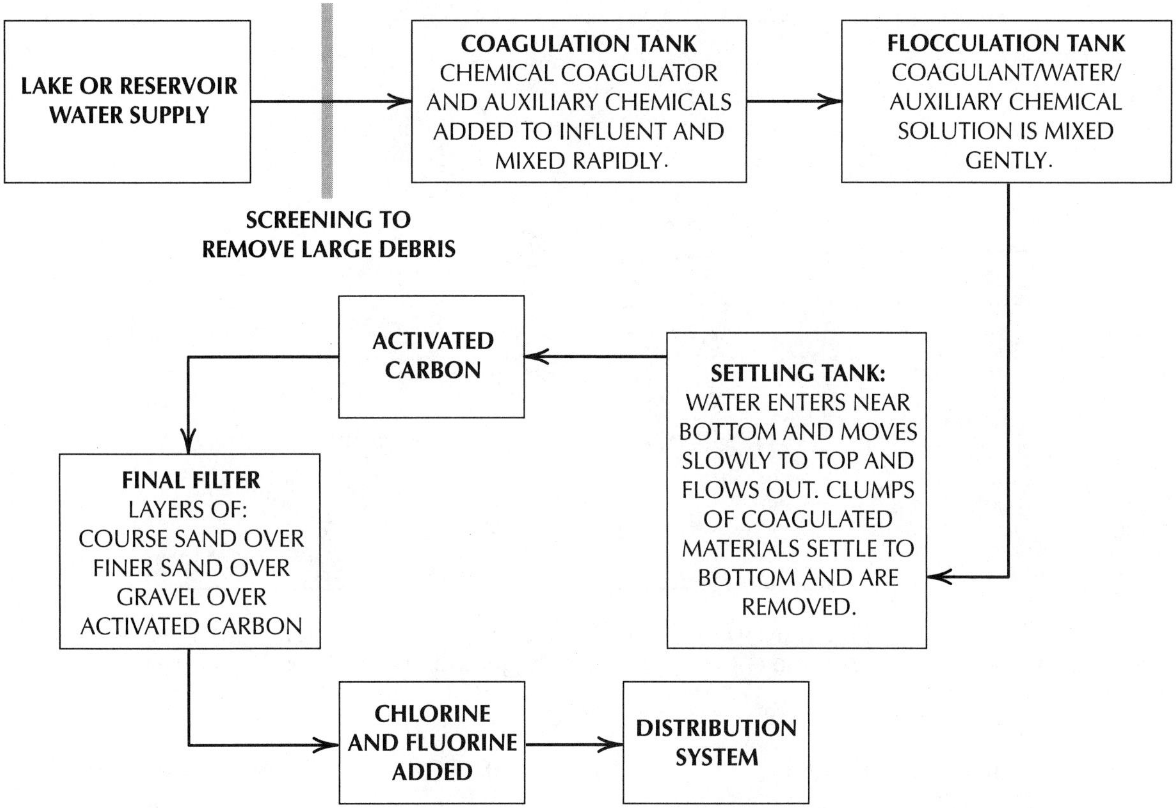

```
┌─────────────────────┐              ┌─────────────────────────┐        ┌─────────────────────────┐
│                     │              │   COAGULATION TANK      │        │   FLOCCULATION TANK     │
│  LAKE OR RESERVOIR  │              │ CHEMICAL COAGULATOR     │        │ COAGULANT/WATER/        │
│   WATER SUPPLY      │───▶          │ AND AUXILIARY CHEMICALS │───▶    │ AUXILIARY CHEMICAL      │
│                     │              │ ADDED TO INFLUENT AND   │        │ SOLUTION IS MIXED       │
│                     │              │ MIXED RAPIDLY.          │        │ GENTLY.                 │
└─────────────────────┘              └─────────────────────────┘        └─────────────────────────┘
```

SCREENING TO REMOVE LARGE DEBRIS

```
            ┌─────────────────┐        ┌─────────────────────────┐
            │    ACTIVATED     │        │  SETTLING TANK:         │
            │    CARBON        │◀───    │ WATER ENTERS NEAR       │
            └─────────────────┘        │ BOTTOM AND MOVES        │
                                       │ SLOWLY TO TOP AND       │
  ┌─────────────────────┐              │ FLOWS OUT. CLUMPS       │
  │   FINAL FILTER      │              │ OF COAGULATED           │
  │   LAYERS OF:        │              │ MATERIALS SETTLE TO     │
  │ COURSE SAND OVER    │              │ BOTTOM AND ARE          │
  │ FINER SAND OVER     │              │ REMOVED.                │◀───
  │ GRAVEL OVER         │              └─────────────────────────┘
  │ ACTIVATED CARBON    │
  └─────────────────────┘

        ┌─────────────────────┐        ┌─────────────────────┐
        │  CHLORINE           │        │  DISTRIBUTION       │
        │  AND FLUORINE       │───▶    │  SYSTEM             │
        │  ADDED              │        │                     │
        └─────────────────────┘        └─────────────────────┘
```

FIGURE 15-5 A schematic diagram of a typical single-stage (relatively clean water source) municipal water treatment system. If the water source is from a river or more contaminated lake, a second co-agulator and settling tank may be included. Such a system would be a two-stage treatment system.

the tank in the form of a sludge that slowly decomposes. The process produces a foul smelling gas (you will know it if you smell it.) The gas rises to the surface, carrying with it oils and greases. The oil and grease form a thick scum at the surface of the liquid. The tank contains a whole array of microorganisms (bacteria, fungi, and others) that live on the materials in the scum and sludge, slowly digesting them. (See Figure 15-7.)

You will notice from looking at the diagram, the tank outlet is located between these layers of scum and sludge. It is in this layer that the clearest liquid is found. The liquid drains out of the tank and moves by gravity to the distribution box. If the **field lines** are downhill from that point, the distribution box simply serves as a holding tank in which still more biological decomposition can occur and from which the liquid flows downhill to the **absorption field**. If the field lines are above either the septic tank or distribution box, a pump will be necessary to raise the liquid to the next component.

Once the liquid reaches the absorption field, it flows into the field lines. The field lines consist of tiles or pipes with spaces between the tiles or holes in the pipes. The liquid seeps

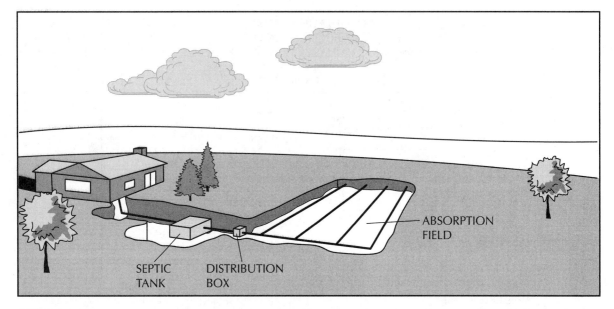

FIGURE 15-6 A typical septic system consists of four components: septic tank, distribution box, field lines, and absorption field. The entire system is underground. *(Drawing courtesy of the United States Environmental Protection Agency)*

SECTION VIEW OF SEPTIC TANK

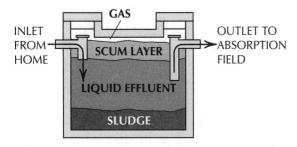

FIGURE 15-7 The septic tank holds effluent from the home while microorganisms digest the organic matter and neutralize most of the pathogenic organisms so that the effluent can be moved into the soil through the absorption field. *(Drawing courtesy of the United States Environmental Protection Agency)*

through the cracks or holes and is absorbed into the surrounding soil through a bed of gravel. (See Figure 15-8.)

SECTION VIEW OF ABSORPTION TRENCH

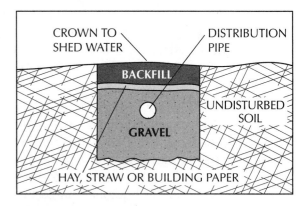

FIGURE 15-8 The field lines slope downhill from the distribution box. Each field line is laid in a trench with a bed of gravel around it. In previous years, the line would be made of tile with cracks left between the tiles. Today, the lines are usually made of synthetic material such as polyethylene with perforations (holes) for the liquid to seep through. *(Drawing courtesy of the United States Environmental Protection Agency)*

With sewage flowing into it daily, it would seem that the septic system would fill up quickly. The liquid flows quickly through the system and into the soil, where it is attenuated naturally. The microorganisms break down the solids into water soluble material, liquids, and gases. That material then flows out with the water into the absorption field and also seeps into the ground. If careful planning as to the size of tank and distribution system is done before installation, the tank should never become full.

Municipal Wastewater Treatment Systems

As more and more people moved into metropolitan areas, septic systems could not handle the increased wastes. The space needed for the decomposition of wastes was not available. Wastewater treatment systems were developed to render the water nonpolluting. This process is divided into three phases: primary waste treatment, secondary waste treatment, and tertiary waste treatment.

The primary waste treatment system is considered a mechanical system that involves collecting the wastewater and removing the items that settle from it. The **primary treatment** system includes a screening device, grit chamber, settling tank, sludge digester, and drying beds. (See Figure 15-9.) The primary system removes about two thirds of the wastes from the water.

As the wastewater moves through the city's sewer system, it collects large and small materials that might have washed from the streets as well as from housing units. It is collected at the sewage treatment plant. The large objects are screened out first at the treatment plant. This could range from tree limbs to auto parts. The remaining liquid and suspended solids are

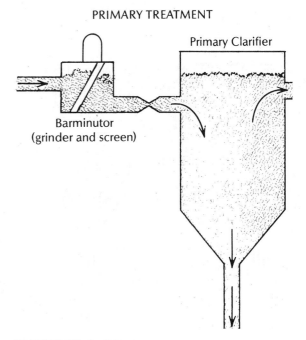

PRIMARY TREATMENT

Primary Clarifier

Barminutor
(grinder and screen)

FIGURE 15-9 Primary treatment system.

then run through a grit chamber. In the grit chamber, the wastes are ground into finer particles to speed up the decomposition process. Once the wastes have been ground, they are allowed to settle out in the sedimentation tank. Here, the soil, grease, and other suspended wastes are removed and placed in the sludge digester. The water then returns to the lake or stream if this is the only system the city has. The solids in the digester are decomposed by anaerobic (non–oxygen-using) bacteria, and the by-products of the digester are sludge and methane gas. The sludge can be sterilized and used as a soil amendment for potting plants as well as a source of plant nutrients for forest or field crops. There is much debate over the safety of using sludge to produce food for animal or human use. The methane gas may be used to heat the treatment plant.

The secondary waste treatment system involves biological processing of sewage.

SECONDARY TREATMENT

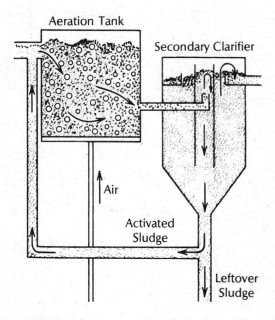

FIGURE 15-10 Secondary treatment system.

(See Figure 15-10.) Over one half of all city treatment plants contain this system. This system will remove about 85 percent of the organic wastes in the water. The **secondary treatment** system is an extension of the primary system. The wastewater is piped through an aeration tank to increase the oxygen content in the sewage water. With the increase in oxygen, the number of aerobic (oxygen-using) bacteria is increased, thus increasing the efficiency of the waste treatment. The water is then pumped into another sedimentation tank for additional settling. The remaining water is discharged after the addition of chlorine. The settled sludge, called activated sludge, is sent through the aeration tank again to allow further bacterial decomposition. This system removes about 90 percent of the organic wastes. Chemical compounds, such as nitrates and phosphates, remain and require additional treatment to be removed.

The third system, tertiary waste treatment, is the chemical processing of sewage wastewater. The **tertiary treatment** system involves removing from the water phosphates and nitrates, which could cause a nutrient-rich situation if the sewage effluent is dumped into a stream. This system is more costly than the secondary system and is used only when the water is going to be reused. The most common procedure is to add lime to remove the phosphates and run the effluent through a stripping tower to remove the nitrates. A stripping tower is simply a vertical tank through which the water moves slowly while the lime and water are in contact, allowing the chemical reaction that removes the excess nitrates.

Although the main goal of sewage treatment is to return the water to a nonpolluting state, a great deal of sludge is produced. In a city of 1 million people, it is estimated that about 35,000 tons of sludge are produced per year. This material can be just as polluting as the wastewater. On the other hand, the sludge can be beneficial as a soil builder and fertilizer. If the sludge is to be used in the production of food products, it must be sterilized to ensure the destruction of disease-producing organisms. Many experts believe that this process renders the sludge safe, but others are unsure.

SUMMARY

Only a tiny portion of the water in the annual hydrologic cycle is ever used by humans. The vast majority of the water we use is actually only diverted temporarily to power or hydroelectric plants and for other uses. That water does not require any sort of treatment before we return it to the hydrologic cycle.

Some of the water will be used for human consumption. In nature, water is never completely pure or clean. Humans require water that is relatively clean before we can use it safely. To take water from the hydrologic cycle and prepare it for human consumption requires some degree of treatment. Water that is relatively clean may only require pumping from the ground. Water in a lake or river must be treated before it is safe to consume.

Once we have used water in our homes, businesses, and factories, it contains many kinds of contaminants. We must remove most of those contaminants before we can safely return the water to the hydrologic cycle. In individual homes, simple septic systems are adequate for that purpose. In larger settings such as towns and cities, wastewater treatment facilities are required.

DISCUSSION QUESTIONS

1. What is the relationship between the natural water (hydrologic) cycle and the human water cycle?

2. What kinds of contaminants do we normally find in water in our lakes and rivers?

3. Describe the mechanical and chemical processes required to prepare water for human consumption.

4. Explain how a septic system works.

5. Explain how a municipal wastewater treatment system works.

SUGGESTED ACTIVITIES

1. To illustrate the effect of suspension, put a small amount of soil into a glass jar or beaker. Pour three to four times as much water over the soil. Put a drop of dish detergent into the mixture. Cap the jar or beaker and shake it vigorously. Observe what happens.

2. Take a class field trip to a wastewater treatment plant in your local community. If that is not possible, invite the plant manager or supervisor to come to your class to explain how the plant works.

3. Set up a sand, gravel, and ground charcoal filtration system in a five-gallon can. Drill holes in the bottom. Pour turbid water in the top and see what comes out of the bottom.

16 WATER-USE PLANNING

OBJECTIVES

After reading this chapter, you should be able to

- explain the principal water management techniques
- explain how to remove salt from water
- identify ways to reuse water
- explain how water runoff can be controlled in urban areas

TERMS TO KNOW		
dams	membrane process	vapor compression distillation
reservoirs	electrodialysis	
ponds	distillation process	flash distillation
desalination	long-tube distillation	

In Chapter 13 we examined water suppliers and water users and saw that water is important to many people and industries. The question soon arises as to whether there will be enough usable water to go around. Through careful conservation practices and water-use planning, each water need should be met. The principal water management techniques include

- dams, reservoirs, and ponds
- waterway utilization
- desalination
- reuse of water
- weather modification and control
- water runoff control in urban areas

DAMS, RESERVOIRS, AND PONDS

Dams, **reservoirs**, and **ponds**, an important part of the water management picture, have two basic uses. The first is to reduce flooding hazard. Additionally, if water can be collected and stored from a watershed area during peak rainfalls, it will be available when rainfall is not as abundant.

Most water-holding structures, whether dams, reservoirs, or ponds, serve several additional uses. They can provide recreation (such as swimming, boating, and fishing), supply water to a town or a city, make water available to industry, or provide irrigation water to farms. Some dams can also provide power to nearby communities or even larger areas. (See Figure 16-1.)

FIGURE 16-1 Dams provide many opportunities for water-use planning. *(Courtesy of Photo-Disc)*

There are also disadvantages associated with the construction of dams and reservoirs. When earth must be moved and concrete poured, great expense is incurred. The cost of the dam must be outweighed either by the value of crops that would be lost, property damage that would occur during the flood season, or by the value of its stored water for recreation, industrial, urban, or agricultural use.

If the main objective of the construction of a dam or reservoir is flood control, the water must be lowered before the rainy season begins. A fluctuating water level can cause serious problems with fish populations. It is usually the goal of the water manager, however, to balance both flood control and fish populations.

Consider the vast landmass covered by the water when dams are built. It seems ironic that we cover thousands of acres of land with water and then pump that same water to irrigate other lands. Nature very rarely balances the amount of water available to all parts of the land and throughout the year. In general, smaller upstream reservoirs are more valuable and less detrimental than huge downstream dams. For whatever purpose the reservoir is constructed, the watershed areas need to remain protected from erosion as much as possible. The life of the reservoirs or dams will be greatly reduced if silt resulting from soil erosion is allowed to pour into the basin. For a discussion on siltation, see Chapter 5.

WATERWAY USE

Another water management practice is the use of water to move commodities. The power required to move goods on water is less than that required to move them on land. Maintenance costs on ships are less than on rail, truck, or air equipment. The capacity of water-transportation systems is generally larger than other methods. The most common water-transported items are petroleum and coal products.

Waterways can also be helpful in our national defense programs. The ability to move cargo on many different avenues may prove advantageous during war or other national emergencies. During World War II, merchant ships used the intracoastal waterway to avoid German submarines.

DESALINATION

The largest bodies of water, which cover over 70 percent of the earth's surface, are the oceans. They contain a large array of wildlife, as will be discussed in Chapter 25. If freshwater supplies lessen to a critical stage, will we be able to use the water of the oceans? Owing to its high salt content, ocean water is not useful in its present state. To tap the ocean water resource, we must first remove the salt. This process of salt extraction is called **desalination**.

Much research has been conducted in recent years on both desalination processes and the economic feasibility of removing the salts. The research has been coordinated by the Office of Saline Water (OSW) with the help of universities, colleges, and the Atomic Energy Commission. The main goals of the OSW are (1) to provide desalination technology to supply immediate water to areas where water is in short supply, (2) to develop hardware for desalination processes, and (3) to develop new processes of desalination. Much work has been done, and test plants have been constructed in the city of Los Angeles with several more planned.

The desalination of ocean water has been accomplished with success by several processes. The two techniques examined in this chapter are membrane and distillation desalination. Other processes being researched include freezing action, treating with chemicals, and humidification. Development of all these processes must be continued to ensure an adequate supply of fresh water.

The **membrane process** of desalination is used mainly on inland brackish water where the salt content is lower. The membranes are filters that allow water to pass through but keep the salts out. The main membrane methods are reverse osmosis and electrodialysis. Reverse osmosis uses the natural principle of liquids flowing through a semipermeable membrane. (See Figure 16-2.) Under natural conditions, water will flow from a fresh to a salt solution. The opposite situation is what we want to happen; if we apply pressure to the salt water, this occurs. This process needs very little energy and no heat, and it is hoped that this method can be developed to become more efficient for the future.

The second membrane process is that of **electrodialysis**. This process uses the fact that ocean salt water contains sodium and chlorine. When subjected to an electric current, these two elements separate and pass through different filters. What is left behind is fresh water, which is pumped away. This process requires a great deal of electric power and uses expensive filters. Unless the costs of

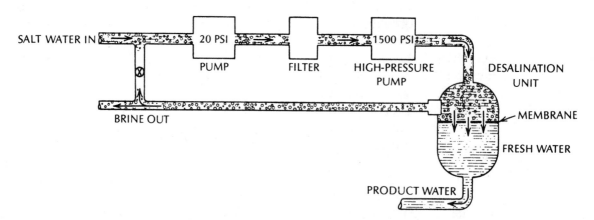

FIGURE 16-2 Membrane desalination method.

both are lowered, the use of electrodialysis will remain limited.

The **distillation process** of desalination is common to anyone who has taken high school chemistry. Nature uses this process in the hydrologic cycle. Basically, when water is heated, it evaporates and leaves behind any solid impurities, namely, the salt. All you have to do is collect and condense the vapor and you have fresh water. The three main distillation processes developed are long tube, vapor compression, and flash distillation.

In **long-tube distillation**, salt water is sent through tubes where it is heated by steam. The vapor (or steam) travels through a series of tubes to where it is finally cooled and collected. The steam cools by transferring its heat to the incoming cold salt water needing to be distilled. This conserves energy and lowers the cost of distillation. (See Figure 16-3.)

The **vapor compression distillation** process involves placing the water under pressure when turned into steam. This steam starts a chain reaction that heats other water. When some of the steam condenses, it falls to a bottom tank and is then collected as fresh water. The energy used in this process is mainly that needed to start the process and to run the compressor motor.

Flash distillation uses the fact that water boils at a lower temperature when it is at a lower pressure. This process involves heating the water and allowing it to flow into a chamber. A partial vacuum is produced causing the water to immediately start to boil or "flash" into steam. The steam is then cooled and collected as fresh water. To save energy, the steam is cooled by the incoming salt water, thus raising its temperature and requiring less heat energy to cause it to flash.

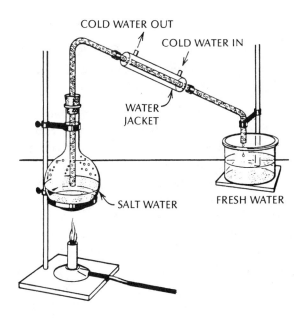

FIGURE 16-3 Distillation method of desalination.

USE AND REUSE OF WATER

What is the quality of water we use? Most commonly the water we use for washing our clothes, cooking, and bathing is the same quality we use for drinking purposes. If water becomes extremely short in supply, it would be wise to reuse or even reclassify the water we use. Why use excellent quality drinking water to flush our toilets, wash our cars, and water our lawns? We could reserve the top-quality water for drinking purposes or for cooking.

One proposal, which would not be popular, is to recycle all water, even sewage water. Although the suggestion seems repulsive at first, many towns and cities are already doing just that. Consider a series of towns and cities located along a common waterway. The first town draws its water from an upstream source and releases its waste effluents downstream. The next town downstream draws its water from an upstream source, which contains the sewage from the first town. This procedure may continue for hundreds of miles and involve many towns. Each town is recycling water from another town. The main problem facing experts in recycling is the removal of odor, tastes, and salts from the water. Also, if pathogens have entered the water, disease such as infectious hepatitis could become a problem. Nobody wants to think of drinking sewage water, but many people do just that every day.

As we discussed in several earlier chapters, industry is a large user of water, and has the most to lose if it is not conserved. The federal government is placing tighter pollution standards on industrial water supplies. It is forcing industry to develop ways to use water over and over.

WEATHER CONTROL

Have you ever gone skiing when they were not calling for snow? That is no big problem. The resort area will just make some new snow for the ski slopes. We have attempted to control the weather for centuries. For hundreds of years, weather control or modification has been considered as a source of water. The Native Americans had rain gods and rain dances. When the West went through a drought, rainmakers would show up claiming they could make it rain. Modern technology has led to the limited use of weather control and weather modification.

The most common and oldest method of making rain is to seed the clouds with silver iodide crystals. These crystals give the water droplets in a cloud something to which to cling. As the droplets come together and become heavy enough, they fall to the earth in the form of rain. Many problems exist in cloud seeding. One common problem is to attempt to find the correct cloud type. At best, we can only increase rainfall by about 20 percent. Another problem with seeding the clouds is to get the rain to fall where it is needed. It takes time for the water droplets to form before they become heavy enough to fall as raindrops. If the lag time is not correctly calculated, the rain will fall at the wrong time and in the wrong place. Another problem is that seeding a cloud in Kansas removes water that might have fallen as rain in Missouri.

Being researched at the present time is the modification of tornados and hurricanes. The experiments have mainly involved lowering the speed of wind associated with the storms. The seeding of the eye of a hurricane with silver iodide has lowered the speed of the hurricane somewhat. Work on tornado modification continues at the present time.

WATER RUNOFF CONTROL IN URBAN AREAS

The runoff from city streets and roads is a sizable amount of water. A 1-inch rain dumps 43,560 gallons of water on each acre of land. With the construction of sidewalks, roadways, and buildings that do not soak up the rainfall, most of the water runs off. City gutters and storm sewers transport the water to streams or rivers where it is dumped. At times this large amount of water causes floods downstream from the city. With possible water shortages, cities might be wise to collect this runoff for a time when water will be in short supply.

SUMMARY

Water-use planning is a very important part of ensuring that an adequate water supply is available for use. The common techniques used in water planning are the construction of dams, reservoirs, and ponds; waterway utilization; desalination; water use and reuse; weather control and modification; and runoff control in urban areas. Each technique has its advantages as well as its shortcomings.

DISCUSSION QUESTIONS

1. How are dams and reservoirs used in water planning?
2. What is desalination?
3. Explain the two techniques of desalination.
4. How does the distillation process work?
5. How can the weather be controlled to increase our water supply?
6. How does the control of water runoff in urban areas increase our water supply?

SUGGESTED ACTIVITIES

1. Visit the water treatment facility serving your community. Questions to ask include: What is the primary water source? What is the annual purification capacity? What are the allowable water quality standards?
2. Set up a small distillation apparatus. Your science or chemistry teacher should have all the equipment needed. Use several sources of water, including some that may not be very clean, to distill some drinkable water. How does distilled water taste? Why?

17 CAREERS IN WATER MANAGEMENT

OBJECTIVES

After reading this chapter, you should be able to

- discuss the jobs, working conditions, and salaries of a career in water management

TERMS TO KNOW

geologist	meteorologist	wastewater treatment plant operator
geophysicist	water treatment plant operator	
hydrologist		
oceanographer		

Water is all around us and affects us in a variety of ways. Careers in water management range from predicting when and where water will fall, to careers in finding out where it went. There are careers requiring much education and careers requiring very little formal education. Usually, as with most career areas, as the requirement for education decreases, so do the salaries received. The four main environmental science occupations are geologists, geophysicists, aquatic scientists, and meteorologists. There are also careers as waste treatment plant operators and many supervised labor positions.

GEOLOGISTS, GEOPHYSICISTS, AND AQUATIC SCIENTISTS

The United States Department of Labor groups geologists, geophysicists, and oceanographers together for reporting purposes. In 1998, according to their estimates, 44,000 people were employed in those three specialties. In addition, many more people are working in colleges and universities as researchers and professors in geology, geophysics, and oceanography. The latest estimate available for numbers in this group working in colleges and universities was 8,500 in 1990.

Nearly one third of the people employed in these professions worked in engineering and management services, and another one sixth were employed by petroleum and mining companies. The federal government employed about 5,800 geologists, geophysicists, and oceanographers in 1998. Over half of those were employed by the Department of Interior, most in the U.S. Geological Survey. An additional 3,000 were employed by state agencies across the country.

Geologists

Nature of Work

Geologists study the structure, composition, and history of the earth's surface. The geologists core into the earth examining and mapping samples. They search for minerals as well as water.

Geologists use tools such as hammers, chisels, levels, transits, cameras, and compasses, and they examine chemical and physical properties of samples under laboratory conditions. Geologists advise construction companies and governmental agencies on the suitability of constructing dams, buildings, and highways; manage research projects; or teach in universities or colleges. Geologists may specialize in three areas: earth material, earth processes, or earth history. A groundwater geologist studies the movement and reserves of underground water resources. This geologist works closely with the hydrologist in seeking new water resources.

Working Conditions

The working conditions of a geologist are varied. Some work in foreign lands looking for resources. They travel in helicopters, by jeep, or on foot. Many times the geologist is underground, but some work in well-equipped laboratories.

Training and Qualifications

A bachelor's degree in geology or a related field is a minimum for entry into the field. Advanced education is necessary for college teaching and research positions.

Geologists usually begin their careers in field exploration or as research assistants in laboratories. With promotions come the positions of project leader, program manager, or other management responsibility.

Economic Outlook

Employment opportunities are good for those with degrees. Demands for geologists are expected to increase as the demand for petroleum products increases. Geologists are also needed to develop adequate water supplies and waste disposal methods as well as to do site evaluations for construction activities.

Earnings

According to the United States Department of Labor, new geology bachelor's degree graduates earned an average starting salary of about $30,900 in the fall of 1997. You should understand that starting salaries vary greatly in any profession. Much depends on the location of the job. Jobs in an expensive area like Washington, D.C. require higher salaries than similar jobs in less expensive places to live. Much also depends on whether the job is with a governmental agency or with private industry. Graduates with previous experience and higher grades can expect higher starting salaries than graduates with no experience or lower grades in school.

In 1996, the American Association of Petroleum Geologists reported an average salary of $48,400 for geoscientists with a bachelor's degree and under two years of experience. Petroleum and mining industries often provide higher salaries, but the job security is much lower in than in other industries and governmental agencies. The federal government reported average salaries in 1997 of $59,700 for geologists, $67,100 for geophysicists, $54,800 for hydrologists, and $62,700 for oceanographers.

Geophysicists

Nature of Work

Geophysics is a catch-all term for the scientific study of the composition and physical aspect of the earth. To conduct their work, **geophysicists** call on other fields, such as meteorology, geology, oceanography, mathematics, and chemistry. Geophysicists concerned with water are called **hydrologists**. Hydrologists study surface and underground waters in the land masses of the earth. They locate underground water sources and estimate their potential. Hydrologists study rainfall, its rate of filtration into the soil, and its return to the ocean. Some work with water supplies, irrigation, flood control, and soil erosion.

Working Conditions

Geophysicists work both outdoors and in the laboratory. They may be stationed aboard ships or on aircraft. The laboratories are usually well equipped and well lighted.

Training

For an entry-level geophysical position, a bachelor's degree is required, and courses in physics, geology, mathematics, chemistry, and engineering are also required. A Ph.D. degree is required for those wanting to work in colleges or with research projects. About forty colleges award bachelor's degrees, and about forty award higher degrees.

Most geophysicists work in teams and start by producing field maps or working on exploration teams. Many assist experienced geophysicists in the laboratory.

Employment Outlook

The geophysics field is a small one. Those with advanced degrees will have the most opportunity. The employment in this field is greatly dependent on federal government funding.

Earnings

Geophysicists have high salaries, with average earnings more than twice those of nonsupervisory workers in private industry. For more specific salary information, see the section on geologist's earnings earlier in this chapter.

For additional information, you should contact

American Geophysical Union
2000 Florida Avenue, NW
Washington, DC 20009

Society of Exploration Geophysicists
P.O. Box 702740
Tulsa, OK 74170-2740
Telephone: 918-497-5500
Fax: 918-497-5557
Email: web@seg.org
Web site: http://www.edge-online.org

For information on federal employment in this area, contact:

U.S. Office of Personnel Management
Telephone: 919-757-3000
Web site: http://www.usajobs.opm.org

Aquatic Scientists

Nature of Work

Water covers over 70 percent of the surface of the planet. Careers in the aquatic sciences require specialized knowledge of the principles and techniques of the natural sciences, mathematics, and engineering.

Aquatic science is the study of our planet's surface waters, including ocean and freshwater environments. According to the American Society of Limnology and Oceanography,

"Oceanography is the study of the biological, chemical, geological, optical and physical characteristics of oceans and estuaries, while limnology involves the study of these same characteristics in inland systems (lakes, rivers, streams, ponds, and wetland) including both fresh and salt waters."

Aquatic scientists study practically everything to do with surface waters. (See Figure 17-1.) **Oceanographers** do research in developing fisheries and in mining the ocean in sustainable ways. They study tides, ocean currents, the ocean floor, marine life, and marine ecosystems. Limnologists study the same kinds of things for inland water systems. Limnologists would also study the application of biological and engineering principles to water treatment for human consumption.

Some examples of research conduced by oceanographers and limnologists include:

- organic, inorganic, and trace-metal chemistry in freshwater and sea water
- water currents ranging from global circulation patterns to mixing patterns in lakes and streams
- the geology of ocean bottoms
- how geothermal energy and other geological processes affect the oceans
- what ocean and lake bottom sediments can tell us about the history of the planet
- the transmission of light through water
- marine and freshwater life from individual species to whole ecosystems

Working Conditions

Most people employed in the aquatic sciences spend part of their time working in offices and the remainder of the time in the field. For oceanographers, fieldwork means getting out

FIGURE 17-1 Technician Jeff Nichols collects a water sample from the Walnut Creek watershed in Ames, Iowa. Samples are collected weekly from this area and surrounding watersheds to study the effects farming practices have on water quality. *(Photo by Keith Weller. Courtesy Agricultural Research Service, United States Department of Agriculture)*

onto the ocean: collecting, measuring, recording, and tagging fish, crustaceans, marine mammals, and other ocean creatures. Most oceanography data are collected on relatively long expeditions involving from one week to as much as three months. Limnologists can expect data collection expeditions of shorter duration lasting only a few days at a time.

Training

A minimum of a bachelor's degree is required for work at the professional level in oceanography. Naturally, all professions require support technicians and laborers with lower levels of education. If you want to move into research in the aquatic sciences, a Ph.D. is required, particularly if you want to work at a college or university.

For more information on careers in the Aquatic Sciences, contact:

American Society of Limnology and
    Oceanography
1444 Eye Street, NW, Suite 200
Washington, DC 20005
Telephone: 202-289-1972, Ext. 249
Fax: 202-628-1509
Web site: http://aslo.org

The Marine Technology Society
1828 L Street, Suite 906
Washington, D.C. 20036
Web site: http://www.mtsociety.org

International Oceanographic Foundation
3979 Rickenbacker Causeway
Miami, Florida 33149
Web site: http://www.rsmas.miami.edu/iof

Marine Technology Society
1825 K Street, NW, Suite 203
Washington, DC 20006
Web site: http://www.mtsociety.org

METEOROLOGISTS

Nature of Work

Meteorology involves studying the air that surrounds the earth. The best-known applica-

FIGURE 17-2 Weather balloons are among the arsenal of tools used to monitor air quality. *(Courtesy Agricultural Research Service, United States Department of Agriculture)*

tion is weather forecasting. (See Figure 17-2.) Trying to predict when it will rain and how much rain will fall is the job of the **meteorologist**. A meteorologist who deals with water supplies, sunshine, and temperature is called a climatologist. Climatologists may collect and give data to companies needing to design heating and cooling systems.

Working Conditions

The weather is a seven-day-a-week job. Meteorologists are expected to work day, night, or on some type of rotation plan. They may be located in an isolated building or be a part of a team.

Places of Employment

Atmospheric scientists, more commonly known as meteorologists, held about 8,400 jobs in the United States in 1998. The largest employer of meteorologists was the federal government, which employed 2,600 in the National Oceanic and Atmospheric Administration (NOAA) and another 280 in the Defense Department. Each of the military services has military specialties as meteorologists who help maintain the latest weather information for military operations. Military meteorologists are not included in this number. In addition, several hundred atmospheric scientists are employed as professors and researchers in colleges and universities.

Training

Entry-level jobs in weather forecasting require a minimum of a bachelor's degree in meteorology. Research and college teaching positions require advanced degrees with courses in physics, mathematics, and chemistry as well as meteorology. Beginning meteorologists often start in jobs involving the collection of data and data analysis. With experience, they may work up to supervisory positions.

Earnings

Atmospheric scientists had a median annual income of $54,430 in 1998. The income of the middle half ranged from $38,570 to $75,260. The lowest 10 percent of meteorologists earned an average of $27,250 while the top 10 percent averaged over $87,760. New meteorologists hired by the federal government with a bachelor's degree and no experience averaged from $20,600 to $25,500 depending on the grades they earned in college.

Employment Outlook

The field of meteorology is small. Most openings develop in the radio and television fields. Colleges and universities offer some opportunities for those with advanced degrees. The federal government is not expected to increase its employment of civilian meteorologists. The earnings of a meteorologist are similar to those of a geologist.

For additional information on meteorology write to

American Meteorological Society
45 Beacon Street
Boston, MA 62108

National Weather Service
Manpower Utilization Staff
Gramax Building
8060 13th Street
Silver Spring, MD 20910

WATER AND WASTEWATER TREATMENT PLANT OPERATORS

Nature of Work

Clean water is absolutely essential to the smooth functioning of our society. **Water treatment plant operators** treat water so that it is safe for human consumption. **Wastewater treatment plant operators** treat sewage and other wastewater to remove pollutants and neutralize pathogens so that the water is safe to return to the hydrologic cycle.

As our population grows, the pressure to provide adequate supplies of clean potable water increases. That makes the job of water treatment plant operator more critical every year. As the pressure increases to treat wastewater more efficiently and more thoroughly, the job of the wastewater treatment plant operator also becomes more important. The operators control equipment, read meters, and adjust chlorine in the water. Operators also make minor repairs on valves, pumps, and equipment.

Both water treatment plant operators and wastewater treatment plant operators may work in industrial plants, small town plants, or large city plants. In the smaller plants, the operator may be the only employee, performing all the work, including records and maintenance of equipment. In larger plants, the staff may include chemists, engineers, and assistants. With the passage of the Clean Water Act of 1972, water pollution standards have become stronger. The Safe Drinking Water Act of 1974 established standards for water for human consumption.

Operators must be ready to operate sophisticated systems. In both water treatment plants and wastewater treatment plants, the actual operation of the equipment is becoming increasingly computerized, but the operator must understand the processes completely so that he or she can monitor the operations carefully.

Working Conditions

Water and wastewater plant operators work both indoors and outdoors. They must work around noisy equipment and machinery and with unpleasant odors. Operators may have to work any or all shifts. Overtime is common,

and operators should not be afraid to work in unsanitary conditions.

Places of Employment

There were about 98,000 people working as full-time water treatment plant operators and wastewater treatment plant operators in 1998. About half worked in water treatment plants and half worked in wastewater treatment plants. The vast majority of these were employed by local governments, although many worked in private industry and in federal installations. Most of them were employed full time, although in some small towns, the treatment plant operators might only work part time. Also in many small towns, both treatment plants may be operated by the same person.

Training

Although a formal degree is not a requirement for plant operators, 49 states require operators to pass a certification examination. There are two-year programs that lead to an associate degree in wastewater technology. The courses cover principles in sludge digestion, odors and their control, chlorination, sedimentation, biological oxidation, and flow measurements. Operators may be promoted to supervisors or plant superintendents. Plant supervisors in large municipal systems usually hold a bachelor's degree in biology, chemistry, or environmental science.

Employment Outlook

The employment of both water treatment plant operators and wastewater treatment plant operators is expected to increase mainly due to the increase in the construction of new plants. Also, as plants modernize their facilities, trained operators will be in high demand. People who enter this field should have steady employment in the years ahead. Moreover, because the number of applications for this field tends to be low, job prospects for qualified applicants with appropriate education and training are very good.

Earnings

The average salary of wastewater treatment plant operators was $29,660 in 1998. Salaries of the middle 50 percent ranged from $23,210 to $36,680. The bottom 10 percent averaged less than $18,500 while the top 10 percent averaged over $44,710.

For more information write to:

Water Pollution Control Federation
2626 Pennsylvania Avenue, NW
Washington, DC 20037

Manpower Planning and Training Branch
Office of Water Program Operations
Environmental Protection Agency
Washington, DC 20460

SUMMARY

Water and its management hold several opportunities to the person who wants to earn a college degree. Most of the high-salary careers require a minimum of a bachelor's degree. Much of the work is done in research at colleges or universities.

There are careers in water management available to those without a college degree. Workers in landfills, city sanitation departments, and waste collection companies require a minimum of education. These people, however, should expect the lowest wages in the industry.

DISCUSSION QUESTIONS

1. What do geologists do? How many geologists are there in the United States?

2. What training is required for geophysicists? What salary can they expect to earn?

3. How many oceanographers are there in the United States? meteorologists? wastewater treatment plant operators? water treatment plant operators?

4. What do meteorologists do? What are the requirements to become a meteorologist?

5. What do wastewater treatment plant operators do? What are the requirements to become a wastewater treatment plant operator?

6. What do water treatment plant operators do? What are the requirements to become a water treatment plant operator?

SUGGESTED ACTIVITIES

1. Visit a meteorological laboratory at a television studio. How is rainfall predicted?

2. Have a speaker from the United States Weather Service come to the class. Have the speaker explain how data for the weather forecasts are collected.

Section III — CASE STUDY

It's in the Water ...

A fisherman in Chesapeake Bay takes in his catch and sees what he didn't want to. His catch is covered in lesions. (See Figure III-A.) On the Texas coast a red tide is coming in. What is happening to our national coastal fisheries? The answer lies in the water.

Harmful algal blooms (HABs) have been identified as the cause of fish lesions, fish kills, and red tides throughout our coastal fisheries. HABs cause deaths in large numbers of fish and have been known to cause skin irritations and lesions on humans as well. (See Figure III-B.)

Of primary concern is the *Pfiesteria piscicida* species that produces high amounts of toxins. (See Figure III-C.) Persons along the North Carolina and Virginia coasts in recent years have been bombarded by frightening news reports. The reports have warned swim-

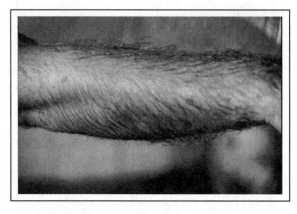

Figure III-B Lesions on a man's arm resulting from contact with *Pfiesteria piscicida* in the coastal waters of North Carolina. *(Photo courtesy of North Carolina State University, Aquatic Botany Laboratory)*

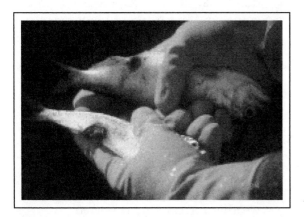

Figure III-A Dead fish showing large bloody lesions that were related to their deaths. *(Photo courtesy of North Carolina State University, Aquatic Botany Laboratory)*

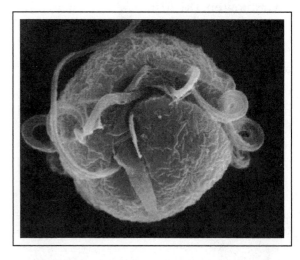

Figure III-C One of the life stages of *Pfiesteria piscicida*. *(Photo courtesy of North Carolina State University, Aquatic Botany Laboratory)*

mers and boaters to stay out of the water because of danger from *Pfiesteria.*

Although it is impossible to determine exactly when one of these outbreaks may occur, the following environmental conditions usually occur in conjunction with HAB outbreaks:

- A general state of high eutrophication is present.

- Systems are warm, are shallow, and have poor circulation.

- Fish are present.

These conditions point to agricultural runoff as a main cause of these outbreaks. Agriculturalists are not so sure. Consider this timeline of events in Chesapeake Bay:

1992	• Jenkins Creek, Choptank River: positive ID of *Pfiesteria.*
Spring 1994 and Summer 1996	• Laboratory fish kills on Paxtuxent River, Maryland—attributed to *Pfiesteria* that probably came in through the seawater flow through system.
Summer 1996	• Striped Bass deaths on Manokin River, Maryland—*Pfiesteria* present along with other dinoflagellates.
1996	• Pocomoke River—fish lesions—*Pfiesteria* tests are negative.
Spring 1997	• Pocomoke River—fish lesions—*Pfiesteria* tests are negative.
May 1997	• Washington area TV station samples show positive *Pfiesteria* findings in Dr. JoAnn Burkholders Lab—above average precipitation in 1996 attributed to shifts in water chemistry.
June 1997	• Lesion fish reports—analysis reveals no chemical contamination.
July 1997	• EPA and NOAA provide $500,000 to support investigations.
July 25, 1997	• Maryland DNR collects sediment samples looking for signs of chemical contaminants.
August 1–3, 1997	• Maryland's Technical Advisory committee determines that many things can be associated with fish kills including: physical irritation from microbial infections of stressed fish; harmful chemicals; secondary infection by bacteria, viruses, and fungi; *Pfiesteria* and other microorganisms.
August 6, 1997	• Hundreds of fish dead on Pocomoke River—later attributed to *Pfiesteria.*
August 26, 1997	• Another fish kill on Pocomoke—still attributed to *Pfiesteria.*
Sept. 10, 1997	• *Pfiesteria* findings confirmed.
Sept. 14, 1997	• DNR submits a model that suggests that nutrient enrichment is the cause of *Pfiesteria* outbreaks.
Throughout the next year	• Multiple commissions, conferences, and meetings held to address the problem. Becomes a multiagency and multistate mission to stop the outbreaks. *Pfiesteria* microbes still found in areas with fish lesions.
October 1999	• *Kudoa* parasite found in the James River.
November 1999	• *Richmond Times Dispatch* reports that *Kudoa* may be responsible for some of the earlier fish problems associated with *Pfiesteria.*

Timeline provided by the North Carolina State University Aquatic Botany Laboratory.

This issue is not clear-cut. Despite numerous tests and management strategies, fish are still dying. Although high eutrophication is associated with the kills and agricultural runoff is an important player in increasing eutrophication, do we know enough to support highly conservative actions such as the closing of agricultural facilities?

What should agriculturalists and state policymakers do? Several options appear to be available:

■ Do nothing.

■ Admit that we cannot prove what really causes these outbreaks. Continue to conduct more research to determine the real culprit.

■ Take a firm stand and place severe limitations on industries associated with high nutrient loads such as poultry and pig farmers.

■ Continue to allow high-nutrient loading but notify coastal residents when dangerous situations develop.

■ Focus on technological improvements that could help prevent nutrient loading.

■ Something else might be possible that we have not thought of yet.

What should we do? You decide.

Section IV
Forest Resources

18 OUR FORESTS AND THEIR PRODUCTS

OBJECTIVES

After reading this chapter, you should be able to

- explain the differences between commercial and noncommercial forests—between growing and mature forests

- list and describe the major forest regions of the United States

- identify the parts of a tree and describe the functions of each part

- differentiate between pure and mixed forests—between even-aged and all-aged forests

- define forest canopy and explain the importance of shade tolerance in the canopy

TERMS TO KNOW

commercial forest	trunk	even aged
certified tree farm	annual ring	all aged
forest regions	crown	canopy
root system	photosynthesis	lumber grades
root hairs	shade tolerant	converted wood

According to legend, a Viking, Leif Ericson, landed in North America about 1,000 years ago. If the legend is true, and most historians think it is, he was probably the first European to set foot on this continent. He found a land "of rolling hills and plenty of trees." The Viking and his crews used wood from the forest to repair their boats and took samples of timber with them. He later returned to an area that is now a part of Canada and established logging camps. The workers there harvested timber to send home by ship. Thus, 600 years before the settlers landed at Jamestown, there was a forestry industry in North America.

Our forests and their products and by-products have always been important. From timber for our homes and factories to the pleasure of a walk in the woods, we gain many things from the forest. The best thing about the forests is that we can keep our forests and use them too! That is because forests are one of our renewable resources. After all, trees are just a very special kind of plant. Just like other plants, trees can be harvested and then replaced by new trees.

Unlike most other plants, though, trees take much longer to grow to maturity.

OUR FORESTS

A forest is a very complex community of associated trees, shrubs, other plants, and animals. In this community, all the members interact with each other and play some part in the complex system.

You could compare a forest to a city. A city is made up of homes, businesses, factories, streets, services, and many other things. A single house alongside a road does not make a city. Even a cluster of homes or stores does not make a city.

By the same token, a tree on a street corner does not form a forest. Even a group of trees planted in a park do not make a forest. A forest is constantly renewing itself. In a forest, food is being produced through photosynthesis every day. Plant parts decay and release their nutrients into the sod where they can be reused. Animals consume the plants or other animals. The animals, too, die and return their nutrients to the sod for reuse. Thus, each member of the forest community takes its needs from the system and then, in return, contributes to the system. This balanced ecosystem is a forest.

As this country was being settled, forests were both an enemy and a friend to the settlers. Let us examine both roles.

THE FOREST AS ENEMY

When the first permanent English settlement in America was established at Jamestown, Virginia, trees were everywhere. Dense forests covered the hills and the valleys. They formed an almost impenetrable barrier to the English settlers. Only a certain amount of food could be shipped from Europe to the colony. Certainly, not enough food could be provided by ship to feed a colony more than temporarily.

Obviously, the first task of the settlers was to provide protection and shelter for themselves. But once that task was begun, the need to establish a reliable food supply became paramount. That meant clearing the forests—getting rid of the trees! Trees covered everything, and nothing could be done until the forests were pushed back.

Early colonists would not have been likely to think of the forests as beautiful or peaceful. (See Figure 18-1.) They would more likely describe the forests as "solemn, interminable, barbaric, harsh. . . ." Trees were the enemy and had to be felled. Until they were gone, there could be neither crops nor fruits: no safety, no ease, no civilization.

THE FOREST AS FRIEND

Even though the forests were in the way of farm production, they were most valuable in many ways. Trees were felled to allow for crop production, and the wood was used to build homes and other structures. It provided furniture. The trees providing a hiding place for potential enemies could be felled to provide material for fortifications. The forest provided summer cooling, and its wood provided heating during the harsh winter. (See Figure 18-2.) Trees were cut for export in the form of beams and finished lumber.

The ship building industry of Europe came to depend on the North American colonies for tall, straight trees for their ships. The naval stores industry grew up around pitch harvesting from pine trees in the southern forest

FIGURE 18-1 In the early history of our country, dense forests presented a formidable problem.

FIGURE 18-2 Trees were important in providing warmth, shelter, and safety. This is a colonial fort re-created at Jamestown, Virginia.

region. Soon, most of the world's production of naval stores was centered in North America.

The forests provided wild game, as well as many fruits and berries. In fact, in the very beginning the settlers, like the Native Americans, relied on forest animals and wild fruit and berries for their very survival.

It is difficult to imagine the vastness of the forests covering this country when European settlers first landed. It would have been possible to walk from Maine to Texas, or from Florida to Minnesota, without ever leaving the forest except to cross rivers or lakes. The plains states

and much of the southwestern and western parts of the country were too dry to support forests, but the eastern and northwestern parts of the continental United States, as well as much of Hawaii and Alaska, were almost entirely covered by dense forests. It has been estimated that the entire area of what is now Arkansas held only about 1,000 acres of open land. The rest was solid forest. It was said that it would have been possible for a squirrel to cross the whole state of Ohio leaping from tree to tree.

FORESTS IN AMERICA TODAY

How Much Is There?

How much land we have in forests depends on how you define it. According to the World Resources Institute (WRI), if you include just areas that are fairly densely forested areas, there were about 8,535 billion acres worldwide, with about 525 billion acres in the United States in 1995. Counting "other woodlands" that are less densely forested, but that have more trees than brush and grasses, WRI indicated that there were about 10,322 billion acres in 1995, with about 731 billion acres in the United States. Worldwide, the trend in forested area is downward. The total worldwide forested area was down 2.2 percent from 1984 to 1994. In the United States, the trend is the opposite. Between 1984 and 1994 it was up about 1.5 percent. That trend has been true for many years and there seems to be no reason to expect it to change in the near future.

Interestingly, the average annual production of wood increased worldwide by 8 percent, and U.S. production was up 11 percent during the same 10-year period. Worldwide, part of the increase in wood harvested might be blamed for the decrease in forested land area. We clear a forest to harvest the wood and convert the land to growing crops or to cities. But, in the United States, production is up substantially more than for the rest of the world, yet the area of forested land in the United States has not increased over that period.

Today, we have about 70 percent as much forestland as when Leif Ericson—and later Christopher Columbus—landed on this continent.

Roughly one third of U.S. forest and woodland is referred to as noncommercial forestland. This area of about 250 million acres is made up of land that is generally not suitable for economical tree production. Included in this are swamps, very rough mountainous land, and other areas where trees grow but cannot be economically harvested. Also included are parks, wilderness preservation areas, game refuges, and other areas where timber harvesting is not allowed.

The other two thirds, about 481 million acres, is generally referred to as commercial forest; but that name is very misleading. The term **commercial forest** simply means that the land is capable of producing economically useful forest. It does not mean that the land is actually being used that way. Almost one half of this area is not actually available for forest production. Much of it is in small parcels: near homes or in suburban areas, or along highways, streams, or lakes. Still more of this so-called commer- cial forest is simply not being managed for wood production.

That leaves only about 250 million acres of actual commercial forestland. That makes about one acre of commercially producing forestland for each man, woman, and child in the United States. Just as the available farmland is being taken over for urban and industrial expansion, so is our forestland. About 12 million acres of U.S. forestland were converted from growing timber to other uses during a recent seven-year period.

Who owns this forestland? Where is it located? Many people believe the large timber companies own most of the forest, but, while these large companies do own vast areas of forest, that reputation is not totally deserved.

Of the 481 million commercial forest acres, private individuals (some 4 million of them) own 58 percent, or 279 million acres. Much of the land classified as commercial but not being used for wood production falls into this category.

The next largest holder of forestland is government. Federal, state, and local governments own about 137 million acres (or about 28 percent) of the commercial forest. As before, much of this is not being managed for forest production.

The forest products industries—Weyerhauser, Kraft, Union-Camp, and other corporations—hold about 68 million acres. That represents only about 14 percent of the total, but over 27 percent of the actual producing commercial forestland.

Every state has at least some forestland. To see how much forestland is in your state, refer to Table 18-1.

While the figures in Table 18-1 represent total forested area, there are at least some truly commercial forestlands in all 50 states. In fact there are nearly 70,000 certified tree farms in this country. Together, they encompass over 85 million acres of forestland. The typical tree farm is a single-family operation of less than 200 acres in size; clearly, on most of these, timber production is not the primary enterprise or source of farm income.

Tree farmers manage their forestland to produce timber, but this often also means better wildlife habitat as well as soil and water management. If you own ten or more acres of forestland, you may be interested

TABLE 18-1 Forestland by State (in million acres)

State	Total Area	Forestland Area	% Forested	State	Total Area	Forestland Area	% Forested
Alabama	32.7	22.0	67.2	Nebraska	49.0	1.0	1.9
Alaska	365.3	129.0	35.3	Nevada	70.3	6.9	9.8
Arizona	72.7	10.5	14.5	New Hampshire	5.8	4.9	84.4
Arkansas	22.6	18.8	55.9	New Jersey	4.8	1.9	39.2
California	100.2	37.3	37.0	New Mexico	77.7	18.1	23.3
Colorado	66.5	21.3	32.0	New York	30.7	16.0	52.1
Connecticut	3.1	1.9	59.4	North Carolina	31.4	19.3	61.4
Delaware	1.3	0.4	30.8	North Dakota	44.5	0.7	1.5
Florida	34.7	16.2	47.7	Ohio	26.2	7.9	30.2
Georgia	37.3	24.1	64.7	Oklahoma	44.1	5.4	12.3
Hawaii	4.1	2.0	48.8	Oregon	61.6	28.0	45.5
Idaho	52.9	18.5	35.0	Pennsylvania	28.8	16.9	58.7
Illinois	35.8	4.3	11.9	Rhode Island	0.7	0.4	60.4
Indiana	23.2	4.4	19.2	South Carolina	19.2	12.7	65.3
Iowa	35.9	2.0	5.7	South Dakota	41.5	0.3	0.7
Kansas	52.5	1.6	2.9	Tennessee	26.7	13.3	50.9
Kentucky	25.5	12.5	49.1	Texas	168.2	19.2	11.4
Louisiana	28.9	13.8	47.8	Utah	52.7	15.7	29.8
Maine	19.9	17.7	89.2	Vermont	5.9	4.4	74.7
Maryland	6.3	2.7	42.7	Virginia	25.5	16.0	62.9
Massachusetts	5.0	3.0	69.7	Washington	42.6	20.5	48.0
Michigan	36.5	19.3	52.8	West Virginia	15.4	12.0	78.1
Minnesota	51.2	16.7	32.6	Wisconsin	35.0	16.0	45.6
Mississippi	30.2	18.6	61.5	Wyoming	62.3	3.2	5.1
Missouri	44.3	14.0	31.6				
Montana	93.1	22.6	24.3	U.S. Totals	2,252.3	715.9	31.8

Sources: American Forest and Paper Association, and American Forest Institute, 1980. (*Note:* These totals differ slightly from the World Resources Institute's estimates for forested area in the United States discussed earlier.)

in establishing a **certified tree farm**. If so, contact

American Forest Foundation
1111 19th Street, NW
Washington, DC 20036

Forest Regions in the United States

In the United States today there are 860 species of trees. Many of these trees will grow only in certain parts of the country. Climate, soil type, and altitude affect the kinds of trees that will grow in a given location. In the continental United States, there are six major **forest regions**.

West Coast forests lie along the Pacific Ocean from central California to the Canadian border. Major species include Douglas fir, coast redwood, western red cedar, Sitka spruce, sugar pine, lodgepole pine, incense cedar, Port oxford cedar, white fir, red aider, and bigleaf maple. Much of our country's lumber and softwood plywood comes from this region.

Western forests are found generally in the mountainous regions from southwestern Texas to Wyoming, to central Washington, and to northern California. Principal species include ponderosa pine, Idaho white pine, sugar pine, Douglas fir, Engelmann spruce, western larch, white fir, incense cedar, lodgepole pine, western red cedar, and aspen.

Central hardwood forests run from parts of New York State to northern Georgia, west to Texas, and north to Minnesota. Prominent tree species include shortleaf pine, Virginia pine, eastern white pine, red cedar, birch, northern red oak, white oak, hickory, elm, white ash, black walnut, sycamore, cottonwood, yellow poplar, black gum, red maple, and sweet gum.

Tropical forests in this country are limited to the southern tips of Florida and Texas. Leading species are mahogany, mangrove, and bay tree.

Northern forests reach from Maine south along the mountains to Georgia. They also cover parts of northern Michigan and Minnesota. Major tree species are eastern white pine, red spruce, black spruce, white spruce, Norway pine, jack pine, balsam fir, white cedar, tamarack, eastern hemlock, aspen, beech, red oak, white oak, yellow birch, paper birch, black birch, black walnut, black cherry, black gum, white ash, basswood, and sugar maple.

Southern forests run from the coast of Virginia to eastern Texas and north to Missouri. The main trees are loblolly pine, longleaf pine, shortleaf pine, slash pine, bald cypress, sweet gum, black gum, Southern red oak, white oak, pin oak, live oak, willow, yellow poplar, cottonwood, white ash, hickory, and pecan.

In addition to the six forest regions in the continental United States, there are four forest regions in Alaska and Hawaii.

Coast forests lie along the southern coast of Alaska. Species include western hemlock, Sitka spruce, western red cedar, and Alaska yellow cedar.

Interior forests of Alaska include white spruce, black spruce, white birch, aspen, and several poplars.

West forests are found in the interior of the larger islands of Hawaii. Species are ohia, roa, tree fern, ruikui, mamani eucalyptus, and tropical ash.

Dry forest lands also appear on several of the larger islands of Hawaii. Major species are alga-roba, roa haole, wiliwili, and monkeypod.

TREES AND THEIR GROWTH

Parts of a Tree

Trees are woody plants with single stems. They generally consist of three major parts: roots, trunk, and crown. (See Figure 18-3.)

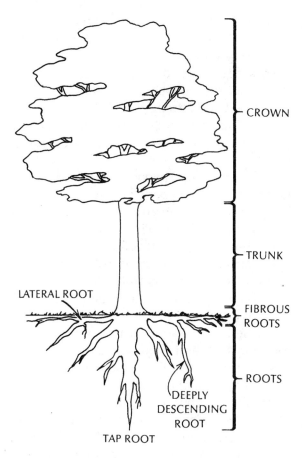

FIGURE 18-3 Parts of a tree.

Roots

The **root system** serves to anchor and support the tree. It takes in water and nutrients from the soil and passes them upward to the rest of the tree. It also serves to store some manufactured food for the rest of the tree. In addition to these functions, the root systems of trees help to hold the soil in place and improve the structure, water-absorbing, and water-holding capacities of the soil.

There are four types of roots in a complete root system: tap root, lateral roots, fibrous roots, and deeply descending roots. (See Figure 18-3.)

Roots grow both in length and in diameter. Growth in length is accomplished by an area

of rapid cell division just behind the root cap on each root. (See Figure 18-4.)

Just behind the growing tip of the root is an area covered by very fine, delicate **root hairs**. (See Figure 18-4.) It is the root hairs that absorb the water and nutrients taken in by the root system.

Lateral growth of the root (diameter) is accomplished in the same way as lateral growth of the trunk. This is discussed in the next section.

Trunk

The **trunk** supports the crown of the tree. It also transports the sap (water and dissolved nutrients) upward from the root system to the crown. The internal structure of the trunk, of branches in the crown, and of roots in the root system is generally alike. Thus, the following discussion applies to the woody parts of all three systems.

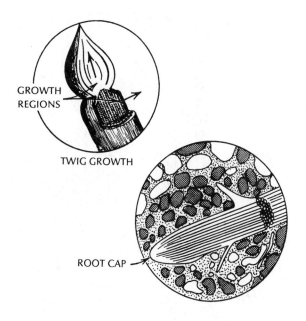

FIGURE 18-4 The length of tree branches and roots increases by growth at the tips.

A mature tree trunk consists of five parts. Starting at the center of the trunk and moving outward they are heartwood, sapwood, cambium, inner bark, and outer bark. (See Figure 18-5.)

The tree grows in length by means of rapidly dividing young cells in the terminal bud on each branch and twig. If there is a central or dominant branch, in many species of trees (such as Douglas fir) the trunk itself will lengthen. If the crown branches, the trunk may not extend in length; but branches will continue to lengthen, as in most oak trees.

The heartwood consists of woody cells that are inactive. In other words, they are dead. This part of the trunk is usually darker and less moist that the sapwood. In fact, heartwood begins as sapwood, and as the tree diameter increases, the sapwood cells near the center of the trunk die and become heartwood. The heartwood adds strength and stiffness to the tree. In mature trees, the heartwood makes it possible for the tree to remain upright. (See Figure 18-5.)

Outside the heartwood is the sapwood. The sapwood is more properly known as the xylem layer. It is lighter colored than heartwood, and the cells are still living. The sapwood serves two major functions. First it is the part of the tree that transports water and dissolved nutrients upward from the roots to the leaves. Second, it provides a storage area for much of the food that has been made in the leaves through photosynthesis. The sapwood feels moist to the touch. It is soft and flexible, and it contributes little to the tree's ability to remain upright.

Surrounding the sapwood layer is the cambium. The cambium is a thin layer of active cells that divides to produce new cells. New cells on the inside of the cambium layer remain in place and become sapwood. New cells produced on the outside of the cambium layer are pushed outward to become inner bark. Other new cells become new cambium as the cambium layer moved outward as a part of the growth process.

Outside the cambium layer is the inner bark. This layer is also known as the phloem. The inner bark consists of living cells. The cells are relatively moist and soft, and they provide the mechanism by which the food produced in the leaves moves downward to the roots and to the rest of the tree.

As the inner bark cells are pushed further outward by new cells being formed under them, they die and become outer bark. This layer consists of harder, drier bark cells. It is

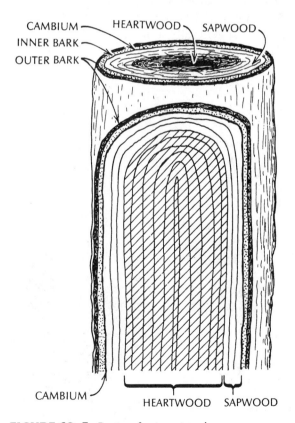

CAMBIUM — HEARTWOOD — SAPWOOD
INNER BARK
OUTER BARK

CAMBIUM

HEARTWOOD SAPWOOD

FIGURE 18-5 Parts of a tree trunk.

inactive, but it provides protection from drying to the inner bark. It also protects the tree from invaders such as disease organisms and insect pests.

In the spring, new sapwood cells are large and soft walled as the tree grows rapidly. They tend to be very light in color and are called springwood. Later in the summer, as the temperatures tend to rise and the amount of rain tends to decrease, these new cells are smaller and denser and a bit darker in color. These cells are known as summerwood. The result of the rapid, soft springwood growth followed by the slow, dense summerwood growth is a ring around the trunk that is visible to the naked eye. This is known as an **annual ring**, and the age of the tree an be determined by counting the annual rings.

During a year with an extremely good growing season, the annual ring will be very wide. During dry years, or when disease, fires, insects, or other problems arise, the rings may be extremely narrow. The result is a record of the history of the tree that can be read much like a book. In fact, it has been possible to follow the annual ring patterns of growth back many centuries. In some areas of the country, in fact, the record of annual growth rings is so complete that tree rings are used to date wooden artifacts recovered at archeological sites. By studying these rings, it is even possible to determine the years during which great droughts or fires occurred. Changes in the climate can be pinpointed by this technique.

Crown

The **crown** of the tree includes branches, twigs, buds, and leaves. It is here that the tree manufactures its food through photosynthesis. It is here also that seeds, fruits, nuts, and berries are produced.

Growth in the crown takes place in two ways. Each year's new growth of twigs is produced to increase the length of the branches, and each branch and twig grow in diameter as described under the previous section on the trunk.

The leaves of green plants, such as trees, are the greatest chemical factory on earth. Each tree leaf accepts water conducted upward from the roots by the trunk and branches. It takes in carbon dioxide (CO_2) from the atmosphere. Then, in the presence of sunlight, it converts all these things into sugar in a process known as **photosynthesis**. It is this miracle of photosynthesis that makes possible all life as we know it.

Excess water from the tree is allowed to evaporate through openings in the leaves. This process is known as transpiration. It is also the process that makes the air in a forest cooler than the air around the forest during summer.

To understand how this works, try an experiment. Place your hand in warm water then remove it from the water. Fan your hand around. As the water on your hand evaporates, your hand gives up warmth to the water vapor and feels cooler. The same thing happens in transpiration. As the water evaporates, it takes warmth from the air and the air feels cooler.

Light Requirements

Different species of trees require different amounts of light. Trees may be classified as either shade tolerant or shade intolerant.

Shade-tolerant trees are those that grow satisfactorily without complete direct sunlight. Some trees do fairly well even in almost total shade. Dogwoods, for instance, grow best in direct sunlight; but they also survive and grow underneath other trees. Most maple varieties are also shade tolerant.

Trees that require some direct sunlight but can also grow in partial shade are known as *moderately shade tolerant*. Ponderosa pine and eastern white pine are moderately shade tolerant as young seedlings, but they become less shade tolerant as they grow larger.

Shade-intolerant trees do not fare well without direct sunlight for at least part of the day. An example is the Douglas fir. It requires some shade for good seedling growth. Thereafter, it can survive even in dense stands; but each fir must have its share of direct sunlight to survive and grow.

Thus, trees growing in the understory are almost always shade-tolerant species. In forests of only shade-intolerant species, the forest floor will have few, if any, growing saplings.

The Forest Canopy

A forest is either pure or mixed. A pure forest has mostly a single species of tree. A mixed forest has more than one species of tree. The stand of trees can be either **even aged** or **all aged**. An even aged forest has mostly trees of a single age and size. An all aged forest will include trees in two or more of the following age/size groupings: seedlings, saplings, pole sized, mature, and veteran.

The ceiling of the forest is called the **canopy** and is made up of the crowns of the taller trees. In relation to the canopy, as shown in Figure 18-6, a tree may be

- dominant (with a crown extending above the general canopy level)
- codominant (with a crown in the general canopy)
- intermediate (with a crown in the general canopy but with little direct light from the side)
- suppressed (with a poorly developed crown below the canopy)

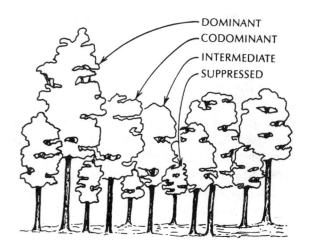

FIGURE 18-6 Members of the forest canopy.

FOREST PRODUCTS

Wood

The most obvious products of the forest are wood. Wood products take many forms.

There are various **lumber grades**. Lumber can be shop and factory grade. This includes wood for furniture, barrels, boxes, cabinets, flooring, millwork, and so forth. The second grade of lumber is structural. Structural lumber includes joists, planks, and laminated wood. The third grade of lumber is yard wood. This includes boards, dimension, and finish lumber.

Ties are used for railroad construction and repair. Poles, posts, and piles are other intact wood products of the forest. Firewood is becoming more and more important.

Converted Wood and Other Forest Products

Once the wood is mechanically or chemically changed, it is referred to as **converted wood**.

Converted wood includes products such as paper, pulp, wood fiber, charcoal, explosives, and plastics. Other wood products include rayon, cellulose, photographic film, sausage cases, sponges, simulated leather, artificial hair, alcohol, shatterproof glass, imitation vanilla, and thousands of other everyday products. It would be almost impossible to spend a day in this country without touching, reading, sitting on, eating, wearing, and looking through a wood product.

OTHER FOREST-PRODUCED BENEFITS

Climate Moderation

Transpiration taking place through forest leaves is the world's largest air conditioner. The temperature in a forest may be as much as 8°F cooler than the air around the forest on a hot summer day. The canopy of the forest blocks sunlight from the ground and keeps the soil from becoming too hot for plant growth. The forest provides shade for wildlife, livestock, and people.

During cold weather, the forest serves to break harsh winds near the surface. We have all experienced the wind-chill effect on a cold, breezy day. Trees make excellent shelters against the cooling or drying effects of the wind.

Forests help quiet loud sounds and harsh noises. They help filter dust from the air. Trees give off pleasant odors, such as the smell of flowering dogwoods or the scent of a pine forest.

Because trees are very complex organisms, many species are affected by air pollution. Thus, forests may provide very graphic warnings of serious pollution problems.

Carbon Dioxide/Oxygen Exchange

The vast blanket of green on the earth's surface consists of many kinds of plants. Forests make up a major part of this vegetative cover. The process of photosynthesis is the major mechanism by which carbon dioxide is removed from the air. It is also the main way fresh oxygen is returned to the atmosphere. Thus, forests are a major contributor to the balancing of the oxygen and carbon dioxide in the atmosphere.

Water and Soil Conservation

As discussed in Chapter 6, forests are the most effective vegetative cover for soil and water control. The canopy of the forest breaks the speed of falling water. The root systems hold soil particles in place. The layer of decaying plant matter on the forest floor holds moisture, speeds absorption, and delays evaporation. No other covering, including concrete, is so protective of the soil in the long run. No other mechanism available promotes a more complete entry of surface water into the soil than a forest.

Forests provide clean, reliable surface water for streams and lakes. At the same time, they help prevent flooding, sedimentation, and surface damage from erosion.

Wildlife and Recreation

Forests provide homes and food for wildlife, including game animals, fur-bearing animals, fish, and game birds. They provide recreation, rest, solace, quiet, and nature study for people.

SUMMARY

The first European-based forest industry on this continent was probably a Viking logging operation about 1,000 years ago. As permanent colonies developed in North America, the forests came to be both friend and enemy to the settlers.

A forest is a complex ecosystem. It consists of an interrelated association of trees, shrubs, other plants, and animals. Forests cover about 731 million acres of this country. One third of that is not economically useful for economic forest production. Another one third is currently not being used for commercial forest production even though it could be. The remaining one third, about 250 million acres, is available for supplying this country with its forest products.

In the continental United States, there are six major forest regions. Alaska and Hawaii have an additional four forest regions.

A tree is made up of three main parts: roots, trunk, and crown. It grows both in length and in diameter by means of cell division in root tips, in buds at the ends of stems, and in the cambium layer. Trees may be classified by their light requirement as shade tolerant and intolerant. They may also be classified by their individual positions in the forest canopy. Forests may be pure or mixed. They may be even aged or all aged.

Forest products include wood, converted wood, and other forest products. The list of forest products is almost endless. Our forest product industry is so extensive that life as we know it would be impossible without it. In addition to the industrial products coming directly from the forest, we receive many forest-related benefits.

Soil and water management depends heavily on the forests. Our recreation often centers around forests. Wildlife depend on them for homes and food. Our climate is moderated and our atmosphere is cleaned by the forests.

DISCUSSION QUESTIONS

1. How was the forest both an enemy and a friend to the early colonists in America?

2. What is a forest? Why is a clump of trees in a park not a forest?

3. How much forestland is there in the United States today? How much is commercial forest?

4. How can commercial forest not really be "commercial"?

5. What are the three main parts of a tree? What does each part do?

6. How does a tree grow in length? in diameter?

7. What are annual rings, and how can scientists "read" them?

8. Define the following terms:
   - shade tolerant
   - shade intolerant
   - dominant
   - codominant
   - intermediate
   - suppressed
   - pure forest
   - mixed forest
   - even-aged forest
   - all-aged forest

9. Look around and make a list of as many forest products as you can find.

10. In addition to actual products, what other benefits do people get from forests?

SUGGESTED ACTIVITIES

1. Measure the temperature on a warm, still, clear day in your school's parking lot. Then measure the temperature in a heavily wooded area. How much cooler is the forest than the parking lot?

2. Find a tree trunk or log and examine its annual rings. Are there any particularly narrow or wide rings? What do they probably mean?

3. Take a field trip to a mixed-species forest. Take a tree identification book with you and try to identify as many different trees as you can.

4. Visit a lumber sales yard and develop a list of all the different types of lumber and other wood products available.

19 WOODLAND MANAGEMENT

OBJECTIVES

After reading this chapter, you should be able to

- define the most common ways to measure wood
- describe the different methods of harvesting a stand of trees and explain the advantages and disadvantages of each
- explain why good woodland management is important to (1) the forest owner, (2) the neighbors of the forest owner, (3) the economy as a whole, and (4) you and me
- explain how a forest can grow faster if the trees are harvested
- describe the main methods used in forest regeneration

TERMS TO KNOW		
board foot	hypsometer	sanitation cutting
cubic foot	cruising	salvage cutting
cord	intermediate cuttings	harvest cutting
gross weight	cleanings	seed tree
dbh	improvement cuttings	clear cutting
dendrometer	thinning	
tree height	liberation	

Forest resources are of great importance to America in many ways—for aesthetic values, environmental improvement, soil and water conservation, and certainly economic production. Forests cover vast areas of land in all of our states.

With such a vast area in trees, you might think the forests could take care of themselves. After all, trees start out as seedlings, grow to maturity, and are cut down. But it is not that simple. Our population continues to grow; our cities, industries, and residential areas continue to expand. Our forested land continues to diminish. And, at the same time, we still insist on using more and more wood and other forest products each year.

Compared to the year 1900, we are using 800 percent more paper and 70 percent more lumber and other wood products in America today; and this must be produced on less forestland. Without the use of good forest management techniques, this increased production from a decreased area would not be possible.

MEASURING THE FOREST

Units of Measure

The most basic consideration in modern forest management is the production of wood. Foresters measure the amount of wood in a given area of forest after the trees are cut. They can estimate the amount of wood before the trees are cut. The following sections define the most common units of measure for wood and describe how estimates are made while the trees are still standing.

Board Foot

The standard unit of measure for most lumber is the **board foot**. Typically boards, planks, timber, or other pieces of lumber are sold at a price per board foot or hundred board feet or thousand board feet. In other words, the number of board feet in a board is determined, then that is multiplied by the price per board foot to determine the total cost.

A board foot is defined as a piece of rough wood 1 foot long, 1 foot wide, and 1 inch thick. Once the wood is surfaced and finished, those measurements will be reduced; but the measurements used are always those before surfacing is done. A board with one board foot need not be exactly this shape, but the volume is always 144 cubic inches (before surfacing). For example, a 2 × 4-inch board that is 18 inches long contains 1 board foot of lumber. A formula for computing board feet is

$$\text{Board feet} = \frac{\text{length} \times \text{width} \times \text{thickness}}{144}$$

in which all measurements are in inches.

Cubic Foot

A unit of measure becoming more popular is the **cubic foot**. A cubic foot of wood is the amount of wood that would be required to fill a space 1 foot wide, 1 foot thick, and 1 foot high. There are 12 board feet in 1 cubic foot or 1,728 cubic inches. A formula for computing the cubic feet in a piece of wood is

$$\text{Cubic feet} = \frac{\text{length} \times \text{width} \times \text{thickness}}{1,728}$$

in which all measurements are in inches.

Cord

At one time, pulpwood was sold by the **cord**. Today, this is usually not the case. However, firewood is still often sold by the cord. A standard cord of wood is defined as a stack of wood 4 feet × 4 feet × 8 feet. This converts to 128 cubic feet. Such a stack of wood, however, is not all wood. It always has spaces between the pieces. Usually a cord of wood has between 75 and 100 cubic feet of actual wood. The rest is empty space; so the cord is not a very accurate unit of measure. A formula for computing the cords in a stack of pulpwood or firewood is

$$\text{Cords} = \frac{\text{length} \times \text{width} \times \text{height}}{128}$$

in which all measurements are in feet.

Gross Weight

Gross weight is by far the most common method of measuring pulpwood today at the wood-buying centers. It is much simpler and quicker than volume measures and, because green wood is heavier than dry wood, it ensures the buyer fresh wood. The pulp mill takes the logs, converts them to chips, and then processes the fibers for use in paper and other products. Fresh logs are easier to chip and process.

Measuring Tree Diameter

The volume of wood in a log is determined by its diameter and length. For a tree that is still standing, the diameter is measured "at breast height." The diameter at breast height (**dbh**) is the thickness across the tree trunk at $4\frac{1}{2}$ feet above the average ground level. Tree dbh is usually recorded in 2-inch increments—6, 8, 10, 12, 14 inches, and so on. The dbh is rounded to the nearest 2-inch class. For instance, a 10.9-inch tree is recorded as 10 inches, whereas an 11.1-inch tree is recorded as 12 inches dbh.

Tree diameters are measured by **dendrometers**. The three most common dendrometers are

1. the Biltmore stick
2. tree calipers
3. diameter tape

The Biltmore stick is the least accurate but quickest way to estimate a tree's dbh. It is somewhat like a yardstick. The stick is held 25 inches from one eye and against the tree at breast height. The left end of the stick is placed (visually) in line with the left edge of the tree, using only one eye for sighting. The same eye is turned to read the approximate tree diameter where the line of sight to the right side of the tree crosses the stick. You must be careful to move only your eye, and not your whole head. That sounds much more complicated than it really is. (See Figure 19-1.)

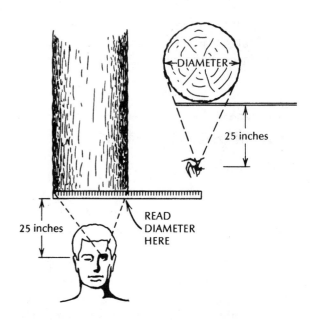

FIGURE 19-1 Using a Biltmore stick to estimate tree diameter breast height (dbh).

The tree caliper is a simple bar of metal or wood. It too is much like a very heavy-duty yardstick. On the zero end is a leg that is placed against one side of the tree. A second, movable leg is moved along the bar to the other side of the tree. The dbh is then read directly from the bar. (See Figure 19-2.)

The diameter tape is much like a steel tape rule. The zero end of the tape is placed on the tree at $4\frac{1}{2}$ feet. The tape is then pulled around the tree and back to the zero end of the tape. The tree's dbh can be read directly from the tape. (See Figure 19-3.) The dbh of the tree is recorded; and once the tree's usable height is determined, we can estimate the timber volume of the tree before it is cut.

Measuring Tree Height

When we speak of measuring **tree height**, we do not mean the height from the base to the

FIGURE 19-2 Using a tree caliper to measure tree dbh.

FIGURE 19-3 Measuring dbh using a diameter tape.

very top of the tree: we are discussing the usable length of the trunk. For sawtimber, this will probably be expressed in terms of 16-foot logs. In other words, the tree grower or buyer is interested in the number of saleable 16-foot logs that can be cut from the tree. Sometimes, 8-foot or 32-foot logs are the length used.

Although logs are measured in 8 ft. or 16 ft. lengths for estimating purposes, sawmills may actually cut lumber in 10 ft., 12 ft., 14 ft., or other lengths for sale.

For pulpwood, the length is in bolts. A bolt is the length to which each pulpwood log

is cut. Usual bolt lengths are 4 feet, 5 feet, and 5 feet 3 inches, depending on the pulpmill's desires.

Tree height is measured from the height of the stump to the point on the trunk where the cutoff diameter is estimated. The usual cutoff diameter is 6 to 8 inches for sawtimber and 4 inches for pulpwood. The height may be estimated in feet or meters and converted to logs or bolts, or it may be estimated directly in logs or bolts. (See Figure 19-4.)

Tree height is measured by a **hypsometer**. The most common hypsometers are

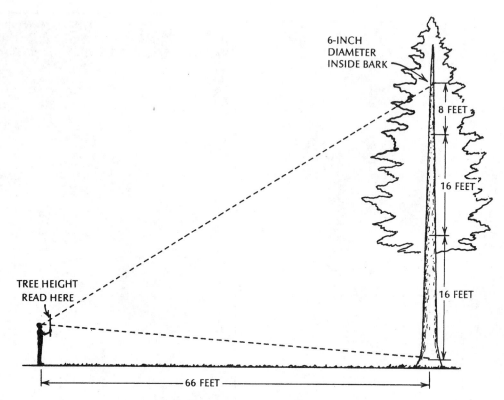

6-INCH
DIAMETER
INSIDE BARK

8 FEET

16 FEET

16 FEET

TREE HEIGHT
READ HERE

66 FEET

FIGURE 19-4 Estimating tree height using the Merritt scale on a Biltmore stick.

1. the Biltmore stick
2. the Abney level
3. the Haga altimeter
4. the Suunto clinometer

Of the four hypsometers, the Biltmore stick is the oldest and the least expensive. It is also easy to fabricate in mechanics laboratories. Because of its low cost, more local schools still use the Biltmore stick than any other method. In commercial forestry, this device is no longer in wide-scale use because it tends to be somewhat less accurate than the more technologically advanced devices. Nevertheless, the concept of estimating standing tree volume is easily taught with any of the four. In recognition of its lower cost and wider avail-

ability in the schools, we discuss only the use of the Biltmore stick here. If your school has one of the other devices, it should come with a set of instructions, and once you have the number of logs estimated and their diameter, the rest of the process is the same.

To estimate tree height with a Biltmore stick, first locate the log scale, or Merritt scale, on the stick. Step away from the tree exactly 66 feet. Hold the stick 25 inches from one eye and line up the base of the stick with the stump height of the tree. Without moving your head, look upward along the trunk and estimate where the trunk gets to the cutoff diameter you are using. Line that point up on the log scale. Read the number (to the nearest one-half log) of logs directly from the scale. (See Figure 19-4.)

Foresters today seldom use Biltmore sticks. But if you learn to use a Biltmore stick, you will understand how tree volume estimation works by the newer methods.

Estimating Timber Volume in a Tree

Once we know the dbh, the small-end diameter, and the number of logs or bolts, we can estimate the board feet or the pulpwood volume of the tree. This is done by means of one of fifty or more so-called "log rules" or tables constructed for that purpose. Some of the most common of these are the Doyle, Scribner, and International rules.

An example, the Doyle rule, is based on a formula

Doyle Volume (board feet) = $(D - 4) \times 2 \times L$

in which D is the diameter at the small end of the log and L is the length of the log in feet. The Doyle rule gives an extremely low estimate of the volume of the log, but it is very simple to use and understand. The Scribner rule is more complex and uses dbh. It gives a higher estimate, often too high on large logs. The International rule is more complex still but gives a more accurate estimate in most cases.

Estimating Standing Timber Volume

All right, so we know how to measure dbh and tree height, and we know how to determine the volume of a log. So what? To the forest landowner who is selling timber, it is important to know how much timber is on the land before the decision is made to cut. In buying or selling timberland, the approximate value of standing trees is important. For the timber buyer, accurate estimates of standing timber are critical. For the forest manager, growth rates in terms of the increase in board feet or cubic feet are essential in management decision making. For the forest researcher, timber volume is often the final result needed.

Estimating standing timber volume is known as **cruising** timber. One way to cruise the timber on a forested area would be to measure dbh and tree height for every tree. The volumes of each tree could then be found by use of the appropriate rule or table and then the volumes of all the trees could be added together. This would be very time-consuming.

A more reasonable approach is to use a sampling of the trees. There are many ways to select the trees to be included in the sample, but the final result is an estimate of the total board feet or cubic feet of timber that can be cut from a stand of trees—while they are still standing.

INTERMEDIATE CUTTINGS

What and Why?

From reproduction to harvest of a forest may take 20 to 100 years. It all depends on the climate, tree species, objective of the manager or owner, and many other considerations. Cuttings made during that period are called **intermediate cuttings**.

Intermediate cuttings to assist young seedlings or saplings are called **cleanings**. In older trees, intermediate cuttings are called **improvement cuttings**.

Thinning

As you will recall from the previous chapter, most trees tend to be more or less shade intolerant. This means each tree must have its share of sunlight to grow properly. A stand with too many trees too close together will not grow well. As a result, the total timber volume produced by such a stand can be increased by removing some of the trees, a procedure called **thinning**.

Once the excess trees are removed, the remaining trees do not have as much competition for light, plant nutrients, and moisture. They can then grow taller, straighter, and faster. The ideal distance between trees varies with the size, species, and objectives of the manager. Thinning may be done by removing individual trees or by cutting alternate rows of trees in a thickly planted stand.

Liberations

In an all-aged forest, the dominant trees may damage the growth of the overall stand. When this is the case, removal of those taller trees, a process called **liberation**, becomes economically desirable. At first, it doesn't seem logical to cut the biggest trees to improve the stand; but the normal objective is to maximize the total timber volume produced in a given period of time. A liberation usually involves cutting the trees, but basal poisoning or girdling is sometimes done.

Sanitation Cuttings

A **sanitation cutting** removes injured, diseased, or insect-infested trees. It leaves only healthy trees, which can then grow without competition from damaged trees. In addition, insects or disease can often spread from infested trees to healthy trees. In the case of insects or diseases, the trees may need to be burned or otherwise destroyed.

Salvage Cuttings

Disease, insects, ice storms, wind storms, and other factors (remember Mount St. Helens?) can kill or damage trees in large numbers. Whenever there are enough trees affected to make a harvest economical, a **salvage cutting** is done. The damaged trees are harvested and sold for wood.

Removal of Undesirable Trees

If the objectives of the forest manager call for an all-pine or all-hardwood forest, other species are undesirable. Even trees that can be very desirable are like "weeds" when they grow in the wrong stand. In addition, trees with forked or crooked trunks may occupy space that better-shaped trees could use. Some species of trees simply grow too slowly or do not produce a good quality of wood. (See Figure 19-5.)

Whatever the reason, these trees take sunlight, nutrients, water, and space from more productive trees. They can be removed by cutting, girdling, or basal herbicide spraying.

On younger stands of trees, removal of faster-growing undesirable trees, shrubs, and vines may be necessary. For instance, pine seedlings can be quickly overtopped by hardwoods so they will grow very slowly, if at all. Removal of the overtopping plants becomes necessary. Cleaning may be done by cutting or spraying the undesirable plants with herbicides.

Prescription Burning

Carefully controlled burning of the undergrowth in a pine forest of saplings to mature trees is normally the most economical form of cleaning when it is appropriate. Prescription burning will be covered in much greater detail in Chapter 21.

FIGURE 19-5 Scrub/wolf tree: undesirable vegetation that should be removed.

HARVEST CUTTINGS

Selective Cutting

Selective cutting, which is usually done in an all-aged forest, is a form of sustained production forestry. (Various **harvest cutting** methods are shown in Figure 19-6.) Using this technique, the forest manager selects individual trees for harvesting. Trees are selected based on maturity, size, species, growth rate, or other factors. Selection is done in such a way that the forest stand continues to grow and still pro-

duces annual (or at least periodic) income. Ideally, the selection process would remove no more wood from the forest than is produced.

In small tracts, selection harvesting can be done for the entire area each year (or other time period), or larger areas may be subdivided with cuttings on one or more subtracts for each harvest.

One problem with selective cutting involves the quality of the trees cut. The tendency is to select the tallest, straightest, most saleable trees each year. If only the best are cut each year, the remaining forest may decline in quality each year. Trees should be selected by considering long-term production as well as short-term profit.

There are two goals in selective cutting. The immediate goal is to produce an income from the harvest. The more important goal is to produce a better yield from the stand.

Shelterwood Cutting

A mature forest is harvested in two or three stages. If the harvest is in two stages, about one-half the trees will be cut at first. The rest will be left to produce seeds. When the seedfall begins in a good seed year, the rest are cut.

The purposes of this harvesting system are (1) to produce a large, immediate income; (2) to provide for the natural reseeding of the area; and (3) to provide partial shelter for the seedlings as they begin to germinate.

Seed-Tree Cutting

In seed-tree cutting, basically the entire stand of trees is removed at harvest, and only a few of the best trees are left standing. These trees, which are selected for their quality and seed-producing ability, are called **seed trees**. After the rest of the stand is harvested, they are left

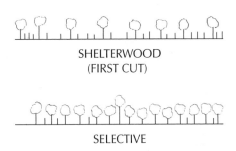

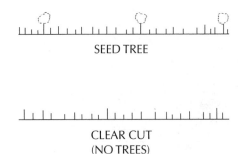

FIGURE 19-6 Four different harvesting systems.

to produce seeds. Normally, four to ten large trees per acre will be enough to regenerate the stand.

Once the new growth begins, the seed trees need to be cut; otherwise their shade would eventually tend to suppress the new trees. Only trees that produce large quantities of seeds are suitable for this method of reproduction. Also, in extremely dry years, the germination and survival rates may be very low for the seeds they produce.

Coppice Cutting

Coppice cuttings resemble seed-tree cuttings. Most of the trees are cut at the initial harvest, but a few a good quality trees are left. These are called standards. They are left not to produce seeds but to provide shelter.

Many hardwood trees will produce sprouts or suckers from their stumps or roots after they have been cut, and this is called coppice growth. (See Figure 19-7.) Coppice growth is very rapid for the first few years of the "new" plant's life as it uses food that was produced by its parent plant and stored in the root system.

There are some disadvantages to this system. First, trees produced from stump sprouts are more susceptible for many years to wind and freezing rain damage. Second, the stand

may become extremely thick in a very short period of time. This necessitates early thinning and, thus, requires substantial labor in stand improvement.

Clear Cutting

Probably no other forestry practice has produced as much controversy as **clear cutting**. (See Figure 19-8.) In some cases, the public outcries have been justified. In recent years, more often than not, the outcries have been unjustified.

Clear cutting is just what it sounds like. All the trees on a tract are harvested in a single operation. This is obviously the most economical harvesting method. No selection is involved. No care is required to leave undamaged seed trees or standards. All the trees are taken and, with the newer harvesting machinery, almost all the tree parts can be taken, too.

The results of clear cutting are extremely visible. An entire area is changed from a mature forest to a collection of ugly tree stumps all in one operation. That is where the controversy comes from.

Many environmentalists argue that clear cutting ruins an area for an entire lifetime. They argue that it causes erosion and upsets the balance of nature for an area.

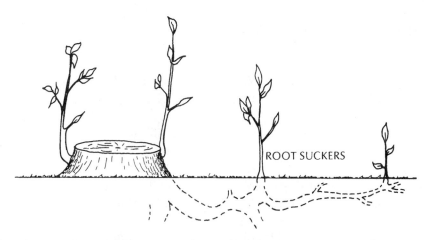

FIGURE 19-7 Using coppice growth to reproduce a forest.

FIGURE 19-8 Even when it is done carefully, clear cutting "appears" to be very damaging to the environment. Clear cutting is a very emotional issue for many people. *(Photo image provided by the Forest Science Data Bank, a partnership between the Department of Forest Science, Oregon State University, and the U.S. Forest Service Pacific Northwest Research Station, Corvallis, Oregon. Significant funding for these images was provided by the National Science Foundation Long-Term Ecological Research program. NSF Grant numbers BSR-90-11663 and DEB-96-32921.)*

Many foresters argue that it produces more wood more efficiently than any other method—both today and when the forest is clear cut again in 20 to 100 years. They point out that wildlife flourish in clear-cut patches, and the resulting regeneration produces an even-aged forest that is more efficient in wood growth.

This argument is unfortunate. Both sides are probably right. Certainly, few things are uglier than a hillside covered with nothing but stumps. Obviously, some increased erosion will occur. At the same time, we need wood for our homes and factories. We want it to be inexpensive and readily available. Most wildlife experts agree with the foresters that clear cutting produces larger populations of healthier wildlife. On the other hand, soil erosion and sedimentation problems may be harmful to fish populations.

REPRODUCING THE FOREST

Natural Seeding

To produce seeds, trees, like other plants, produce some type of flower. They may look like familiar flowers (the magnolia, tulip poplar, and dogwood, for example), or they may not look like flowers at all (examples are pines, firs, and oaks).

Seeds from trees take many forms. They come in all sizes and are transported in many ways. When the seeds end up in the right location with enough moisture and warmth and light, they can germinate. For example, shortleaf pine seeds are released in large numbers from their cones. They are winged and very lightweight. The wind carries them easily, sometimes for great distances.

One successful management technique for regenerating a stand of trees is natural reseeding. The previous section pointed out how the seedtree and shelterwood methods of harvesting promoted natural reseeding. After clear cutting or a wildfire, natural reseeding of an area will eventually occur. It may take a long time if the cleared or burned area is very large.

There is only fair control over the species that will be reseeded even if seed trees are selected for that purpose. Unfortunately, other nearby trees of different species produce seeds too. For instance, cottonwood seeds may be carried by wind for miles. On the other hand, natural reseeding requires little labor and expense; if adequate seed trees are left, a full stand will result in a few years.

Direct Seeding

Tree seeds can be applied directly to an area where trees are desired. They can be broadcast by hand. They can be applied by seed spreaders carried by hand or operated behind tractors, or by tractor-powered grain drills. On very large or remote areas, they can be spread from airplanes or helicopters. A small airplane can seed as many as 1,500 acres per day, and a helicopter as many as 2,500 acres per day.

Direct seeding is usually less expensive than planting seedlings; but, except for the grain drill, direct seeding produces stands of trees without rows. Planting in rows helps in many management steps and gives better control of spacing.

As you can see, direct seeding may involve simply broadcasting seeds on unprepared soil or it may involve planting seeds on prepared seedbeds with grain drills. As you can guess, the results differ widely, depending on the method. And, as you can also guess, the best results come from the most expensive methods of seeding.

Cuttings (Sticks)

One very interesting method of tree reproduction is by means of hardwood cuttings. Some tree species, such as willows and cottonwoods, grow readily from cuttings. Cuttings are usually taken during winter and planted before warm weather comes. Such a cutting looks just like a stick. It must be kept moist so the cambium layer does not dry out.

A hole is punched or dug in the sod with a metal rod. The bottom end of the cutting is placed in the hole, and the soil is firmed around the stick. In the spring, the underground end develops roots and the aboveground end develops leaves and twigs. Now it is a complete, young tree.

Many farmers have used freshly cut hardwood posts in building fences. Sometimes these posts act like cuttings. They root and sprout and begin to grow.

Plant Seedlings

The most certain, quickest way to get a full stand of trees is to plant nursery-produced seedlings. This also requires more labor than the other methods.

Advantages of planting seedlings over other methods are

- complete control over the species

- easy selection of different varieties for special purposes

- introduction of a type of tree that does not already grow in an area

- replacement of the trees in less time than direct seeding or natural seeding

- fast erosion control

- control of tree spacing

- creation of pure and even-aged forests

- availability of healthy, genetically superior nursery seedlings

Planting has disadvantages too:

- introduced trees may not do as well as native varieties

- the initial investment is high, both in terms of labor and seedling costs

- because all the trees are alike, a disease or insect problem is potentially more devastating than for naturally reseeded trees

Seedlings may be obtained at very low cost through most state forestry agents. Commercial tree nurseries produce and sell large quantities and many varieties of trees for forestry use. Seedlings are usually secured and planted during the winter. They may be planted by hand using hand tools or by workers riding on tractor-drawn equipment.

STEPS IN FOREST MANAGEMENT

The American Forest Institute says there are four steps in developing a good forest management program:

1. Analyze the situation.
2. Decide on your objectives.
3. Develop a plan.
4. Measure the results.

The first step is analyzing the situation. Take a close look at the forestry resources that are to be managed. What is the size of the area? What kinds of trees are currently on it? What are their ages, sizes, and quality? Are there any particular problems, such as disease, insects, low-grade growth, severe erosion danger, and so forth? Determine all the factors that will affect the forest use in the short run.

Step two is to decide on the objectives. For instance, if cash is needed right away and the stand is even aged and mature, a clear cut might be indicated. If the sale of timber from the stand is planned for 10 or 20 years from now, a different plan will result. What are the goals? How are the trees on the forestland to help reach those goals?

How long is the manager willing and able to wait? The objectives should be both short range and long range.

The third step is to develop a management plan. Analysis of the situation will indicate what is possible, and the objectives will determine where the trees fit into the picture. The management plan is simply the plan for reaching the objectives with the resources available. It should take into account both short- and long-range objectives. A good forest management plan considers the effects of actions on the land, forest, neighbors, and watershed, as well as on the objectives.

Step four is to measure the results. This is how the manager keeps track of the plan's progress. What were the costs of the harvest and reforestation? What revenues were generated? Was there a profit? How much new growth are you getting per acre each year? When will the next harvest be ready?

SUMMARY

Since 1900, American consumption of wood and wood products has increased tremendously, but we have less forestland now than then. This strange situation is made possible by two things: (1) more wood is produced per acre on our managed forestlands, and (2) we use the harvested trees more fully and more efficiently.

Wood is measured in many ways, such as board feet, cubic feet, cords, and gross weight. It is possible to estimate the amount of wood that can be obtained from a log before it is cut. It is even possible to estimate the amount of wood in a single tree or an entire forest before the first tree is cut.

If you can estimate the standing volume of one tree, you can do the same for a whole forest. This process is called timber cruising. Landowners, forest managers, potential buyers, and forest researchers need to know how much wood there is in a growing tract of trees for many reasons. Timber cruising allows us to estimate value and growth rates of trees without cutting them.

There are two broad categories of cuttings: intermediate and harvest. Intermediate cuttings are done either to clean out the forest floor so that young trees can grow or to improve the stand of trees. Harvest cuttings are done to produce wood for sale and may be selective, shelterwood, seedtree or coppice, or clear-cut harvestings. The clear cut is the most economical method, but it is also the most controversial.

Forests can be reproduced by natural seeding, direct seeding, coppice growth, or planting seedlings. The method with the least direct cost is natural seeding, but it is also the slowest and the least reliable. The most expensive, fastest, and most reliable method is planting seedlings.

Forest management can be thought of as a four-step process: analyzing the situation, deciding on the objectives, developing a plan, and measuring the results. The objectives should take into account not only the immediate but also the long-term future. The plan should consider the responsible use of the forest resource and its proper regeneration.

DISCUSSION QUESTIONS

1. How can we be getting more forest products from less forest and no more cut trees than we did in 1900?

2. What is a board foot? cubic foot? cord?

3. Why is gross weight probably a better measure for pulpwood than the cord?

4. What is dbh? How is it measured?

5. Why is the height of a tree for sawtimber measured in logs instead of feet?

6. What are intermediate cuttings? What is their purpose?

7. List five types of harvest cuttings.

8. Do you favor or oppose clear cutting? Why? (Why not organize a class debate on this issue?)

9. List and describe four methods of forest reproduction. What are some advantages and disadvantages of each?

10. List and discuss the four steps in developing a forest management program.

SUGGESTED ACTIVITIES

1. Visit a tree farm and discuss the management techniques being used. Find out how the trees were established and what harvesting method is being planned or used.

2. Visit a pulpwood or lumber-purchasing yard. What kinds of trees are purchased? What units of measure are used? What prices are paid to the grower? How and where are the logs sent for processing? What wood products will result from the processing?

3. Arrange for a forester to visit your class to discuss forest management techniques most appropriate for your area.

20 FOREST ENEMIES AND THEIR CONTROL

OBJECTIVES

After reading this chapter, you should be able to

- describe the major insect pests of our forests

- describe the most important disease problems of our forests

- outline other enemies of the forests

- outline woodland management techniques for controlling forest insect problems, disease problems, and problems caused by other forest enemies

TERMS TO KNOW		
forest enemies	biological insect control	forest pathology
bark borers	pathogen	noninfectious diseases
defoliating insects	insect resistance	infectious diseases

In terms of visual impact, there is no comparison between a forest wildfire and an infestation of beetles eating the cambium layers of the trees in a forest. Clearly a wildfire is dramatic. We spend an entire chapter discussing forest fires, both wildfire and fire used as a forest management tool. Yet, far more damaging to our forests, both in terms of trees killed and lumber lost, are the forest's other enemies: insects, disease, wildlife, overgrazing, and weather damage.

What do we mean by **forest enemies**? Trees, like other plants and animals, begin life as embryos. They grow in size and eventually reach maturity. Trees are adapted to grow best in certain climates and under certain conditions. For the purposes of this chapter, forest enemies include anything that interferes with the normal reproduction and growth of forest trees, given favorable climate and growing conditions.

As previous chapters pointed out, our population is growing, our forestland is shrinking, and we need to continue to produce more wood and other forest products on less land area. Accomplishing this means continuing to improve our woodland management, maintaining our wildfire protection programs, continuing research into genetic improvements, and improving the efficiency with which our mills use the trees we do produce. But as important as any of these is the protection of our forests against the enemies this chapter discusses.

INSECTS

Types of Insects

According to some authorities, insects kill more trees than any other enemy of the forest! Forest insects may infest a single growing or mature tree, or they may attack small areas within a forest. They may be specific to one or only a few species, or they may attack a wide range of tree species. In major outbreaks, whole forests can be damaged or killed.

The western pine beetle may have destroyed as much as 25 billion board feet of timber in a 25-year period starting in the mid-1920s. Gypsy moth outbreaks in the northeastern United States continue to worry foresters.

Insect damage to trees can be of several types. The following paragraphs summarize the main forms of insect damage.

Bark Borers

These tiny beetles attack trees by tunneling underneath the bark. (See Figure 20-1.) The soft cambium and newly produced wood and bark cells are destroyed, effectively girdling the tree. In addition, the beetle's damage allows other destructive agents (such as fungus) to enter the tree. When the beetle attack on a tree is severe, the tree dies. It is easy to identify the damage done by **bark borers**. Bark beetles are particularly damaging in pine forests.

The Southern Pine Bark Beetle, *Dendroctonus frontalis*, is one of the most destructive enemies of forests in the South. The first part of the name, *Dendroctonus*, means "tree killer," which is certainly an appropriate name. The Southern Pine Bark Beetle burrows through the outer bark and inner bark and eats the inner

(a)

(b)

FIGURE 20-1 (a) An adult Southern Pine Bark Beetle feeding underneath the bark of a pine tree. (b) Beetle damage in a southern forest. This is typical damage resulting from bark beetle infestation. The tree can be completely girdled, and thus killed. *(Photos courtesy of United States Department of Agriculture, U.S. Forest Service)*

bark and cambium layer of cells. After a short period the beetle can tunnel completely around the tree, girdling it. When a tree is girdled, it can no longer send food from the leaves to the roots. The roots die and then the entire tree dies. (See Figure 20-1.)

The Southern Pine Bark Beetle is one of only a few insects that can invade healthy trees. Most others require injuries to branches or bark in order to get into the tree. During an outbreak from 1973 to 1977, this beetle killed trees totaling about 4.5 billion board feet of pine timber. According to a report by the Florida Division of Forestry in 1996 an outbreak of Southern Pine Bark Beetle along the Suwannee River invaded a number of large pine plantations. The report went on to say that the Southern Pine Bark Beetle had caused $900 million in damage to U.S. pine forests from 1960 through 1990.

Defoliators

Some insects damage trees by feeding on the leaves or needles. In most cases, damage is done by insect larvae instead of the adults. It is easy to identify the damage of **defoliating insects**, especially when the attack is severe. Leaves are partially or completely eaten. The midribs may be left. In severe attacks, the whole tree may be stripped of leaves. Caterpillars (insect larvae) may be visible on the tree.

Wood Borers

Wood-boring insects eat their way through the sapwood and heartwood of the tree. They weaken and damage the wood. They may attack live trees, or they may attack logs or wood after it is processed. Termites are not the only wood borers, but most people probably think of this pest first.

Tip Feeders

Insects that attack young twigs, stems, or buds are tip feeders. These are the most tender and moist parts of the tree. A tree that forks too near the ground may not be useful for timber. Twists and bends in a tree trunk lower its value. These deformities may have resulted from tip-feeding insects. Repeated and severe tip damage may kill the tree.

Sap Suckers

Insects with penetrating mouthparts may feed on tree sap. Aphids and scales are examples. To understand this group of insects, think of an example with which you are familiar. A mosquito that "bites" you is much like an aphid that "bites" a tree leaf. Sap suckers usually only weaken trees and slow their growth.

Galling Insects

Some insects damage trees by producing galls on twigs, limbs, trunks, or leaves. Galls are abnormal growth and usually cause little serious damage.

Seed Eaters

Trees reproduce by several methods as explained in Chapter 18. One important way is by seeds. Some insects live by eating tree nuts, fruits, and seeds. If the seed is damaged by insect feeding, it cannot produce a tree.

Root Feeders

Insects in the sod may feed on tree roots. Particularly susceptible are tree seedlings. The growing tips of any plant root, however, may make a meal for white grubs, wireworms, and other root feeders.

Control of Insects

Natural Controls

People are always talking about "the balance of nature," and in nature there are factors that tend to limit the size of insect populations in the long run. Extremely high or low temperatures can kill insects or, at least, limit insect activity.

Insects have natural enemies. Insects that eat tree seeds may be very appetizing to other types of insects, birds, or animals. Insects, like other animals, are subject to diseases caused by fungi, bacteria, and viruses. As the damaging insect population grows, they provide a more abundant food supply for their predators. They may also become more susceptible to diseases.

Such natural controls are slow. In addition, as humans change the forest to improve its production, these natural controls may no longer work very well.

Biological Controls

Biological controls exist in nature. Predators and diseases of insects were discussed previously. Whenever foresters alter these natural controls, we say they are using **biological insect controls**.

An example of a biological control would be the introduction of a new predator to an area. A bird that is particularly fond of a problem insect may not be native to an infested forest, and introducing the bird into the problem area would be a biological control.

Another example is the introduction of pathogens. A **pathogen** is any disease-causing organism. A pathogen, bacterium, or fungus that attacks specific insects may be available and can be introduced into the insect population.

Management Controls

Infested trees and tree parts are removed during thinning or selective harvest cuttings. They may be cut and burned or otherwise destroyed. Tent caterpillars produce large, silk-like tents. These can be removed by hand and burned.

Trees that have been damaged are susceptible to insect attack. Ice storms, wind storms, or other sources of injury may cause such damage. Damaged trees should be harvested or destroyed because (1) they will not be efficient and valuable wood producers, (2) they take up space, water, and light needed by other trees, and (3) their presence can promote insect or disease problems.

Researchers have been working for many years to develop better varieties of trees, and one aspect they have worked on is **insect resistance**. Tree varieties are now available that are less likely to have serious insect problems. These varieties can be selected for reforesting harvested areas.

When trees are too crowded, several things happen. Their growth is slowed. More trees in a given area do not mean more wood. In addition, the trees are not as strong or healthy. They are more susceptible to insect and disease problems than are healthy trees.

Prescribed burns are used to control heavy undergrowth in some areas. One side benefit of such burns is the reduction of insect populations.

Chemical Controls

At one time it was thought that chemicals would solve most of our forest insect problems. A very widely used chemical in the early 1960s was DDT. It was very effective on a wide range of insects and was relatively inexpensive. Unfortunately, it had many undesirable side effects.

Today, extensive research must be done before a pesticide can be used in this country. The pesticide must be proven reliable and safe; only then will it be approved by the Environmental Protection Agency (EPA) and sold in this country.

Such regulation makes pesticides safer—but also more expensive and more difficult to get and use. In short, chemicals apparently will not provide the solution foresters once hoped for. In severe cases, however, chemicals can provide a quick and effective short-term solution.

DISEASES

Forest Pathology

Forest pathology is the study of tree diseases—their characteristics, causes, prevention, and treatment. Forest pathologists generally classify forest diseases as either infectious or non-infectious.

Noninfectious diseases are usually caused by environmental problems. Air pollution, acid rain, sunscald, drought damage, flood damage, ice damage, and wind damage are examples. Noninfectious diseases are discussed in other parts of this chapter.

Infectious diseases are usually caused by parasites. Parasites are organisms that rely on other organisms for their food. In a very real sense, humans are parasites of trees. The parasites of interest here are those living in infected trees. In general, there are five major groups of disease-causing organisms for trees: fungi, bacteria, nematodes, viruses, and mistletoes. Of these five, the most important diseases by far are those caused by fungi.

Fungus-Caused Diseases

Fungi are always present in large quantities and in wide varieties in the forest. Fungi reproduce and spread by tiny spores. The spores may move directly from an infected tree and infect another tree, or they may require an "intermediate" host. Spores may be spread by the wind, insects, or other wildlife.

Fungus diseases may attack the leaves, stems, or roots of the tree. Of these three, foliage diseases are the least dangerous. Leaf-attacking fungi cause dead spots on the leaves or kill entire leaves. This damage may be unsightly, but it probably will not kill the tree. At worst, a reduction in growth rate will result.

Stem diseases are more important. Stem diseases may block the flow of moisture and nutrients upward from the roots, or they may block the downward flow of sap. The heartwood may be destroyed altogether, and the tree may be damaged by strong winds.

The best-known fungus-caused stem diseases probably are Dutch elm disease, chestnut blight, fusiform rust, and various types of heart rot. (See Figure 20-2.)

Root-rotting fungi are generally less important than the stem diseases, with one possible exception. Fomes root disease weakens the root system making the tree less resistant to wind damage. Root rot may destroy the entire root system. In severe cases, roots become soft, spongy, and stringy; the tree, if not killed, will be badly damaged. Spores of the organism (*Fomes annosis*) enter stumps from harvested trees, allowing the disease to spread from infected trees to new areas. Once the annosis fungus is in the soil, it spreads from one tree to the next. It often causes very high death rates in infected pine plantations.

FIGURE 20-2 This tree is suffering from heart rot.

Control of Forest Diseases

The most effective control of forest diseases is good management. A well-managed forest is usually healthy and almost disease free.

Fomes rot can be prevented from entering an uninfected forest. When trees are cut for any reason, the stumps are dusted with borax or creosote. For an infected area, little can be done realistically. A common flower, marigold, and several other plants appear to check the spread of the disease. The only workable alternatives are either to accept some losses or to change the stand to a nonsusceptible tree species.

For stem fungus diseases, good management techniques usually work. Diseased trees must be removed and used or destroyed. Infected branches are pruned and destroyed. Branches over 3 inches should not be pruned; they should be removed.

Damaged trees are more subject to disease. When a tree is damaged, it should be salvaged or destroyed. Crowded trees are more subject to disease. Forest stands should be thinned as needed for best tree density.

Forest researchers have been working to help here too. Disease-resistant varieties have been (and are being) developed for many species of commercial trees. Trees resistant to southern fusiform rust, chestnut blight, Dutch elm disease, and white pine blister rust are already available or soon will be.

In some cases, diseases are spread by insects. Insecticides to kill the carriers are often effective in preventing their spread. In other cases, intermediate hosts can be removed. For instance, white pine blister rust must have currant or gooseberry bushes in its life cycle. The disease can be prevented in a stand of trees by removing all currant and gooseberry bushes within 900 to 1,000 feet of the forest. Infected trees are removed and salvaged or destroyed to prevent infection of new trees.

Once a tree is diseased, chemical treatment is not usually done. Prevention is the answer, and good management is the key.

WILDLIFE

In general, any animal living in the forest gets its food from the forest. Squirrels eat tree seeds, acorns, nuts, and so forth. Birds eat seeds and young seedlings. Rabbits eat seeds, seedlings, and the young bark of seedlings and saplings. Deer feed on seedlings and saplings—mostly hardwoods. Such browsing generally favors conifer production because conifers (pines, fir, cedar, etc.) are not a favored food but hardwoods are. Browsing animals eat the young hardwoods and tend to leave conifers alone.

A number of birds and mammals feed on tree shoots, twigs, and buds. Such feeding stunts tree growth and causes deformities even if the tree survives. Beaver actually fell smaller trees (usually less than 4 inches in diameter). However, these toothy creatures have been known to cut down trees up to 18 inches across.

When the wildlife population of an area is at normal levels, forest damage is not a big problem; if populations increase, damage can result. When the squirrel population becomes too dense, all the seed crop will be destroyed. Not only that, but buds and twigs will be eaten. The same holds true for other animals too. Deer usually prefer other foods to tree bark; but when the deer population becomes large, forest damage results.

When the wildlife population of an area becomes too great, not only the forest suffers; the animals also suffer. Mass starvations and fatal diseases often result. The best solution to this is a well-managed hunting and trapping program. For more information on game management, see Chapter 24.

GRAZING

In many places, particularly in the southern and western United States, livestock are grazed in the forests. This has several benefits. Dense undergrowth is controlled because the goats, hogs, or cattle eat or trample the bushes or vines. Also, hardwoods are eaten, which helps in establishing pure pine forests. Insect and disease problems also may be helped because the undergrowth is controlled.

At the same time, the forest soil is compacted. Drainage may become a problem. Root development is hindered, and trees may be stunted or even killed. In the West, the grasses may be important in erosion control. For further discussion on this, see Chapters 5 and 6.

Controlled grazing is one thing; overgrazing is different. Overgrazing in the forest can actually lower the long-term income from the land. It can also cause long-term damage to the forest and to the soil.

ENVIRONMENT

Sheet ice storms and heavy snows break limbs and even whole trees. (See Figure 20-3.) Wind can bend, break, or uproot trees; heavy rains cause flooding and drainage problems.

Extremely hot weather can damage new seedlings. An unusually warm spell in the winter can cause early budding and leaf and stem growth. If this is followed by a very cold spell, this new growth often dies. Excessively hot weather and direct sunlight can cause sunscald on exposed tree bark, especially on young trees.

Environmental damage can range from minor to extreme. In extreme cases, a whole forest can be killed by environmental conditions such as a tornado. In less extreme cases, individual trees can be slightly damaged. Regardless of the degree of damage, weakened trees are much more likely to suffer still more damage from insects and disease later. A tree in which a branch is broken by ice may not

FIGURE 20-3 Trees can be bent or broken by wind and ice.

seem to be damaged severely, yet the tree may die several years later. It is likely that the tree was actually killed by insects or disease pathogens that invaded it through the break in the outer bark resulting from the ice storm. Trees can be weakened by long drought, bending from heavy ice, unusually cold or unusually hot temperatures, and many other environmental factors.

The best way to minimize these types of weather damage is good forest management. Well-managed forestry reduces the amount of damage the forest will suffer. In a well-man-

aged forest, trees are properly spaced, strong, and healthy. They do not bend or break as easily, and they are not easily uprooted. Strong branches support a good amount of snow and ice. Strong trunks resist strong winds. A solid canopy does not allow enough direct sunlight to scald tree trunks. A border of bushes around a forest prevents wind from getting under the canopy.

Selection of tree varieties is important. If eastern white pine does not normally grow in southern Florida, there may be a reason. Perhaps the weather is too hot for it. If slash pine does not normally grow in North Dakota, perhaps there is a reason. Maybe it does not do well under heavy ice or snow weights or in cold weather.

Bringing a new tree species or variety into an area can be a risky business. If it is suited to the climate, soil, altitude, latitude, and so forth, it may do well. If it is unsuited, it may not survive there. Worse still, it may live for several years only to be killed by a weather extreme later.

Once a tree has been damaged by weather extremes, it probably should be removed. A bent tree will be bent forever. Straightening a young tree in your yard may be possible, but such operations are not economical in the forest. A large tree damaged by lightning can often be saved, but the cost of doing so is too great for trees being grown for wood.

Once ice, snow, wind, or other weather problems occur, the normal step is salvage cutting. If only a few trees are damaged, selective cutting is appropriate. The cut trees may be simply destroyed if there are too few to justify sale. On the other hand, many or all of the trees in an area may be damaged. When that happens, a full harvest may be necessary. If the

FIGURE 20-4 These trees were toppled by winds from a hurricane. Salvage cuttings will allow the wood to be harvested.

trees are too small for lumber, they may be cut for pulpwood. (See Figure 20-4.)

SUMMARY

Many things can affect the growth rate and overall health of the forest. Even with a favorable climate and good growing conditions, trees face many enemies. This chapter covered the major enemies of the forest (except wildfire, which is covered in Chapter 21).

Many people believe that of all the forest's enemies, insects are the most damaging. Insect damage can be as small as a minor discoloration on a few leaves, or it can be as large as a whole forest completely destroyed. Major pests are bark-boring, defoliating, wood-boring, tip-feeding, sap-sucking, galling, seed-eating, and root-feeding insects. Insect controls are gener-

ally grouped into four categories—natural, biological, management, and chemical controls.

Tree diseases are another major enemy of the forest. Forest pathologists classify diseases as either infectious or noninfectious. Noninfectious diseases are caused by environmental problems. Infectious diseases are caused by parasites. The major parasites that attack forest trees are fungi, bacteria, nematodes, viruses, and mistletoes. Of these five, fungi are by far the most important. This chapter discussed only fungus-caused infectious diseases.

Wildlife can also damage the forest. When wildlife populations are at normal levels, there is little problem; but populations of deer, rabbit, squirrel, beaver, or some other animal can become too large. When this happens, the forest suffers from unusual or excessive feeding. Prevention of wildlife damage depends on good wildlife management techniques.

Grazing sometimes becomes a problem. Many farmers graze livestock in the forests of the South and West. This alone is not usually a problem; but when too many animals are pastured there, overgrazing can be a problem. The results can be damage to trees, soil compaction, drainage problems, or severe erosion. The solution is to control grazing at a level that does not cause such damage.

Environmental damages are considered to be a noninfectious form of disease. Environmental diseases discussed included ice storm damage, wind damage, snowstorm damage, sunscald, and others. Again, the best solution to these problems is good forest management. Proper variety selection, spacing, thinning, and removal of damaged trees are important. Environmental damage can sometimes be very severe and sudden.

7. What management steps can a forest owner take to help prevent forest disease problems?

8. What are some specific ways to combat root rot?

9. How do wildlife damage the forest?

10. What can we do to control wildlife damage to our forests?

11. When is grazing a problem in the forest?

12. List some environmental enemies of the forest.

13. How can we help cut down on environmental damage to forestland?

14. Once a commercial forest tree has been damaged, what should probably be done?

DISCUSSION QUESTIONS

1. What do we mean by "enemies" of the forest?

2. What is the single most destructive enemy of the forest? What makes them so very destructive?

3. List and describe the types of damage insects cause to trees.

4. List and describe the four categories of forest insect control measures.

5. How do infectious and noninfectious forest diseases differ? How are they alike?

6. What are the five major disease-causing organisms? Of these, which causes the most serious problem?

SUGGESTED ACTIVITIES

1. Visit an unmanaged forest of mixed species. Try to find examples of tree damage from insects, diseases, environmental, and other problems. Identify trees that would need to be removed because of those reasons if the forest was being managed for wood production.

2. Visit a tree farm. What management techniques are being used to control insects, diseases, and other forest enemies?

3. Arrange for a forester to visit your class to discuss locally important forest diseases, insect pests, and other problems—as well as their control.

21 FIRE!

OBJECTIVES

After reading this chapter, you should be able to

- list and describe some of the most destructive forest fires in U.S. history

- draw and explain the fire triangle

- explain how fire can be used as a positive tool in woodland management

- describe the anatomy of a typical forest wildfire

- explain how firefighters find and attack a forest wildfire

TERMS TO KNOW		
prescribed fire	ignition temperature	alidade
Great Chicago Fire	fire triangle	triangulation
Peshtigo fire	incendiary fire	fire suppression
Miramichi fire	ground fire	fire barrier
fuel	surface fire	backfire
oxygen	crown fire	

The camera sweeps across a burned hillside. Smoke seeps upward from still-smoldering tree trunks. A small animal killed by the fire lies on the ground. A deep voice reminds, "Remember, only you can prevent forest fires." Smokey Bear, the Keep America Green programs, and other efforts aimed at forest fire prevention are familiar to most Americans. Television commercials, radio announcements, comic books, highway billboards, and other things aimed at forest fire prevention have been around since World War II.

The original Smokey Bear was actually a survivor of a forest fire. Badly burned by a wildfire, the bear was rescued, nursed back to health, and moved to the National Zoo in Washington, D.C. There, Smokey remained a living symbol for many years. At his death, he was replaced by another bear that carried on the tradition.

Nobody knows how many acres and lives have been saved because of these education programs; but, since they began in the 1940s, the number of wildfires reported each year has dropped dramatically.

FIRE AS A FOREST MANAGEMENT TOOL

Prescribed Fire

Before European settlers came to America, Native Americans used fire as a way to clear the land and to improve hunting. The early settlers also soon learned to use fire as a management tool. These intentional fires often got out of hand, causing wildfires. Combining those wildfires with natural and other wildfires soon produced a serious problem in this country. As a result, the use of fire as a management tool fell into disfavor.

By the beginning of the twentieth century, foresters were beginning once again to advocate the use of fire as a management tool in southern slash pine forests. By 1907, forest owners were beginning to use planned fires to reduce the litter on the forest floor. By removing this fuel, the likelihood of wildfire decreased. Such a managed, intentional fire is called a **prescribed fire**.

Purposes of a Prescribed Fire

A prescribed fire produces many benefits for the forest, wildlife, and people. A few of the benefits follow:

- reducing the hazard of wildfire by removing fuel from the forest floor
- getting sites ready for seeding and planting
- improving wildlife habitats
- removing undesirable trees and brush cluttering the forest understory
- helping control forest diseases
- improving the quality of grass for grazing by removing brush and dried weeds

- improving the appearance of the forest by cleaning out weeds and brush
- making access to the forest easier

How Does a Prescribed Fire Work?

First, it must be controlled very closely. Only a small area is burned at a time. The humidity and moisture content in the forest must not be too low, and there should be only a slight breeze. Many other conditions must exist for the prescribed fire to be done safely.

Second, the fire must not be allowed to get too hot because fire damages trees in several ways. If the fire is too hot, the crown can wilt, causing damage to the leaves. The heat can become so intense that the cambium layer under the bark literally cooks. In extreme cases, the crown and even the trunk can actually catch on fire. If the prescribed fire is planned and controlled properly, the heat will be low enough to prevent such damage.

Third, the fire must not be allowed to get out of control. If a prescribed fire breaks out, it becomes a wildfire and can cause great damage.

As you can see, using a prescribed fire is not a simple task. Only a trained forester should attempt this technique.

WILDFIRE IN AMERICA

Probably the most famous fire in this country's history is the **Great Chicago Fire** of October 8-10, 1871. Most students of American folklore know the story of the cow that supposedly kicked over the lantern and ignited that blaze. When the flames were finally out, much of Chicago was in ashes and an estimated 300 people had died.

At almost the same time as the Chicago fire, a group of loggers were burning slash near Peshtigo, Wisconsin. The fires were not fully extinguished, and high winds whipped them out of control. The summer had been very dry and hot. The pine forests were dry and very susceptible to fire.

The fire storm, driven by the wind, raged across the dry forest. Burning cinders ignited many spot fires, often miles ahead of the main fire. The flames moved so fast and in such great leaps that escape was not possible for over a thousand people and millions of wild animals.

By the time the great **Peshtigo fire** was out, the damage was far greater than that of the more famous Chicago fire. In all, 1.3 million forest acres in Wisconsin and 2.5 million forest acres in Michigan were destroyed. Think of that—3.8 million acres of forest destroyed in a single fire. In addition, an estimated 1,638 lives were lost. Nobody really knows for sure; such a large area was burned that only a rough guess was possible. Whole villages were wiped out.

The second largest fire in U.S. history was the **Miramichi fire** of 1825. In just a few days, 3 million acres of forests in Maine and New Brunswick, Canada, were destroyed. The fire claimed at least 160 lives.

During the decade of 1986-95, the summer weather in the far western states of California, Idaho, Utah, Washington, Montana, and Oregon tended to be drier than in many previous years. The drying trend is thought to be related to a series of unusually strong and frequent warm surface currents (called El Niño currents) in the Pacific Ocean. Regardless of the cause, the result was a rash of unusually large and violent forest wildfires in those states in 1994. On one day in mid-August 1994, at least 370,000 acres were burning in

those states, in 39 major fires, and involving over 24,000 firefighters. You may remember the live television coverage of the fires. You may also recall that 10 U.S. Forest Service firefighters were killed in a forest wildfire in Colorado that same summer. The efforts at extinguishing those western forest fires involved over 150 helicopters, 40 air tankers, and over 24,000 firefighters, including several thousand military personnel. (See Figure 21-1.)

Even though live television coverage of fires in the Northwest and in California makes it seem that wildfires are concentrated in those regions, actually the reverse is true. Surprisingly, over the years, the greatest numbers of wildfires as well as the greatest overall damage have been concentrated in the southern United States. (See Table 21-1.)

Television news reports also give the impression that forest wildfires are a growing problem. That also is untrue, except for the unusually destructive 1994 fire season when almost twice as many acres were burned. Forest fire prevention, detection, and suppression programs in the United States have been among the most successful environmental or conservation efforts of all time. The federal budget for forest fire prevention, detection, and control exceeded $1 billion in 1994. Such federal, state, and local expenditures over the years have produced a wildfire management program that may be too successful. In fact, the forest fire management program in the United States has been so successful that our forests probably need far more forest fires than are occurring!

A fully mature forest may be beautiful, but it is very unproductive from many perspectives. Little wildlife live in a mature forest, particularly on the forest floor, because little vegetation grows there. Most of the sunlight is captured by the tall crowns of the mature

FIGURE 21-1 Wildfire in Colorado. *(Photo courtesy United States Department of Agriculture, U.S. Forest Service, Montrose Interagency Dispatch Center)*

TABLE 21-1 Total Number of Forest Fires in the United States Compared with Those in the Southern United States

Year	Number of U.S. Fires	Number of U.S. Acres	Number of Southern Fires	Number of Southern Acres
1942	208,218	31,854,963	153,089	29,773,110
1952	188,277	14,187,325	134,255	11,498,476
1962	115,345	4,078,894	72,444	3,064,285
1972	124,554	2,641,166	54,228	894,627
1984*	118,636	2,266,134	63,937	617,091
1992	87,394	2,069,926	43,227	498,348

* 1982 data not available

trees and little is left to produce food for ground-dwelling animals and predatory birds. Also, in a mature forest, the growth of new trees and the production of new wood are balanced by the death of old trees and the decomposition of old wood.

There are only three ways a forest can remain healthy and growing. Periodically, it must be burned, harvested, or destroyed by insects, disease, or other calamity. One result of the Forest Service's success in fire management is that many of our national forests and parks have more trees per acre than can remain healthy. With increasingly successful efforts by more and more environmental groups and wildlife activists to prevent logging, many of those forests have become "ripe" for disaster. Heavier-than-normal undergrowth promotes disease, insect infestations, and more wildfire.

As a result of this paradox, many environmentalists and forest conservationists are now advocating a more "hands-off" approach to forest wildfires. They contend:

1. Wildfires are a part of the natural system.

2. National efforts at preventing and controlling forest fires have resulted in unnaturally heavy production of undergrowth and trees weak from overcompetition.

3. As a result, there is a growing danger of major forest fires, forest disease, and insect infestations.

4. To correct that problem, we should decrease our future attempts at preventing forest fires, and limit our efforts to preventing damage to humans from future wildfires.

Forest managers would prefer to be allowed to use increased harvesting of mature trees and prescribed burning to correct the problem; but that approach is producing much controversy. Increased harvesting is opposed by preservationists who want to allow the forests to return to so-called "natural" conditions. Harvesting the largest trees is opposed by some biologists. They contend that the tallest and healthiest trees should be left to produce future generations of trees and that the shorter, more undesirable trees should be harvested instead. Prescribed burning is opposed by animal rights activists who fear that existing animals and birds will be harmed by the intentional fires. The forest managers argue that such practices would result in economic losses that would be unacceptable to forest owners and would produce greatly increased prices for wood and wood products.

CAUSES OF FOREST FIRES

The Fire Triangle

Fire is both a chemical and a physical process. Forest fires result from the rapid combination of oxygen with other chemicals contained in wood, leaves, and other material in the forest.

Forest fires always involve heat and usually produce fight and flame; but this is not necessarily the case, as we will see later. Forest fires range from slow, smoldering, very gentle fires to raging infernos of flames, smoke, and high winds.

For a forest fire to occur, three things are required. The first is **fuel**—that is, something to burn. What in a forest can burn? Trees, dead leaves, grasses, forest litter, vines, brush, and many other things in the forest are combustible. Green leaves and wood do not burn as readily as dry leaves and wood, but they will burn when the conditions are right. In short, just about everything that makes up a forest will burn. There is seldom a natural shortage of fuel for a forest fire.

The second requirement for a forest fire is **oxygen**. About 20 percent of our surface atmosphere is oxygen gas. As the oxygen right around a flame is used up, the flame dies down. When most of the oxygen is gone, the flame will go out. Light a match and place it in a small jar, then cover the jar. The flame probably goes out before the match completely burns. If the air were perfectly still around the flames, a forest fire couldn't grow; but, it isn't. Winds, even gentle breezes, fan the flames. Even if there is no breeze, hot air rises. Thus, a flame creates its own "breeze" as heated air uses and fresh air is drawn toward the flame. A large forest fire creates its own wind, even when there is no breeze blowing. When the flames are being blasted by high winds, the fire can spread very rapidly.

The third requirement for a forest fire is heat. For a combustible material to burn, it much reach its **ignition temperature**. Most forest fuels have ignition temperatures of 600° to 880°F (316° to 471°C). This ignition temperature is the same whether the material is wet or dry, but wet leaves and wood do not burn as readily. Why? Because water boils at a temperature well below these common ignition temperatures; wet leaves and wood exposed to an open flame do not get much above the boiling point of water until all the water evaporates.

Thus, green or wet wood or leaves do not catch fire as readily as dry wood. Once the fire begins, the heat produced by the fire dries out nearby wood and leaves. This allows the fire to spread. The fire gains speed and momentum as it goes, drying more fuel with its own heat.

That is why the danger of a forest fire is low during wet weather and high during dry weather. The fuel and oxygen are there all the time, but it takes a much greater amount of heat to evaporate all the water and still reach the forest's ignition temperature.

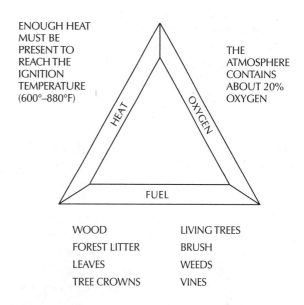

FIGURE 21-2 The fire triangle. All three sides must be present for a forest fire to occur.

This relationship can be illustrated as a triangle—the **fire triangle**. (See Figure 21-2.) To be a fire triangle, all three sides must be present. If one side is removed, it is no longer a triangle. To burn, a forest fire must have all three sides of the fire triangle: fuel, oxygen, and heat. When one side is removed, the fire goes out. This simple idea forms the basis for the very complicated processes of fighting forest wildfires and controlling prescribed burns.

Sources of Fire

Most forest fires are caused by people. Some forest fires are natural. Long before human ancestors learned to control and use fire, forest fires occurred. The giant redwood forests in California have fire scars that date back to the year A.D. 245, but that is not the usual case.

The forest industry has identified the major sources of forest wildfires as follows:

Incendiary fires make up one major problem. This category includes malicious burning or arson. It is not, however, limited to people who set the woods on fire intentionally. It also includes forest owners who burn the forest undergrowth and allow their planned fires to get out of control. Planned fires are done for many reasons. Wise use of fire as a forest management tool is discussed earlier in this chapter.

Debris burning causes many forest fires. Burning trash, brush, tree tops and branches after harvest (slash), and the like, often produces unplanned results. This is one of the biggest causes of wildfire in this country. Many states prohibit this type of burning. Even if it is permitted in your area, you should check with the forest service before burning large quantities of debris.

Smokers were once a much more serious problem than they are today. The education campaigns mentioned earlier have certainly helped. A still-burning match or cigarette provides the heat required to start a forest fire.

Campers have always been a problem. Campfires, hot charcoal, and lanterns must be carefully used in the forest.

Railroads were once a major problem. Old-fashioned steam locomotives produced sparks from their fires. At one time, fire was used as a technique to clear brush and grass from railroad tracks and rights-of-way. This is no longer a major cause of forest fires.

Logging operations, too, are no longer the major cause of forest fires they once were.

Lightning is the main natural cause of forest fires. This is particularly true in the western United States, where dry lightning storms are more frequent than elsewhere. A bolt of lightning produces great heat, but during a rainstorm there is little danger of forest fire. Electrical storms without rain produce up to 9 percent of the fires on protected forest lands in this country.

Children, machinery operation, spontaneous combustion, and many other miscellaneous causes account for the rest of the forest fires.

TYPES OF FOREST FIRES

There are basically three categories of forest fires. The categories are based on the intensity of the fire.

Ground Fire

The **ground fire** is not particularly common outside the very wet, bog-type areas of this country. A ground fire results from spontaneous combustion. It smolders, so there is seldom a flame.

In the thick humus or peat in these forests, decay produces great heat. The bogs, however, are usually damp. The heat becomes intense beneath the surface, but the loss of heat to the air keeps the surface cooler. The ignition point is reached, but the low levels of oxygen and slow air movement prevent any actual flames. On the other hand, actual ignition of the humus by an external source can produce nearly the same result. The humus or peat smolders and burns from the surface downward.

Ground fires produce very intense heat at the soil surface. As a result, plant root systems are killed, and the entire forest is destroyed. Once the fire is out, there is little sign that there has been a fire, except that all the vegetation may be dead.

Surface Fires

Surface fires burn near the ground. The flames generally consume the grasses, brush, and vines on the surface; but the forest canopy remains undamaged. This is the most common kind of forest fire. The most easily combustible fuel in the forest is the dry litter on the ground; that is, the amount of heat required to ignite dry litter is less than for the green leaves higher up.

It is on the surface that almost all forest fires start. In addition, the surface fire is probably the easiest to control.

Crown Fire

The **crown fire** is the most violent and dangerous of the forest fires. Chapter 18 described the parts of a tree. You should recall that the tree's crown includes its branches, twigs, and leaves. In most hardwood trees, the crowns are not particularly flammable. They will burn if the fire is intense enough, but they are not as explosive as conifers.

Cedars, pines, firs, and other conifers have very flammable crowns. To prove this, you need only read the newspapers during the Christmas season. Every year many homes are burned and lives are lost because of Christmas trees that burst into flame. They burn easier and faster when dry but, even when still green, such trees are highly flammable.

Once a surface fire gets into the forest canopy, it becomes a crown fire. It is then that the fire becomes extremely dangerous. It is then that the fire can spread most quickly because the wind can carry burning leaves and sparks for great distances.

DETECTING A WILDFIRE

Once a forest fire begins, there are numerous ways for it to be detected and located.

We have probably all seen forest lookout towers like the one shown in Figure 21-3. During

FIGURE 21-3 Lookout towers are still used during peak fire seasons to locate forest fires.

times when the danger of forest fires is high, these are operated continuously. When the operator in one tower spots a column of suspicious-looking smoke, he or she does two things. The tower operator radios or telephones the fire's approximate location to the district forestry office or similar fire management office. He or she then determines the direction (azimuth) from the tower to the fire with the help of an instrument called an **alidade**. When a second tower operator then locates the fire, the azimuth from that tower is also measured. The forestry office then uses a process called **triangulation** to determine the actual location of the fire. (See Figure 21-4.) Triangulation

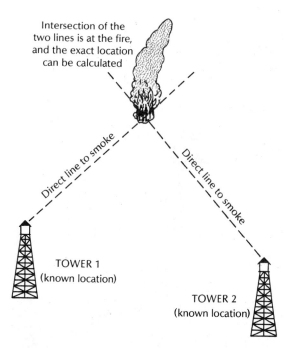

FIGURE 21-4 Two sightings are required to exactly locate a fire by triangulation.

FIGURE 21-5 The district forestry office is responsible for coordinating the forest fire prevention and suppression effort.

offices are responsible for coordinating the efforts to combat forest fires. (See Figure 21-5.)

involves plotting fines connecting each tower (or other known location) with the smoke. The point at which the two lines intersect is the exact location of the fire.

Another way the forestry office locates fires is by telephone reports. Motorists may report fires near the highway, and forest and brush fires in rural areas are frequently spotted and reported by residents before they are spotted by lookout towers.

An increasingly important fire-detection technique is the use of fire watch planes. During high fire hazard periods, the planes continuously patrol very large forest areas. A well-trained pilot can locate and report forest fires very quickly and with fair accuracy. The use of airborne infrared heat detectors, radar, and satellites for tracking electrical storms are becoming more important today. District forestry

THE ANATOMY OF A FIRE

Forest fires are not all alike. Many factors affect the size, shape, direction, speed, and intensity of a particular fire. It is possible, however, to describe a typical fire.

A typical fire starts from a single point. It spreads most quickly in the direction the wind is blowing. It spreads least quickly toward the wind. If there are small fires produced by blowing sparks and debris, they will be downwind. The leading edges of a fire are seldom smooth and straight. These characteristics are illustrated in Figure 21-6.

Many factors affect the anatomy of a particular fire. These include slope, moisture conditions, types of fuel, wind, open areas, roads, streams, lakes, and other natural barriers. Firebreaks placed in forests before a fire occurs supplement the natural barriers. Such natural and artificial barriers not only affect

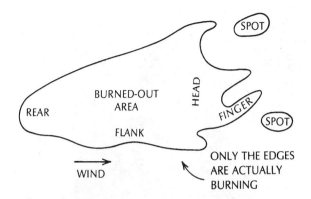

FIGURE 21-6 The anatomy of a forest fire. Actual burning occurs only at the edge of the burned-out area.

the anatomy of the fire, they can also be used as a part of the attack plan against the fire.

PREVENTION AND SUPPRESSION

Prevention

The best way to combat fires is to prevent them from starting in the first place. Efforts at forest fire prevention take many forms.

We have already discussed the Keep America Green program and Smokey Bear campaigns. Such educational efforts have been very successful in decreasing fires caused by smokers, campers, and some other forms of carelessness.

It is possible to predict when fires are most likely to occur. Foresters can estimate the danger of fire risk to a particular area at any given time. Moisture levels, relative humidity, wind speed, and other factors were combined using the National Fire Danger Rating System (NFDRS). The system results in a series of mathematical indices that predict the likelihood and probable severity of forest fires in a given situation. When the burning index is particularly high, extra caution is needed.

In national, state, and large commercial forests, hunters and other forest users can be advised to be extra careful. When the index is very high, the forests can be closed altogether. Logging operations are confined to low-risk times. Lookout towers are operated on a 24-hour basis for all towers, for only selected towers, or not at all depending on the burning index.

Direct Attack

When a fire is small and spreading slowly, the best method of **fire suppression** may be a direct attack. These methods require that the flames themselves be attacked. Use of a shovel or rake to toss soil into the flames is typical. Wet clothes, green branches, or swatters can be used to smother the flames. These techniques remove oxygen from the fuel triangle and the flames are extinguished.

Spraying water from a backpack pump, from a tanker, or from a lake or stream with a pump truck is another technique. The water may temporarily smother the flames (remove oxygen) but, more importantly, it also cools the fuel below its ignition point. (It removes heat.) On larger fires, airplanes or helicopters are used to dump water or chemical fire retardant directly onto the fire. (See Figure 21-7.)

A narrow fire lane can be cut just ahead of the fire. This is done with a shovel, rake, hoe, mattock, axe, or other hand tool. It involves removing all combustible materials down to bare soil in a narrow strip around the fire. The fire burns up to the fire lane and burns itself out. The firefighters must ensure that sparks do not jump across the lane and start fresh fires.

Indirect Attack

An indirect attack on a wildfire involves removing fuel from the fire triangle. There are basically two techniques: barriers and backfires.

FIGURE 21-7 Forest Service aircraft dropping fire retardant chemical on a wildfire in Colorado. *(Photo courtesy United States Department of Agriculture, U.S. Forest Service, Montrose Interagency Dispatch Center)*

A **fire barrier** may be any road, stream, bare area, field, lake, or other natural obstacle to the fire. Natural or existing barriers, if any, are coupled with hastily constructed barriers. Bulldozers and special plows are used to clear fire lanes or firebreaks ahead of the fire. Chainsaws or other saws are used to remove trees that could fall across the barriers.

Barriers are normally constructed or completed first in front of the head of the fire because the fire spreads most quickly in that direction. Once the barriers are complete across the head of the fire, they are extended around its flanks and, finally, its rear. Of course, if homes or businesses were threatened by the flanks or rear, they might be protected first; but normally the head is attacked first, then the flanks, then the rear.

Once a barrier is complete along the head, the decision must be made as to whether to use **backfire**. If the fire is spreading quickly, if the wind is fairly strong, or if the fire is fairly large, it may jump across the narrow fire lanes. Backfire can be used to greatly widen the fire lane.

Remember that a fire spreads most slowly toward the wind. The head of the fire is moving away from the wind and toward the firebreaks. Backfires are smaller fires set along the firebreak on the side toward the fire. They slowly burn toward the wind and the wildfire. As they burn, they use up the fuel and make the firebreaks wider. This removes the fuel from the head of the fire, which then must go out. (See Figure 21-8.)

Backfires, in most situations, are set by firefighters on foot; but in the last few years,

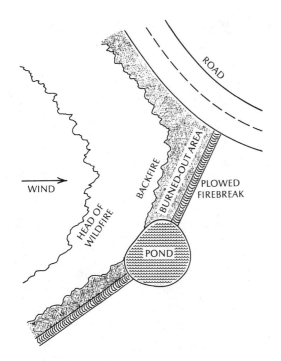

FIGURE 21-8 Backfires and firebreaks are combined to stop a forest wildfire. The burned-out area no longer has fuel.

aerial backfiring techniques have been developed using a spontaneously igniting chemical combination (potassium permanganate and ethylene glycol) in small, plastic balls like ping-pong balls.

Ground crews must constantly patrol the fire line to ensure that the fire does not get across. Spot fires are a normal occurrence. They can be controlled by direct attack if caught soon enough.

Mopping Up

Mopping up occurs after the fire is under control. This operation involves patrolling the fire line until the fire is no longer dangerous. It also includes watching for later spot fires. In addition, it means entering the burned-out area to extinguish any small fires that still remain.

It is not feasible to put out every glowing ember, but care is taken to extinguish any fire that could spread or that would produce blowing embers in the wind.

SUMMARY

Efforts to prevent forest wildfires in this country intensified after World War II. Since that time, educational programs have helped decrease the number of wildfires occurring. Fire detection and suppression techniques have improved with technology to decrease the amount of forestland destroyed each year by fire.

The most destructive fire in American history burned about 3.8 million acres of woodland and killed an estimated 1,638 persons. It began near Peshtigo, Wisconsin, in October 1871. Such huge and spectacular fires are not likely today because of improved roads and fire-fighting equipment; but as recently as 1970 the Laguna fire in California burned 185,000 acres and killed 10 people.

To exist, a forest fire needs three things: fuel, oxygen, and heat. These three things are known as the fire triangle. Once a fire has started, suppression efforts involve removing one or more of the sides of the triangle.

People cause most fires through carelessness. Lightning is the major natural cause of fires in this country. It accounts for about 9 percent of our forest fires.

There are three general types of fires. Ground fires are slow, smoldering fires in the humus or peat layer of the forest floor. Surface fires burn the forest litter and undergrowth of the forest. Crown fires burn the entire tree and destroy the forest canopy. Most fires are surface fires.

Once they begin, fires may be located by reports from motorists or residents in an area, sightings from fire-control (lookout) towers, or fire-watch aircraft. The most accurate and dependable source is the use of two or more lookout towers.

It is very important to understand that not all forest fires are bad. Prescribed burning can be used in forest management to clear undergrowth, improve wildlife conditions, cause seed release, reduce the danger of wildlife, and minimize insect and disease problems. In fact, controlled fires may be one of the most important forest management tools in some parts of the country.

DISCUSSION QUESTIONS

1. What great forest fire occurred the same time as the more famous Chicago fire? How big was it?

2. What are the three sides of the fire triangle? Give examples to illustrate each side.

3. What are the main causes of forest fires in the United States?

4. List and describe the three types of forest fires.

5. What is triangulation in forest fire detection? Describe how it works.

6. Draw and label the parts of a "typical fire."

7. What part of the fire is normally attacked first?

8. What is a direct attack? When would it be used in forest fire fighting?

9. Describe indirect attack methods in forest fire fighting.

10. What are some techniques being used in the prevention of forest fires?

SUGGESTED ACTIVITIES

1. Visit a tree farm. Determine what management techniques are being used to minimize the danger of wildfires. Is prescribed burning used?

2. Has there been a forest wildfire in your area in the last year or two? If so, visit the site and examine the nature and extent of the damage. Has reforestation started yet? If so, how is it being accomplished?

3. Invite a forester to your class to discuss the use of controlled fire as a management tool as well as in prevention and suppression of wildfire.

4. Prepare a class report on fire as a tool in the forest industry.

OBJECTIVES

After reading this chapter, you should be able to

- list some important careers in the forest industry

- describe the training required for entry into these careers

- describe some of the tasks performed by workers in these careers

TERMS TO KNOW		
forester	undercut	skidding
forestry technician	backcut	postsecondary school
logger	bucked	

The forest and forest products industries in this country make up a multibillion dollar part of our economy. Experts estimate that the forest industry produces about 6 percent of the gross national product in the United States. Almost 5 percent of all jobs in the United States are in the forestry and forest products industries. Well over 3 million American jobs come from production, harvest, processing, manufacturing, and marketing of forest products.

Clearly, this is big business; but in selecting forestry as a career, other factors must be considered. Most of those jobs are in the manufacturing, processing, and marketing ends of the industry. Most do not even resemble the job of forester; secretaries, accountants, production workers, truckers, and hundreds of other positions are examples. Even though these workers are in the forest industry, they do not generally receive training in forestry. Moreover, they are not directly involved in working with the forests themselves.

This chapter discusses some of the major occupations dealing directly with forestry. If you are interested in forestry as a career, one of these occupations is probably what you are thinking about.

FORESTER

Nature of Work

The occupation of **forester** makes up the professional center of the forestry industry. Foresters are the professionals who help manage, direct, and protect our nation's forest resources.

Foresters are responsible for surveying existing forest resources. They estimate standing timber volume, predict future growth, identify tree species, and judge the health and usefulness of individual trees.

They are responsible for protection of the forest. Foresters must watch for and predict insect and disease problems as well as wildlife and livestock damage. When potential problems begin to emerge, the forester must detect them and plan ways to solve them.

This country's massive forest fire prevention and protection system is the responsibility of professional foresters. They plan fire prevention programs and determine fire danger ratings throughout the country on a continuing basis. They schedule the location and operation of forest-fire lookout towers, spotting planes, and other detection techniques. Once fires are detected, planning and managing the fire-fighting effort is the responsibility of still other foresters.

Additional responsibilities include planning and managing timber sales. Foresters may oversee logging and replanting operations. They may be responsible for camping, hiking, picnicking, fishing, boating, and other outdoor recreation on forestland.

Working conditions for foresters vary widely. Much of the time a forester may be in his or her office handling paperwork. On the other hand, time will also be spent in the forests. Some foresters work in or near large cities; others work in remote and rugged areas.

Foresters use computers and photographs taken by satellites. They must be able to use sophisticated surveying instruments as well as simple tools like Biltmore sticks.

Beginning foresters often spend most of their time out of doors. As they gain experience, they may seek more office-oriented or people-oriented work.

Job Outlook

In 1997, there were about 37,000 professional foresters and other conservationists in the United States. About 25 percent worked for private industries; companies such as Weyerhaeuser, Union-Camp, Georgia-Pacific, and others employ many professional foresters.

Another 30 percent worked for the federal government, and another 30 percent worked for state and local governments. Some worked for the United States Park Service in the Department of the Interior but were employed by the Forest Service and other agencies in the Department of Agriculture.

The remaining foresters (about 15 percent) occupied a wide range of positions. Some were employed by the various state forestry agencies. Others worked for local governments, colleges, extension services, or managed their own forestlands. Still others served as consultants to other individuals or agencies.

Employment opportunities for professional foresters are expected to grow slowly throughout the rest of the century. For many years, the number of forestry graduates has exceeded the number of openings. Forest products industries, however, are able to employ the surplus graduates in sales, management, and related jobs.

In 1997, newly hired foresters with a bachelor's degree started in the U.S. Forest Service at annual salaries of $19,500 or $24,200 depending on academic achievement (so work on those grades—they really do make a difference.) With a master's degree, beginning foresters could expect starting salaries of $24,200 or $29,600, again depending on academic achievement. Beginning salaries in the federal government are slightly higher in certain "high-cost" geographic areas. Overall, the beginning forestry school graduate with a bachelor's degree started at $24,800 in 1997. As a general rule, foresters entering private industry earned about the same salaries as beginning federal employees with the same educational levels and academic achievement. The average salary for professional foresters in nonmanagement positions was about $47,600 in 1997. Forest product technologists, foresters with additional graduate preparation, had average salaries of about $62,000 that year.

Training and Qualifications

To become a professional forester, at least a bachelor's degree, probably in forestry, is necessary. Because the competition for good forestry positions is so keen, an advanced degree may be needed.

Practically all land grant institutions have departments, schools, or colleges of forestry. In 1999, the Society of American Foresters accredited 48 of those programs. Professional preparation programs such as these stress the sciences (chemistry, biology, and physics), mathematics, communication skills (written and oral), and computer applications. Courses on wildlife science, water quality, dendrology, and entomology are becoming increasingly important at these schools.

Accredited programs (*Source:* American Society of Foresters, Bethesda, MD) include the following:

1. Alabama: Auburn University, School of Forestry, Auburn, AL 36849-5418

2. Alaska: University of Alaska, Department of Forest Sciences, Fairbanks, AK 99775

3. Arizona: Northern Arizona University, School of Forestry, Flagstaff, AZ 86011

4. Arkansas: University of Arkansas-Monticello, School of Forest Resources, Monticello, AR 71655

5. California: California Polytechnic State University, Natural Resources Management Department, San Luis Obispo, CA 93407

6. University of California-Berkeley, College of Natural Resources, Berkeley, CA 94720

7. Humboldt State University, Department of Forestry, Arcata, CA 95521. 1979.

8. Colorado: Colorado State University, Department of Forest Resources, Fort Collins, CO 80523-1401

9. Connecticut: Yale University, School of Forestry and Environmental Studies, New Haven, CT 06511

10. Florida: University of Florida, School of Forest Resources and Conservation, Gainesville, FL 32611-0410

11. Georgia: University of Georgia, School of Forest Resources, Athens, GA 30602-2152. 1938, 1991.

12. Idaho: University of Idaho, Department of Forest Resources, Moscow, ID 83844-1133

13. Illinois: University of Illinois, Department of Natural Resources and Environmental Sciences, Urbana, IL 61801

14. Southern Illinois University, Department of Forestry, Carbondale, IL 62901-4411

15. Indiana: Purdue University, Department of Forestry and Natural Resources, West Lafayette, IN 47907-1159

16. Iowa: Iowa State University, Department of Forestry, Ames, IA 50011

17. Kentucky: University of Kentucky, Department of Forestry, Lexington, KY 40546-0073

18. Louisiana: Louisiana State University, School of Forestry, Wildlife and Fisheries, Baton Rouge, LA 70803-6200

19. Louisiana Tech University, School of Forestry, Ruston, LA 71272

20. Maine: University of Maine, College of Natural Sciences, Forestry and Agriculture, Orono, ME 04469

21. Massachusetts: University of Massachusetts, Department of Natural Resources Conservation, Amherst, MA 01003-4210

22. Michigan: Michigan State University, Department of Forestry, East Lansing, MI 48824-1222

23. Michigan Technological University, School of Forestry and Wood Products, Houghton, MI 49931

24. University of Michigan, School of Natural Resources and Environment, Ann Arbor, MI 48109-1115

25. Minnesota: University of Minnesota, College of Natural Resources, St. Paul, MN 55108-1030

26. Mississippi: Mississippi State University, College of Forest Resources, Mississippi State, MS 39762-9680

27. Missouri: University of Missouri, School of Natural Resources, Columbia, MO 65211

28. Montana: University of Montana, School of Forestry, Missoula, MT 59812

29. New Hampshire: University of New Hampshire, Department of Natural Resources, Durham, NH 03824-3589

30. New York: State University of New York-ESF, Faculty of Forestry, Syracuse, NY 13210

31. North Carolina: Duke University, Nicholas School of the Environment, Durham, NC 27708-0329

32. North Carolina State University, Department of Forestry, Raleigh, NC 27695-8008

33. Ohio: The Ohio State University, School of Natural Resources, Columbus, OH 43210. 1993.

34. Oklahoma: Oklahoma State University, Department of Forestry, Stillwater, OK 74078. 1971.

35. Oregon: Oregon State University, College of Forestry, Corvallis, OR 97331-5704

36. Pennsylvania: The Pennsylvania State University, School of Forest Resources, University Park, PA 16802-4300

37. South Carolina: Clemson University, Department of Forest Resources, Clemson, SC 29634-0306

38. Tennessee: University of Tennessee, Department of Forestry, Wildlife and Fisheries, Knoxville, TN 37901-1071

39. Texas: Stephen F. Austin State University, Arthur Temple College of Forestry, Nacogdoches, TX 75962-6109

40. Texas A&M University, Department of Forest Science, College Station, TX 77843-2135

41. Utah: Utah State University, Department of Forest Resources, Logan, UT 84322-5215

42. Vermont: University of Vermont, School of Natural Resources, Burlington, VT 05405-0088

43. Virginia: Virginia Polytechnic Institute and State University, College of Natural Resources, Blacksburg, VA 24061-0324

44. Washington: Washington State University, Department of Natural Resource Sciences, Pullman, WA 99164-6410

45. University of Washington, College of Forest Resources, Seattle, WA 98195

46. West Virginia: West Virginia University, Division of Forestry, Morgantown, WV 26506. 1947.

47. Wisconsin: University of Wisconsin-Madison, Department of Forest Ecology and Management, Madison, WI 53706-1598

48. University of Wisconsin-Stevens Point, College of Natural Resources, Stevens Point, WI 54481

Suggested high school courses include chemistry, mathematics, physics, biology, and forestry or agriculture. Also helpful is prior experience in forestry or conservation work. Summer jobs in parks or with forest products industries offer this kind of experience.

Foresters must be prepared for hard work and outdoor life, at least for the first few years, and so must be healthy and strong and like outside work.

More information on the profession can be obtained from any of the following organizations:

Society of American Foresters
5400 Grosvenor Lane
Bethesda, MD 20814

U.S. Forest Service
United States Department of Agriculture
P.O. Box 96090
Washington, DC 20090-6090

FOREST AND CONSERVATION WORKER

Nature of Work

People who work in this category are variously known as **forestry technicians**, forestry aids, conservation technicians, and conservation aids. The work is typically done under the supervision of a professional forester or other conservationist. Most of these positions are with federal, state, or local government agencies. In reality, these workers can accomplish many of the tasks of professional foresters.

Before a logging operation, forestry technicians may assist in estimating the amount of standing timber in a given area and marking trees to be harvested or left standing. (See Figure 22-1.) They may be tasked to observe logging operations to ensure that only the intended trees are harvested and that excessive damage is not done in the harvesting process. They may help in clearing areas for various forestry tasks and they may assist in land measurement and surveying.

FIGURE 22-1 Estimating the amount of lumber in a tree and on a tract of wooded land is an important forest management and marketing skill. *(Courtesy National FFA Organization)*

Forestry technicians occupy fire towers to watch for wildfires, operate radios, dispatch equipment, and provide general assistance in forestry offices. They help identify and remove diseased and damaged trees, clear brush, monitor insect populations, help rescue and care for injured animals and birds, and apply all sorts of additional forest, wildlife, and conservation-related management techniques.

Employment Outlook

In all, about 40,000 forest and conservation workers were employed in 1996. Average yearly earnings for beginning workers in this category were about $19,800 in 1998.

Much work in this category is seasonal. During fire season, the U.S. Forest Service as well as state forest agencies and private forest industries may hire large numbers of temporary workers to assist in firefighting operations.

The work is often strenuous and dirty, but many young people love the out-of-doors, and the opportunity to work away from the smells and noise of factories and service establishments attracts them to this kind of work. The employment outlook is for stable numbers of positions through at least 2006. Opportunities for ambitious young persons to get into this kind of work are good, especially with appropriate postsecondary preparation.

Training and Qualifications

Technicians or aids do not need a bachelor's degree, but postsecondary education is typically required for entry. Such training programs are available at community colleges, vocational-technical centers, through private industry training programs, or through government-run training programs. Most beginning technicians have completed one- or two-year postsecondary programs. Some beginners get started through experience they have gained part-time on firefighting crews, during tree planting season, work in parks or recreation facilities such as campgrounds, or in tree nurseries.

LOGGER

Nature of Work

Before the **loggers** ever get into the forests, many others have already been there. Foresters have decided that the trees are ready for harvest. They, or forestry technicians, have selected the trees and marked the areas for cutting. Timber cruisers have estimated the amount of wood to be cut. Equipment operators have prepared logging roads and trails and have set up loading areas.

The first task of the logger is felling the trees. A tree is cut in such a way that its fall does the least damage. Seed trees to be left should not be damaged. As a tree falls, it should not fall into other trees where it can wedge or hang. Trees can be felled manually, or they can be cut by a very powerful tractor-mounted shear or feller-buncher. Manual felling is done as shown in Figure 22-2.

The **undercut** removes a wedge of wood from the side where the logger wants the tree to fall. Then the **backcut** is made just above the undercut on the opposite side. The weight of wood above the undercut pulls downward, causing the tree to fall in that direction. If the tree's crown is heavier on one side, or if the tree is leaning, its general direction of fall may not be changed. Even then, an experienced logger can guide the tree to an exact location in the general direction.

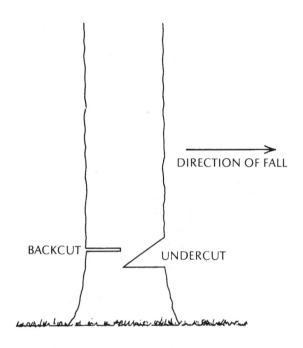

FIGURE 22-2 Manual tree felling.

FIGURE 22-3 Logging today is increasingly done by heavy equipment. *(Photo image provided by the Forest Science Data Bank, a partnership between the Department of Forest Science, Oregon State University, and the U.S. Forest Service Pacific Northwest Research Station, Corvallis, Oregon. Significant funding for these images was provided by the National Science Foundation Long-Term Ecological Research program. NSF Grant numbers BSR-90-11663 and DEB-96-32921.)*

Once the tree has been felled, it is limbed and **bucked**—the limbs are removed and the tree is cut into logs or bolts.

The next step is **skidding**. The logs are dragged to a loading area where they are sorted by size and mechanically loaded onto trucks for hauling. Pulpwood bolts are stacked by hand and then loaded onto trucks by means of hoists or other lifts for hauling.

Logging is a very physically demanding job. Almost all logging work is done outside. The conditions may be very hot or cold. The work involves lifting, pushing, climbing, pulling, and operating chain saws, heavy equipment, and hand tools such as axes, crosscut saws, shovels, and chain hoists. (See Figure 22-3.)

This type of work may sound undesirable, but the chance to work outdoors, away from the smell and routine of factories, makes up for the disadvantages for many people.

Employment Outlook

In 1996, private logging companies in this country employed about

- 33,000 log-handling equipment operators
- 21,000 logging truck and tractor operators
- 17,000 fellers and buckers
- 11,000 other logging support workers

Most of these workers spent the majority of their working hours in the forests actually cutting and moving trees or hauling logs to sawmills and paper mills. Many of them live for extended periods of time in logging camps at remote harvesting sites. An additional 190,000 workers were employed in sawmills and planing mills where they take the logs and convert them into lumber. An additional 70,000 people were estimated to be self-employed as loggers, drivers, and pulpwood harvesters in support of this huge industry.

Jobs in logging are expected to remain steady through 2006, primarily because of mechanization in the logging industry. Manual felling of trees is being increasingly replaced by mechanical felling. In addition, smaller trees make up an increasing share of the harvest. This means that even more trees can be harvested by machine.

On the other hand, the jobs are still very physical and demanding. For those young people who desire very physical outdoor jobs, such work should be available.

The ups and downs of the economy are particularly important in the logging industry. During a recession, few houses are built. With fewer houses, less lumber is used. With a lower demand for lumber, fewer logs are needed. With fewer logs, fewer loggers are needed. By the same token, when the economy is strong, many loggers are needed.

One last note on this occupation is important. Because labor market conditions vary widely and because the work can be so very strenuous and even dangerous, wages of logging workers vary greatly. Logging heavy equipment operators in Alaska can expect to make MUCH more money per hour than chainsaw operators in other states. Fallers and buckers earned between $8.11 and $23.64 per hour, depending on location, in 1997. The work is also highly dependent on the strength of the general economy. During the 1980s a general recession caused new home construction to decline. That meant huge cutbacks in the logging industry because the need for new lumber fell drastically. In the late 1990s a particularly strong economy meant a booming housing industry, which meant, in turn, a great

need for skilled workers in logging, and thus markedly higher wages. Interestingly, that is not expected to translate into more logging jobs, because of the increase in productivity resulting from mechanized tree harvesting.

Training and Qualifications

Traditionally, loggers required little technical training. Physical strength, endurance, and a willingness to work hard under dangerous conditions was all that was needed. With the movement away from the manual felling of very large trees toward the mechanical harvesting of smaller trees, however, this is changing.

The work is not as hard, heavy, or dangerous as it was. Safe and efficient use of heavy equipment requires highly skilled operators. Chainsaws have largely replaced crosscut saws, and the safe and efficient use of chainsaws also requires trained operators. Every time machine replaces muscle, the requirement for training increases. In this regard, logging is like any other occupation. Nevertheless, logging still requires strength and physical endurance.

FORESTRY TEACHING AND RESEARCH

Nature of Work

Forestry subjects are taught at three levels. Entry-level training is offered at the postsecondary level, and professional forestry training is offered at the college and university level.

High school programs are taught through the vocational agriculture program. There are about 10,000 high school agriculture teachers in the United States. Of these, about 20 percent teach some forestry. A smaller number, probably about 10 percent, are primarily natural resources management teachers. These programs are heavily forestry oriented. Graduates of such programs are employed in entry-level jobs in the forest industry. Some may become technicians after experience, and many go on for further training.

High school forestry or natural resources management teaching positions have certain prerequisites. A bachelor's degree in agricultural education is necessary, and courses should emphasize forestry. A year or more of experience in the forest industry, or in some other conservation field, is also needed.

Programs to train forestry technicians and other specialized workers are offered at **postsecondary schools** such as community colleges, technical institutes, and vocational-technical schools. To become a postsecondary forestry instructor, experience and education are necessary. A degree in forestry, or in agricultural education with a forestry emphasis, is very important. In addition, experience in the industry or in high school teaching of forestry is helpful.

Professional foresters are almost all graduates of forestry programs at the college or university level. Forestry is often a part of a school or college of agriculture, or there may be a separate school or college of forestry. In either case, a forestry department or school requires a faculty of professionally trained foresters.

Professors in forestry teach scientific, technical, and professional forestry courses. They also conduct research on how to improve the forest industry, on increasing tree growth rates, on reducing disease problems, or on other problems of interest.

A professor at a forestry college must have a degree in forestry or some closely related area in addition to at least a master's degree, and probably a doctoral degree, in some specialty area. Experience in the forest industry is also helpful. Competition for these jobs has always been and will continue to be very intense.

SUMMARY

The forestry industry in this country is big business. It employs as much as 5 percent of our workers and produces 6 percent of the gross national product. Of the 3 million workers employed in this industry, only a small portion work directly with the production and harvesting of trees. This chapter dealt with the largest of the occupations dealing directly with the forests.

Foresters are the professional center of the forestry program and industry. They manage, supervise, and plan our nation's forests. Foresters must be technically competent in the skills required to manage the forests. They must also understand the forests—how trees grow, what they need, and what can harm them. Foresters almost always have college degrees in forestry.

Forestry technicians, or aides, assist foresters. They must be able to perform almost all the skilled tasks of the forester. The forestry technician's work is usually outdoors. To become a forestry technician, completion of a postsecondary training program in forestry is necessary.

The occupation of logger traditionally has not required as much training in most cases. With the increasing use of mechanical logging equipment, however, the training requirements will continue to increase. At the same time, the number of logging jobs is expected to decline. Even with the use of labor-saving machines, the job is a physically demanding one. It is also subject to weather and economic fluctuations.

Teaching of forestry is done at three levels: high school, postsecondary, and college or university. Teachers in the first two levels must have degrees in agricultural education and/or forestry. Teachers and researchers at the college or university level usually have doctoral degrees.

8. Describe how to fell a tree manually so its direction of fall is controlled.

9. What are the three levels at which forestry training is offered in this country?

10. What training is required to become a forestry technician? A logger?

DISCUSSION QUESTIONS

1. How big is the forest industry in the United States?

2. What are some technical skills a forester must be able to perform?

3. How many foresters and conservationists are there in the United States? Where do they work?

4. What training is required to become a forester?

5. What is a forestry technician? What does he or she do?

6. How many people are employed as forestry technicians? Who employs them?

7. What are the major tasks of a logger? Describe each one.

SUGGESTED ACTIVITIES

1. Write to the dean of agriculture at your state land-grant college or other college of agriculture to request information on their forestry program and careers in the forest industry.

2. Write to the USDA Forest Service and request information on careers available in forestry and the training requirements for these careers.

3. Interview a professional forester to find out what he or she does on the job and what training he or she had to help prepare for the profession. What are the advantages and disadvantages of working as a forester? Do the same thing with a forest technician.

Section IV — CASE STUDY

Taxol: The Miracle Cure?

Every year, 12,000 women face the harsh reality of dying from ovarian cancer. In the 1960s, taxol (a miracle cure for some) was discovered in the bark of the Pacific Yew Tree. It wasn't until 1979 that Dr. Susan Horwitz, a scientist at Albert Einstein College of Medicine in New York City, discovered taxol's cancer-fighting properties. Trial testing throughout the 1980s had patients begging for increased harvesting of the yew. (See Figure IV-A.) As people waited for the harvesting of yew to save their lives, it became apparent that they waited in vain.

Pacific Yew grows primarily in old-growth forests. (See Figure IV-B.) Old-growth forests support a unique blend of plants and animals found nowhere else in the world. People around the world agree that this ecosystem needs protection. Though some logging is allowed, debate continues over this practice.

Until taxol was discovered, the yew was considered a trash/wildlife tree to loggers. When trees were harvested, the tree was left alone or burned depending on harvesting technique. Taxol opened a new and extremely valuable market for these trees. By harvesting the bark of the yew, you kill the tree. Questions were raised over how much was too much. With the number of patients seeking taxol raising on a daily basis, it was feared the tree would be harvested to extinction. It also opened further debate over logging in old growth on a general basis. Those interested in the preservation of these forests lobbied for the halt of harvesting. The absence of yew would create a hole in the old-growth ecosystem that none

FIGURE IV-A Pacific yew (*Taxus brevifolia*), is found primarily in the understory of old growth coniferous forests in the Pacific Northwest. *(Photo courtesy Edward C. Jensen, Oregon State University)*

FIGURE IV-B Aerial view of H.J. Andrews Headquarters compound, nestled in an old-growth Douglas-fir forest near the mouth of Lookout Creek on the Andrews Experimental Forest. *(Photo courtesy U.S. Department of Agriculture, U.S. Forest Service. Photographed by Al Levno)*

would be positive on how to patch. The miracle cure would have to wait while politicians battled over the issue.

The controversy over harvesting put pressure on the scientific community to come up with a synthetic or man-made version of the chemical. People found that the yew could be grown commercially and harvested outside the confines of the old-growth forest. In the early 1990s a synthetic version was created and the issue died. With taxol, we were lucky.

The central point to this whole issue was never settled. What do you think the real question was all about? How do we choose our battles? If the old growth forests needed to be harvested for the miracle cure for AIDS or some other untreatable disease/virus, would we harvest? Would we preserve the old-growth forest as a national treasure even if it meant that people would have to die? You decide.

Section V
Fish and Wildlife Resources

23

FISH AND WILDLIFE IN AMERICA

OBJECTIVES

After reading this chapter, you should be able to

- explain the difference between extinct and endangered species of wildlife

- discuss endangered mammals, birds, and fish species

- explain how various species of animals became extinct

TERMS TO KNOW		
extinct	rare	endangered

Of the many resources that are abundant in America, about half can be considered nonrenewable and the other half renewable. Wise management practices of nonrenewable resources are always stressed because once used, nonrenewable resources do not regenerate. We usually take renewable resources for granted because we have developed the attitude that there will always be more. Animals usually fall into the latter classification because of their ability to reproduce, but even though they reproduce, many species have died out. It quickly becomes evident that management of our renewable resources, such as birds, mammals, and fish, is just as important as management of our coal and oil. This unit looks at the wildlife resources in America as well as possible management procedures. We look at the future to see what can be done, and we examine the past for possible mismanagement procedures.

BACKGROUND INFORMATION

Before beginning to examine our fish and wildlife resources, we must first define to exactly what wildlife refers. A common dictionary definition would define wildlife as living things that are neither human nor domesticated, especially mammals, birds, and fishes. Using this definition, one would have to include both plant and animal life in this broad statement. For the purposes of this unit, we include mainly the higher forms of life, that is, vertebrates, as being wildlife. It is important to note at this point, however, that plant life and lower forms of animal life are also important in maintaining the balance of nature.

The United States is a nation rich in natural resources, especially wildlife. It has been estimated that the continental United States contains over 2,300 different vertebrate species. The game animals make up only a small portion of this number.

It was these mammals, birds, and fish that allowed the early pioneers to survive. The meat supplied food; skins were used for shelter and clothing; oil kept firearms usable and provided light in cabins. Without wildlife resources, the wilderness would have never been conquered.

America was established in the world fur trade business but at the expense of the wildlife. Fur trappers took the animals faster than they could multiply, treating their resources as the crop farmers, fishers, cattle ranchers, and loggers treated theirs.

Many wildlife species were killed because they appeared hostile. A huge bear was certain to be a threat. The wildcat had to be killed to protect the people. Many species were also destroyed because they threatened the safety of domestic animals. Specific animals were often unjustly accused. For example, hawks were thought to kill domestic farm animals such as chickens, and mass destruction of the hawk took place. When the stomachs were examined from a sample of the killed hawks, they were found to contain 40 percent insects, 30 percent frogs and snakes, 23 percent rats and mice, 3.4 percent small birds, 2 percent aquatic wildlife, 0.5 percent game birds, and 0.5 percent rabbits. For some species such as the red-tailed hawk, poultry parts were discovered, but those parts were only a small percentage of the diet. The initial fear was unfounded.

Many species of wildlife have become so few in number they are considered rare or endangered. In 1966, Congress enacted the Endangered Species Conservation Act, amending it in 1969, to protect fish and wildlife on a worldwide basis. The act called for the protection and conservation of species of native fish known to be threatened with extinction. The 1969 amendment dealt with the importation of endangered species into the United States from anywhere in the world. The amendment called for the formation of an endangered species list, which is updated every five years. In 1970, the list contained 133 species of mammals, 124 birds, 24 reptiles, 25 fish, and 1 mollusk.

By 2000, the list had been expanded to include much larger numbers of plants and animals. The United States Fisheries and Wildlife Service provides what is referred to as a TESS (Threatened and Endangered Species System) Box Score. The TESS box score summarizes the status of all species listed as threatened and endangered worldwide. (See Table 23-1.) Part of the reason for the increase in numbers is a result of growing scientific knowledge. A few plants and animals have been added because of loss of habitat, and a few more have become endangered or threatened because of other reasons—many related to increased human population and technology.

EXTINCT, RARE, AND ENDANGERED SPECIES

Extinct Species

Some of our wildlife species are considered rare and some are extinct. What is the difference between the two? **Extinct** species no longer exist outside of museums and photographs. A **rare** or **endangered** species is one that is no longer common and is in danger of becoming extinct. A zoo may hold the last examples of the species.

One of the most noted cases of wildlife extinction is that of the passenger pigeon. At one time, the population of the pigeon was thought to be in the billions. The bird, 16 to 18 inches long with blue and reddish-brown feathers and a white-colored breast, flew in enormous flocks. John James Audubon, the well-known bird expert, estimated one flock

Table 23-1 Threatened and Endangered Species System Box Score

Group	Endangered		Threatened		Total Species	Total Species with Recovery Plans
	U.S.	Foreign	U.S.	Foreign		
Mammals	63	251	9	16	339	47
Birds	77	176	15	6	274	76
Reptiles	14	65	22	14	115	30
Amphibians	10	8	8	1	27	12
Fishes	68	11	44	0	123	90
Clams	61	2	8	0	71	45
Snails	20	1	11	0	32	20
Insects	30	4	8	0	42	27
Arachnids	6	0	0	0	6	5
Crustaceans	18	0	3	0	21	12
ANIMAL SUBTOTAL	367	518	128	37	1,050	364
Flowering Plants	565	1	139	0	705	528
Conifers and Cycads	2	0	1	2	5	2
Ferns and Allies	24	0	2	0	26	26
Lichens	2	0	0	0	2	2
PLANT SUBTOTAL	593	1	142	2	738	558
GRAND TOTAL	960	519	270	39	1,788	922

he saw at over 1 billion birds. When these large flocks reached an area, they would strip all the foliage, leaving the land bare. People declared war on the birds, killing all they could find. The young were captured, killed, and shipped to cities as food. The flocks soon disappeared. States such as New York, Massachusetts, and Pennsylvania passed laws to protect the bird, but they came too late. Those held in captivity would not breed, and the last-known passenger pigeon died in 1914 in a Cincinnati zoo. Her body is now preserved in the Smithsonian Institute in Washington, D.C. (See Figure 23-1.)

Other extinct species include the Carolina parakeet, heath hen, and Labrador duck. The Carolina parakeet was seen in many states. It was able to live in the colder U.S. climates and

FIGURE 23-1 Passenger pigeon.

traveled in huge flocks making a screaming cry. The birds were fond of cockleburs, and sheep farmers liked these birds because they kept the burs from getting into the sheep's wool. The brightly colored feathers caused their downfall. These feathers were prized for decorating women's hats and made the birds popular as pets. Their final extinction also came in 1914.

The heath hen is a relative of the prairie chicken and was used as food by the early settlers. In the early 1900s, people realized that the bird was becoming scarce, and a bird sanctuary was set up. The flock soon grew, but when a fire swept across the sanctuary, only a few males survived. The last bird died in 1932.

The Labrador duck became extinct before anyone realized it was gone. Most of the birds were killed for their feathers, which were used to stuff pillows.

The extinct animals are gone forever. We can never bring them back. Endangered species can be managed, and it should be our goal to make sure that they do not go the way of extinct species.

Endangered Mammals

There are 251 mammals on the endangered list distributed by the United States Department of the Interior. The common endangered mammals include bighorn sheep, polar bears, key deer, wolves, and mountain lions. Most were hunted extensively without considering possible extinction. By closely examining these species, we should find clues as to why they are endangered.

Bighorn Sheep

The bighorn sheep (Figure 23-2) are threatened by extinction from two sides: humans and disease. These large sheep, which are relatives of domestic sheep, live high in the mountains,

FIGURE 23-2 Bighorn sheep.

often well above the tree line. They measure from 6 to 7 feet in length and have long, curved horns.

The bighorn is hunted for trophies. Many carcasses are found with the head removed. Because the sheep is so alert, it is almost impossible to track; therefore, most hunters will find a waterhole frequently used by the sheep and wait in the brush. When the sheep gets within firing range, the slaughter begins. The sheep are no match for high-powered rifles and scopes. The heads are removed for trophies, and the remainder of the bodies are left for the scavengers. The bighorn sheep are under the protection of game laws, but poachers continue to take the sheep. The only hope for the bighorn sheep is to keep them in special reserves with no hunting. Poachers are still a problem, however, and game wardens are always on the lookout for them.

Polar Bears

The population of this large white bear (Figure 23-3) is rapidly diminishing. They are an important source of food for many Eskimos. Their fur is used for clothing, and their meat

FIGURE 23-3 Polar bear.

FIGURE 23-4 American wolf.

is used for food. This by itself would not be so bad. Fur collectors, however, are using technology, such as airplane hunting, to kill vast numbers of the animals. Females produce only two young every other year. They stay with the mother for about 10 months. Canada and the former Soviet Union have passed laws to protect these animals.

Key Deer

The smallest white tail deer is the key deer, which resides in the Florida Keys. They were sought for trophies because of their small size. Today, there are strict laws prohibiting the hunting of the deer. In 1953, the 6,745-acre Key Deer Wildlife Refuge was established, and a population estimated to be between 25 and 80 has increased to between 250 and 300 in 2000. Man and the automobile continue to be the main menaces.

Wolves

For years, stories have been told of wild wolf packs terrorizing homesteads and towns. It was considered a common fact that wolves were dangerous to man.

The wolf (Figure 23-4) resembles a wild dog and hunts at night in packs. Each wolf pack is a single family of adults and young. The female bears five to fourteen pups in a den that is guarded by the male. The male and female mate for life. As humans cleared more and more land, the wolf was forced to feed on domestic livestock. The attacks intensified. There have been bounties on the wolf since 1630, but today bounties are allowed only if the wolf population endangers the deer population.

Mountain Lions

Except for a few animals in southern Florida, the mountain lion has been eliminated in the eastern United States. A short-haired relative of the African lion, the mountain lion hunts at night, feeding mainly on deer. (See Figure 23-5.) It seldom attacks humans, but humans frequently hunt the lion. Most are hunted with dogs that tree the animal; tired from the chase, the lions become very easy targets. The animals are hunted for their skins and their heads. Hunting is not allowed in the national parks,

FIGURE 23-5 Mountain lion.

but mountain lions are still hunted in U.S. forest areas.

Endangered Birds

Just as mammals have been hunted on the ground, birds have been hunted in the air. The sight of birds in flight has compelled many hunters to use them as target practice. As previously mentioned, the United States Department of the Interior has placed 124 birds on the endangered species list. The most common examples are the whooping crane, bald eagle, ivory-billed woodpecker, and prairie chicken.

Whooping Cranes

One of the most beautiful birds is the white whooping crane. This migrating bird has been in danger of extinction for many years. They fly from Texas to Canada each year on the migration, and many hunters have shot these large white targets. The whooping crane nests in Canada where up to two young per nest hatch from eggs. The young fly south with the adults as winter approaches. Each year when they return to Texas at the Arkansas National Wildlife Refuge, they are counted. The count

has steadily increased. As of May 2000, there were an estimated 188 whooping cranes remaining. There is also a proposed plan to establish a refuge in Nebraska where the birds commonly stop on the way back to Texas.

Bald Eagles

It seems ironic that the national bird of the United States was until recently in danger of extinction. This fact should awaken people to wildlife conservation. The bald eagle (Figure 23-6) has a white head and tail and, when viewed at a distance, appears to be bald. Bounties were once placed for the birds' talons, and it has been reported that over 115,000 pairs were brought in for the bounty. The birds were killed mainly because it was believed they fed on salmon. Actually, the birds fed on dead salmon and rarely caught live fish.

The birds are now protected by law. A fine of $500 can be received for tampering with an eagle or its nest. In the first half of this century, when DDT was used widely to kill mosquitos, it also contaminated the fish, causing problems with the eagle. There have been several refuges set aside for eagles to grow and prosper. It takes the young 4 years to reach maturity, but the great bird can live up to 100 years.

As of this writing, the bald eagle remains on the TESS list. In July 1999 President Clinton announced a proposal to remove this species from the list because of its remarkable recovery over the past three decades. A news release prepared by the U.S. Fish and Wildlife Service said:

> As a symbol of freedom, strength, and courage, the bald eagle represents the best of what America has to offer. On the eve of Independence Day weekend, President Clinton marked the culmination of a three-decade effort to protect and recover this majestic bird by announcing a proposal to

FIGURE 23-6 Bald eagle.

remove it from the list of threatened and endangered species.

"The American bald eagle is now back from the brink of extinction, thriving in virtually every state of the union," President Clinton said. "I can think of no better way to honor the birth of our nation than by celebrating the rebirth of our proudest living symbol."

The bald eagle once ranged throughout every state in the Union except Hawaii. When America adopted the bird as its national symbol in 1782, as many as 100,000 nesting bald eagles lived in the continental United States, excluding Alaska. By 1963, only 417 nesting pairs were found in the lower 48.

Today, due to recovery efforts by the Interior Department's U.S. Fish and Wildlife Service in partnership with other federal agencies, tribes, state and local governments, conservation organizations, universities, corporations and thousands of individual Americans, this number has risen to an estimated 5,748 nesting pairs. As a result, biologists believe it may no longer require the special protection of the Endangered Species Act.

By the time you read this, the bald eagle will probably no longer be listed as an endangered species. This is truly a success story for American conservation efforts.

Ivory-Billed Woodpeckers

The ivory-billed woodpecker is the largest woodpecker in North America. It lives in the southern states where it builds nests in tops of old forests. Its principal foods are insects and grubs found in old and living trees. At one time, the birds were thought to be extinct; but, they have reappeared. At the present time, they are being kept in a secret reserve. One day they will be reintroduced to the public.

Prairie Chickens

The prairie chicken is a type of grouse that once populated the great American prairies. Native American dances imitated their mating dances. Texas permitted year-round hunting until 1937. With the increase in oil drilling and years of droughts, the populations started to dwindle. In 1959, the Prairie Chicken Foundation was formed to work for the management of the prairie chicken.

Endangered Fish

Much work has been done with endangered fish. In 1987, there were twenty-five fish on the

TABLE 23-2 List of Endangered Fish Species

Atlantic sturgeon	Gila trout
Lake sturgeon	Greenback cutthroat
Shortnose sturgeon	trout
Longjaw asco	Lahonta trout
Arctic grayling	Montana trout
Atlantic salmon	Piute trout
Apache trout	Rio Grande trout
Colorado River	Little Colorado
squawfish	spinedance
Desert dace	Chi-vii
Humpback club	Comanche pupfish
Pahrump killfish	Blind fish
Gila topminnow	Blue pike

Department of the Interior's endangered species list. (Commercial fishing industries have also worked in this area.) Many of the fish in the list (Table 23-2) will be covered in Chapters 25 and 26. Two species are located in the Mammoth Caves and Death Valley (pupfish and blind fish).

SUMMARY

Fish and wildlife in America have not fared as well as many other resources. Game species have received more attention than nongame species. Most of the management has been through the United States Fish and Wildlife Service and the National Parks System. The time has come, however, when more and more people are becoming conservation minded toward wildlife. It has become evident that we must do everything in our power to conserve what we have. In the next few chapters, we will closely examine game as well as marine and freshwater wildlife. We also look at possible careers in the wildlife area.

DISCUSSION QUESTIONS

1. Define wildlife.

2. Using fur trapping as an example, how have humans caused the extinction of wildlife?

3. Explain how each of the following became endangered:

   a. Passenger pigeon
   b. Carolina parakeet
   c. Heath hen
   d. Labrador duck
   e. Bighorn sheep
   f. Polar bear
   g. Key deer
   h. Wolf
   i. Mountain lion
   j. Whooping crane
   k. Bald eagle
   l. Ivory-billed woodpecker
   m. Prairie chicken

SUGGESTED ACTIVITIES

1. Prepare a report on one endangered animal. How can we prevent this animal from becoming extinct?

2. Have all the students in the class who hunt or fish make a list of the game animals and fish that can be found in or near your community.

24

GAME MANAGEMENT

OBJECTIVES

After reading this chapter, you should be able to

- identify the habitat requirements of wildlife

- discuss the difference between an euryphagous and stenophagous animal

- explain the most commonly accepted methods of game management

- explain how an individual landowner can employ game management techniques

- discuss major legislation affecting game management

TERMS TO KNOW		
habitat	territory	predator control
euryphagous	refuges	artificial stocking
stenophagous	habitat development	population density
home range	hunting	

When European settlers first came to the United States, they were greeted with a vast array of plant and animal life. Trees were everywhere they looked, and the game in the forest was more plentiful than they imagined possible. The idea of considering a program to conserve and manage their newfound resources was not a high priority. The only thoughts they had were for their own survival. Little did they know that, in a short period of time, Americans would be concerned about the decline of wildlife and game and would actually begin to initiate programs to manage them.

What is game management? Aldo Leopold, an early authority on game management, referred to it "as the art of making land produce sustained annual crops of wild game for recreational use." Wildlife scientists have since expanded the definition to mean the science and art of changing characteristics and interactions of habitats, wild animal populations, and humans to achieve specific human goals by means of the wildlife resource. It is important to point out that habitats, populations, and people all play key roles in game management.

HABITAT REQUIREMENTS

The basic requirements of food, cover, water, territory, and home range are all part of an animal's **habitat**. A discussion of each of these requirements follows.

Food

Wildlife can be classified as to both the type and amount of food they consume. The most common classification deals with the type of food they consume. The most common classifications using food types include herbivores (plant eaters), carnivores (meat eaters), insectivores (insect eaters), frugivores (fruit eaters), omnivores (animals that eat many different types of food), and spermivores (seed eaters). The classification of wildlife by food quantity includes euryphagous and stenophagous animals. A **euryphagous** animal is one that consumes great varieties of food. Because its choice of food is usually varied, its chances of survival are usually great. The opossum, which eats fruits, berries, corn, worms, frogs, snakes, and even mice, is an example of a euryphagous animal. A **stenophagous** animal is one that eats a specialized diet. It usually has less chance to adapt to new food sources if its traditional food supply is not available. Because of the failure to adapt, it is more likely to starve in a food-scarce season. The black-footed ferret eats prairie dogs almost exclusively.

Cover

For game to survive harsh weather conditions, they must find cover—a place that will protect them. This cover will also provide a place of protection from their predators. Cover ranges from a thicket or fencerow to water, as in the case of beavers and muskrats.

Water

Water is one of the most important requirements of wildlife. The bodies of most game animals consist of 60 to 80 percent water. Water is important in blood composition, temperature regulation, and nutrient transport. Without a supply of abundant fresh water, an area will soon become desolate of wild game.

Home Range and Territory

The last requirement of wildlife is its home range and territory. Some biologists refer to these areas as breeding sites; however, the game needs more area than just a breeding ground. The area over which the game travels is called its **home range**. This area may be as small as an acre or as large as a township. The area an animal will defend, often to the death, is called its **territory**. It is important to note that many animals' home ranges can overlap, but their territories never will. The only exception is when animals are mating.

GAME MANAGEMENT PROCEDURES

Game management procedures have been attempted with a variety of methods. The most commonly accepted methods include

- game refuges
- habitat development and improvement
- coordination with other resources
- hunting regulations
- predator control
- artificial stocking

Game Refuges

One game management procedure is to set aside land for the protection of wildlife species. These areas are called **refuges**, reserves, and wilderness areas. They provide the basics for survival without the threat of hunters.

The early refuges were for private use. The first state reserve was established in 1870 in California. Since that time, areas for colony-nesting birds, migratory waterfowl, pelicans, and large game animals have been established.

Refuges are not the answer to all wildlife problems. Refuges only protect the wildlife from hunters, not from their natural enemies. The refuge also does not protect the game from hunters once they leave its boundaries. Refuges are only a portion of a total game management plan.

Habitat Development and Improvement

Previously, we have seen that game requires food, water, shelter, and a living area. To increase game populations, the development and improvement of their habitats must be an integral part of game management procedures.

The common methods for **habitat development** are fencerow plantings (Figure 24-1) and woodland management. When the majority of the farming programs consisted of both livestock and crop enterprises, fences were abundant. With the onset of large tillage equipment, along with the broad use of livestock confinement systems, fences became unnecessary and cumbersome. The fencerow area once used by game for cover and food has become nonexistent. Farmers have been asked to set aside a small area, such as an end row, for the purpose of wildlife feeding; and those cooperating with game management personnel see a large increase in game populations in their areas.

One of the easiest ways farmers or landowners can contribute to the management of game is through careful consideration of the way their woodlands are handled. The question of woodland management comes down to grazing versus nongrazing. If the farmer allows livestock to graze in his or her woodland, a definite decrease in the game-carrying potential of that acreage will result. There can be a decrease, if not a complete loss, of small game such as rabbits, squirrels, and quail. By not using the woodland for grazing, farmers can see an increase in these game populations. (See Figure 24-2.)

FIGURE 24-1 Fencerows provide a suitable habitat for many species.

FIGURE 24-2 Crops left in the field will be a valuable food supply for wildlife during the winter. On state-owned land, corn is left for the wildlife population.

Coordination with Other Resources

The management of game as a resource is not independent of other resources. Those involved in game management must be aware of other natural resources and attempt to coordinate their procedures in line with the management of other resources. Occasionally, we find competition among resources. A farmer manages cropland to produce maximum yield. He or she manages soil to prevent soil erosion and nutrient depletion. Convincing the farmer to leave a row of corn or an area of wheat for wild game can be difficult because it may be difficult for the farmer to realize that the game on the farm is an important resource, too.

Hunting Regulations

Game was abundant to the early colonists. Some colonists, however, had the foresight to avoid the depletion of their wildlife resources because of **hunting**. In 1646 Rhode Island became the first state to establish a closed season on game. Although the law was specific for white-tailed deer, it led the way for the other states to follow suit. The first bag limit was initiated in 1878 by the state of Iowa.

One of the most abundant animals on the western plains was the American bison. This animal was the most important resource to the Plains Indians; it provided food, clothing, and shelter products. When the railroad was being constructed through the West, thousands of bison were slaughtered to feed the workers. The railroad hunters cared nothing about the hides Native Americans used for clothing and shelter, or the bones they used as tools. Both were left to decay and waste. Word of the vast and abundant bison herds quickly spread east. This produced the onset of a slaughter as millions more were killed by hunters for such things as their tongue, an Eastern delicacy, or their hides. The entire carcass was not utilized, and often more was left behind than was taken. Bison were also slaughtered as a means to control Native Americans, who traditionally relied on these animals for food, clothing, and shelter. This uncontrolled slaughter brought legislation from Congress because extinction was so near. The bill, however, was vetoed by President Grant. Laws were passed later, but not before this wildlife resource was all but extinct.

Since the bison crisis, responsible solutions have developed to the overhunting problem. Biologists and scientists have started to evaluate the positive and negative aspects of hunting and the pressures it places on game. Hunting is being controlled on both public and private lands. Biologists have banded, tagged, radioisotoped, and tracked wild animals to measure their territories and home range. They have learned that game populations depend on many factors such as reproduction rates, climate, disease, habitat, and predators. Experts have also worked with fluctuations in the year-to-year populations and have determined that procedures such as bag limits, hunting seasons, and closed seasons for a given game species must be changed each year in accordance with the population size. The hunter must be prepared to accept, with low and high population fluctuations, changes in permissible bag limits and season lengths.

Predator Control

One facet of game management is to control the predators on the game rather than controlling the game itself. The control of predators that feed on game animals involves much study and is generally based upon two beliefs:

- Medium-sized and large predators can be dangerous to people.
- Predators can be a threat to domestic livestock and wild game.

Although both beliefs are probably correct, additional questions must be asked and evaluated. Do the predator's activities have any benefits either directly or indirectly? Does the predator keep the game population at too low a level?

Predators can be beneficial to the game management plan. First, by controlling the predator's population either through seasons, bounties, or bag limits, we can also control the populations of the prey species. Second, predators feed on animals that humans consider pests, such as rats and mice. Third, **predator control** practices keep the species in a healthier condition. Fourth, predators help maintain an improved game population by killing the weak and handicapped individuals.

Artificial Stocking

An area of great concern is the practice of **artificial stocking**. (See Figure 24-3.) This can be either the stocking of game natural to the area or the introduction of species new to the area. Bringing in new species is called the "introduction of exotics."

Artificial stocking of exotic species is usually done to introduce new or additional species to an area, or to supply predators for a problem or overpopulated game animal. Introduction of exotics, however, can have detrimental effects. The introduction of the English sparrow in 1865 is a case in point. The English sparrow was able to adapt and multiply in its new environment. At the present time, most

FIGURE 24-3 An artificially stocked farm pond.

people consider the small bird to be a pest. The rabbit, once introduced in Australia, soon became a major pest.

Whenever humans try to manipulate game artificially, two important principles must be carefully examined. The wildlife biologist first must examine the desired population density of the area and, second, the estimated carrying capacity.

Population density refers to the number of game animals in a defined area—for instance, 1 bear per 100,000 acres or 3 quail coveys per 20 acres. If authorities are not careful with the amount of game introduced to a certain area, it will become overpopulated and result in additional environmental problems. In addition, this works to the disadvantage of the species being introduced.

Carrying capacity refers to the amount of game for which a given area will provide the essentials for life. Population density differs from carrying capacity in that the former deals with population numbers, and the latter deals with food supplies and animal needs. If the amounts are not carefully calculated, then too many game could result. This could then lead to starvation, unhealthy animals, and possible disease problems because of weak animals with low resistance.

THE INDIVIDUAL LANDOWNER AND GAME MANAGEMENT

After looking at the various game management techniques that can be employed, most landowners seldom know where to start. They usually want the answers to two questions: What can they do to increase the game on their land, and where do they go for additional technical assistance if they feel uncertain of which technique to use.

The individual landowner can increase the area's population of pheasant, rabbit, and quail by remembering the essentials for game: water, food, and shelter. The landowner may have a low area on his or her property that stays wet much of the time. Through the help of the Natural Resources Conservation Service, the farmer may construct a farm pond. This pond is a practical way to supply water for fish production and game animals, as well as for livestock and fire protection. A pond can be used to attract game birds and migratory waterfowl. Ducks breed by ponds and find food along its banks.

Providing an abundant food supply may also attract animals to an area. (See Figure 24-4.) Fencerows and ditch banks are a starting place for this food. Grain such as sorghum can be a valuable food source. Grain sorghum produces a seed head on top of a stalk similar to corn. When snow covers another food, the heavy, cornlike stalk will hold the seed well above the snow. The seed has about the same nutritive value as corn and is readily accepted by most game animals. Conservation departments in many states make free sorghum seed available to landowners for this purpose. One

FIGURE 24-4 Food plots at Mississenewa Reservoir, Indiana.

pound of sorghum seed will plant approximately 2,700 square feet of land area (about one-sixteenth acre). Conservation organizations such as Quail Unlimited offer seed and other incentives to landowners to plant such crops.

Cover or shelter is as important to game animals as food and water. If the animals do not have adequate shelter to protect them from the elements, their chance for survival will be slim. A landowner can also help game in the attainment of adequate shelter. Examine the fencerow again. Should it be burned down to nothing? If sorghum seed is planted, shelter will be present. What about that small area of woodlot? If that area is allowed to grow, the brush will supply a very good habitat for pheasant, quail, and rabbits. When bad weather comes, those animals will be able to find shelter in the brush.

The local landowner has ample opportunity to help in the game management problem. One does not need a huge acreage or elaborate refuges. If the landowner carefully examines the needs of the game in his or her area, and then inventories the resources available, he or she probably will be surprised at the similarities. One needs to be observant to the needs of the game and ask for assistance when necessary.

Where can the landowner gain technical assistance in game management procedures? There are numerous agencies available. Government agencies such as the United States Forest Service under the direction of the United States Department of Agriculture can help individuals or groups with their woodlot management program. Also, each state has a Department of Natural Resources (although the exact name will vary by state). Within these departments are fish and game divisions employing wildlife biologists trained in game management. The Natural Resources Conservation Service can help in the construction of ponds, development of wildlife land areas, construction of windbreaks, fishstream improvement, or woodland improvement programs. The landowner may also wish to ask for assistance from the many fish and game clubs organized throughout the country. There are many programs available to individuals who wish to improve the game resources in their areas.

LEGISLATION AND GAME MANAGEMENT

Who owns the game animals in a given area? Are they owned by the landowner, the state, or the federal government? As early as 1896, the Supreme Court gave the management of wildlife to the states. The landowner, however, does have property rights; therefore, the laws covering wildlife are few and simple.

The major federal legislation acts governing wildlife are

1. *Lacey Act of 1900.* The Lacey Act of 1900 was the first major law affecting wildlife. The law made it a federal offense to transport illegally taken wildlife across state boundaries. This law was the basis for state and federal cooperation in enforcing state wildlife laws.

2. *Migratory Bird Act of 1929.* The Migratory Bird Act of 1929 was a broader statement of the Lacey Act. The main purpose of this law was to provide refuges for migratory birds.

3. *Migratory Bird Hunting Stamp of 1934.* This law gave the federal government a way of raising revenues for acquiring lands for migratory game birds.

4. *Pittman-Robertson Program and Dingell-Johnson Act of 1950.* These two laws allow an excise tax to be placed on guns, ammunition, and fishing tackle by states. The states then match the funds collected and allocate the total to their fish and game departments.

5. *Lea Act of 1948.* The Lea Act was enacted to help farmers who were experiencing problems with crop damage from ducks and geese. This law provided federal land for use as waterfowl feeding areas to lure these birds from private cropland.

6. *Endangered Species Act of 1966.* This law gave the authority of examining and recommending procedures protecting rare and endangered species to the Fish and Wildlife Service.

SUMMARY

Game management involves careful consideration of all natural resources. It involves people in both the public and private sector. Game management also involves the application of all procedures discussed in this chapter.

In individual cases, one method may be more appropriate than another method. Each game management procedure must be evaluated, its benefits weighed, and its adversities examined, before the best plan or plans are put into operation for conserving this valuable natural resource.

DISCUSSION QUESTIONS

1. What are the habitat requirements of wildlife?

2. Explain the difference between home range and territory.

3. How should a woodland be managed to increase game populations?

4. How is wildlife coordinated with other natural resources?

5. Discuss the advantages and disadvantages of hunting.

6. How does controlling predators manage game?

7. How do carrying capacity and population density differ?

8. What can the private landowner do to increase game in an area?

9. Discuss the agencies the private landowner can contact to get technical assistance in game management procedures.

10. Discuss the six major laws concerning game management. What have these laws accomplished?

SUGGESTED ACTIVITIES

1. Invite a game warden to speak on the value of game management. What is done in your area?

2. Obtain a list of the hunting regulations for your area. Include protected species and bag limits. Why are the bag limits set as they are?

3. Plan a field trip to an area that could be set up as a game protection area. Prepare a proposal of the possible ways the area could be developed.

25

MARINE FISHERIES MANAGEMENT

OBJECTIVES

After reading this chapter, you should be able to

- list and explain the ways the ocean is zoned

- discuss the types of ocean water movements, including waves, tides, and currents

- know the characteristics of marine fish, marine shellfish, and marine mammals

- explain the characteristics of the estuarine ecosystem

- discuss how the ocean can be artificially cultivated

TERMS TO KNOW

ocean zonation	thermocline	long-lining
supratidal	internal wave	purse seining
intertidal	catastrophic wave	prawns
neritic zone	tidal wave	spat
bathyal zone	stationary wave	baleen whale
abyssal zone	tides	whalebone
euphotic zone	currents	rorqual
salinity	plankton	rookeries
temperature stratification	bait fishing	estuary

The ocean can be considered the last frontier on this planet. Due to the ocean's unique physical characteristics, it can support a wide variety of marine plant and animal life. In this chapter, we examine both the physical and biological components of this important natural resource. With knowledge of the composition of the ocean, we can formulate management techniques to more fully utilize our marine resources.

OCEAN PHYSICAL CHARACTERISTICS

The physical characteristics of the ocean can be divided into four major areas:

- zonation
- salinity
- temperature density
- water movements

Zonation

We can classify ocean zones by a variety of methods such as depth measurements, temperature changes, pressure variations, and light penetration regions. **Ocean zonation** by depth and light penetration are the two most common methods used and will be discussed in this chapter.

Eugene P. Odum, a retired ecologist at the University of Georgia, divides the ocean into five zones: supratidal, intertidal, neritic, bathyal, and abyssal. The **supratidal** and **intertidal** areas are above the water level and are omitted in some classifications. The supratidal zone lies above high tide and below the vegetation line. The intertidal zone, sometimes called the littoral zone, is the area between high tide and low tide. At this waterline the **neritic zone** starts, therefore, making this zone the first the marine manager is usually concerned with.

The neritic zone contains more biological substance than any other part of the ocean. This zone is about 10 to 200 miles wide and reaches depths of 200 to 600 feet. The neritic zone stops at the end of the continental shelf and, in the past, has been used as a great dumping ground for wastes from industries and cities.

The second marine zone is the **bathyal zone**. It contains the continental slope and rise and is regarded as a geologically active area with its underwater avalanches and slides.

The **abyssal zone** is considered the ocean deep zone, reaching depths of 5,000 meters. Some trenches may extend to more than 6,000 meters, which is referred to as the hadal zone. Because of the depth, scarcity of food, increased water pressure, and lack of dissolved oxygen, the animal life able to live and grow in this zone must be highly specialized. The marine life must be able to obtain nutrents from the rich sediments lying on the ocean floor.

The light penetration method for zoning the ocean is one of a vertical nature rather than the horizontal method previously examined. The first zone, the **euphotic zone**, is sometimes referred to as the "twilight zone" and is the part of the ocean in which sunlight penetrates the water. This zone supports plant and animal life that requires sunlight to grow. The depth of the euphotic zone is from the surface to about 600 feet beyond the horizontal shelf. Below the euphotic zone is the cold, dark abyssal zone.

Salinity

Salinity refers to the concentration of salts within the ocean water. Although the concentrations and types of salts vary throughout the ocean, the most commonly found salt compounds consist of sodium, chlorine, magnesium, calcium, and potassium. Salinity is defined as *the number of grams of dissolved salt in 1,000 grams of sea water,* and the symbol for expressing salinity is ‰. The range of the ocean's salinity is from 33‰ to 38‰ with an average of about 35‰. If you prefer to think in terms of percentages, that would be a range of 3.3% to 3.8% with an average of 3.5% dissolved salt.

Temperature Density

The ocean can be compared to a giant heat pump that moves and transports heat from the equator to the poles. Temperature changes occur as we move both to the different latitudes of the ocean and to different depths of the ocean. The **temperature stratification** consists of three layers: the mixed surface layer, the middle **thermocline** layer (10 to 1,000 meters), and the cold, bottom, deepwater layer (1,000 to 6,000 meters). The thermocline, which is below the light penetration depth, suggests that there is a transfer of heat vertically as well as horizontally.

Density, defined as mass per unit volume, is dependent upon salinity, temperature, and pressure. The changes in density result from evaporation and heating of the ocean's surface. The higher the temperature, the lower the density for a given salinity and pressure. The density of the ocean increases as the pressure and depth increase.

Water Movements

The fourth major area of the ocean's physical characteristics is that of water movement: waves, tides, and currents. The movement of water dictates such items as temperature, saline readings, and nutrient levels as well as animal and plant life.

Waves can be classified as wind generated, internal, catastrophic, and stationary. The most common wave is that of the wind-generated kind. These waves can be divided into sea, swell, and surf. The wind-generated sea wave is an irregular wave with no systematic pattern. The sea wave travels at different heights and changes directions as it moves. Wind-generated swell waves, on the other hand, are uniform waves with similar dimensions traveling together because of their similar speed. They remain at a constant speed as they travel but decrease in height, sometimes traveling across an entire ocean. The wind-generated surf wave is one occurring close to shore. Water particles move in an orbital motion, but toward the beach. This type of wave has less depth. Therefore its energy is directed toward the shore.

A more difficult wave to detect is the underwater **internal wave**. This wave is found with the temperature changes in the depths of the ocean. It travels more slowly but has a greater height.

The **catastrophic waves** are those with massive power behind them. These waves are caused by storms, hurricanes, and landslides on the shore. A commonly known catastrophic wave is the **tidal wave**.

The last type is the **stationary wave**. These waves occur in bays and calmer waters. The wave itself does not move horizontally, but the water surface moves up and down.

Tides are specialized waves caused by the gravitational attraction of the sun and moon on the earth. The tides occur at very exact times with one-half lunar day (12 hours, 25 minutes) between high tides. Because the lunar day is longer than the solar day, the tides occur 50 minutes later each day. An area will have either one high tide and one low tide each day (diurnal) or two high tides and two low tides each day (semidiurnal). The effect of the sun becomes especially important on the tides. When the sun and the moon line up with the earth, a strong tide is produced; this happens every 14 days at the time of the new and full moons. This exceptionally high tide is called a spring tide. When the sun and the moon are at right angles with each other, which occurs during the half moon, an exceptionally low tide is produced. This tide is called a neap tide.

The final way water is moved is by **currents**. The most common types of currents are surface currents, turbidity currents, and bottom currents. Surface currents, which are caused by the prevailing winds, reach velocities of about 3 knots. When a hurricane reaches shore or an earthquake occurs, underwater landslides can occur; these in turn produce a turbidity current underwater. Currents in the deep water become the bottom currents. They are responsible for moving sediments on the ocean floor and for transporting water from the Atlantic Basin to the southern Atlantic Ocean.

THE BIOLOGICAL OCEAN

Just like its terrestrial counterpart, the ocean is a wonderland of animal and plant life. The ecological principles discussed in Chapter 3 hold equally true in the ocean habitat. The storehouse of nutrients and food is virtually untapped and, with proper management procedures, could provide relief to the world food problem.

Marine animal life can be divided into four major groups: microscopic marine animals, marine fish, marine shellfish, and marine mammals.

Microscopic Marine Animals

The most common microscopic animals are the **plankton**, particularly the zooplankton. This animal life is the staple food for species ranging from tiny fish to whales. Many zooplankton are related to the marine crabs and shrimp. The waters vary in plankton concentration, but on the average, there are about one tenth of a gram per cubic meter of water. Some scientists have suggested the possibility of harvesting plankton because they contain the nutrients and amino acids found in human foods. It would, however, require the filtering of over one million gallons of water with sophisticated nets to collect one pound of dry plankton material. It has been generally concluded that humans should leave the harvesting of plankton to

the fish and collect the larger animals farther up the food chain.

Marine Fish

The four most important marine fish species are salmon, tuna, menhaden, and flounder. These species make up the majority of the world fishing markets. Examples of less economically important species include haddock, herring, cod, and mackerel. (See Figure 25-1.)

Salmon

There are seven main species of salmon: Atlantic, cherry, chinook, chum, coho, pink, and sockeye. They range in size from 16 inches long and weighing 5 pounds to 36 inches long and weighing 25 pounds. They live in both the Atlantic and Pacific Oceans.

(a)

(b)

FIGURE 25-1 Common marine fish: (a) menhaden, (b) flounder. *(Courtesy of Dick Mermon)*

The salmon begins life in fresh water but migrates to the ocean to live and grow. The Atlantic salmon can make the trip repeatedly, but the Pacific species (which include all species but the Atlantic) spawn only once and die soon afterward. The salmon spawns in shallow streams during late summer or early fall. The female finds a rocky stream bed where she digs a saucer-shaped nest with her tail. The male salmon stays close by to guard the female. She then deposits her eggs in the nest and the male fertilizes them. The procedure is then repeated upstream. The female will lay 2,000 to 10,000 eggs during this spawning time.

The salmon eggs hatch in three to four months. The fry retreat to the gravel bottom for several weeks trying to avoid their predators, birds and other fish. The young salmon live off food in the yolk sack attached to their stomachs. Some species of salmon move toward the ocean waters immediately, whereas others stay in fresh water for up to three years. The adult salmon remain in the ocean from six months to five years before returning to the freshwater stream to spawn. Once the salmon reaches fresh water, it stops eating and relies on its body fat for nourishment. The male develops a hooked snout, and some species develop a hump on their backs. Commercial and sports fishers try to make their catch before this time because the salmon meat loses some of its flavor, thus lowering its quality.

Management techniques used to preserve and protect the salmon population include fish ladders, to help salmon over dams; construction of artificial spawning channels, where temperature and water flow are controlled; use of hatcheries to propagate small salmon; and construction of escarpments, which allow a certain number of fish to spawn. Salmon have also been used to control other rapidly producing fish species. For example, the coho salmon was introduced into the Great Lakes to control alewives, a very fast-reproducing fish species.

Commercially, salmon are caught in nets and sold fresh, frozen, smoked, and canned. The main salmon fishing countries include Japan, the United States, and Canada.

Tuna

Tuna is a member of the mackerel family and the leading game fish in the United States. The three most commercially important types are albacore, skipjack, and yellowfin tuna. The tuna species range in mature size from the 10-foot-long, 2,000-pound northern bluefin to the 2-foot-long, 10-pound frigate. Tuna are fast swimmers, reaching speeds of over 45 miles per hour. The tuna does not have the ability to push water through its gills; therefore, it must swim continuously to live.

Tuna are caught by bait fishing, long-lining, and purse seining. **Bait fishing** involves using live bait to attract tuna, then catching them with a hook and line. When **long-lining**, fishers reel out a line up to 75 miles long with as many as 2,000 hooks. **Purse seining** involves using nets, called purse seines, to catch the fish. Porpoises characteristically travel above schools of tuna. Pilots locate the porpoises from the air and radio the fishing boats, which, in turn, net the tuna. A major problem with this netting procedure is the accidental, but unlawful, netting of the porpoises swimming above the tuna.

Menhaden

Menhaden, also known as bony fish or fatbacks, live in the Atlantic Ocean and feed on plankton. They swim in schools close to the ocean's surface, which makes them an easy catch. They are 12 to 18 inches long and weigh about 1 pound at maturity. Most menhaden fish are used in products such as livestock feed, soap, and fertilizer, rather than for human food.

Flounder

Flounder, or plaice, is a saltwater flat fish that lives on the sandy and muddy bottoms of bays. Flounder have a flat body with both eyes on the same side of the head (refer to Figure 25-1). They feed on shrimp and small fish.

Marine Shellfish

Marine shellfish include shrimp, oysters, crab, and lobster.

Shrimp

The most economically important shellfish is the shrimp. Related to crabs and lobsters, shrimp live in salt and fresh water. They are excellent swimmers in both forward and backward directions. Large shrimp, called **prawns**, feed from the ocean floor and grow as large as 12 inches in length.

The most common shrimp used for food is the peneid shrimp, which is hatched from eggs laid in the ocean. The female may lay 500,000 to 1 million eggs. As the young shrimp move toward shore, about 80 percent are lost due to natural predators. Once near shore, the shrimp settle in bays and river mouths until they are five to seven months old. At that time, the shrimp move back to deeper water for breeding. The female dies soon after she lays her eggs.

Modern shrimp-gathering methods use nets called trawls, which are dragged on the ocean floor. Once netted, the shrimp are usually sold frozen or canned. Leading shrimping countries include the United States, India, and Japan.

Oyster

A type of mollusk, the oyster is a shellfish with a two-piece shell protecting a soft inner body. Oysters live in mild or warm climates. The oysters of the Persian Gulf and Pacific Ocean are responsible for making pearls.

The female oyster lays about 500 million eggs each year, spraying them into the water. The young oyster, called a **spat**, hatches 10 hours later and is about the size of the point of a needle. Within 24 hours, the shell starts to form and the young oyster attaches to a rock where it spends the rest of its life. Some oysters have lived up to 20 years. The oyster's greatest enemies include humans, fish, crabs, starfish, and oyster drills.

Crabs

The third economically important marine shellfish is the crab. Crab habitats range from the shallow waters close to shore to the deep waters of the ocean. Over 4,500 different kinds of crabs have been identified. The most common, the Atlantic crab, lives in burrows on saltwater streams. The largest crab is the Alaskan king crab, which weighs up to 12 pounds at maturity and is the most sought-after crab by American and Japanese fishers.

Lobster

The American lobster lives on ocean bottoms near the shore in both the Atlantic and Pacific Oceans. The lobster averages 12 to 24 inches in length and weighs from 1 to 20 pounds. It feeds by burrowing into a hole with only its claw at the edge. As prey comes close to the claw, it quickly becomes the lobster's next meal. The main food of the lobster consists of crabs, snails, small fish, and other lobsters.

The female lobster carries her eggs under her tail for 11 to 12 months. Every 2 years she lays 5,000 to 100,000 eggs by shaking the eggs from eggshells. The eggs will quickly rise to the surface. The young lobster drifts on the surface for three to five weeks; it is easy prey for birds and fish. It then sinks to the ocean floor where it spends its remaining life, which can last up to 15 years.

Lobsters are caught in traps called pots. The pots allow the lobster to enter, but vertical wooden bars confuse the lobsters and prevent them from finding their way out. The fishers must check the traps daily. Two lobsters in confinement will fight until one dies. To prevent possible injury to the lobster crop, a wooden nail is placed on the joint behind the claw, or the claw is immobilized with a rubber band, to disarm it, enabling several lobsters to be kept together for shipment.

Marine Mammals

Marine mammals differ from other marine life in two important ways. First, marine mammals, like land mammals, are warm blooded rather than cold blooded. Second, the mammals have lungs as breathing devices rather than gills. It is necessary for them to surface occasionally to take on a new supply of oxygen. This also allows some marine mammals to survive on land. The main marine mammals include whales, porpoises, walruses, and fur seals.

Whales

Whales include the largest mammals that have ever inhabited the earth. The largest blue whale on record was 100 feet long and weighed 136 metric tons. Whales can be divided into baleen whales and toothed whales.

Baleen whales obtain their food by straining plankton from seawater through tiny plates called **whalebone**. There are ten different kinds of baleen whales in three major groups: right, gray, and rorqual. Right baleen whales include the bowhead, the longest baleen whale; the black right, which averages 60 feet long; and the Pygmy baleen, the smallest baleen whale. Gray baleen whales are black to gray in color and dotted with white blotches. They feed from the ocean floor. The **rorqual** baleen whale is the fastest whale species. This whale was able to flee the whaler's harpoon until the introduction of the fast, diesel-powered whaling ships used today. Rorquals are sometimes referred to as finback whales because of the fin protruding from their back. The most common kinds of rorqual baleen whales are blue, Bryde's, fin, humpback, minke, and sei. The common food for the rorqual is a small, shrimplike animal called krill. Some scientists have experimented with propagating and catching krill to be used in livestock feeds, and this alarms many marine biologists because of the possibility of taking away the main food supply of many whales.

Toothed whales, as their name implies, have a lower set of peglike teeth. There are over 65 different kinds of toothed whales divided into five major groups: sperm, beaked, belugas and narwhales, dolphins and porpoises, and river dolphins. The largest toothed whale is the sperm whale, which reaches about 60 feet in length. Sperm whales are blue-gray to black in color and have an enormous, square-shaped head. Sperm whales live in tropical waters and feed off squid, barracuda, and sharks.

Whales are seasonal breeders, usually mating during the winter months. The male, called a bull, and the female, called a cow, migrate to the equator for mating. Most baleen whales migrate, whereas most toothed whales do not. Pregnancy for the female whale lasts an average of 10 to 12 months. When the calf is born, it weighs nearly 4,000 pounds and is 23 feet long. The cow nurses her offspring for about 7 months. Because of the rich milk, the offspring of some species will gain an average of 200 pounds per day.

Whales travel in herds of 100 to 1,000. An adult male whale will form a harem school consisting of himself, his females, and their young. Nursery groups of females and young and bachelor groups containing young males also travel along with a harem. When migrating, the mothers and young lead the herd, followed by males and non-pregnant females, with pregnant females at the rear. The average lifespan for whales ranges from 15 years (for a porpoise) to 40 years (for a baleen whale).

Whaling was always a dangerous occupation. Matching a small wooden harpoon boat against a giant whale produced constant danger for the fishers. Whaling changed, however, in 1860 with the invention of an exploding harpoon. This new harpoon carried an explosive charge on its end that exploded once in the whale's flesh. Death was certain. Today, modern whalers use heavier and faster boats to improve their catches.

Modern whaling techniques consist of a factory ship and a fleet of high-speed, diesel-powered catcher boats. The whalers use airplanes, helicopters, and sonar devices to locate whale schools. Once the catcher boats make the kill, the dead whales are filled with air to keep them afloat. Markers using radar and flags are placed on the whales, and the catcher boats move on to new territories. Buoy boats are sent to tow the dead whales to the factory ship. Once a dead whale is on board, its blubber is removed and cooked to remove the oil. The meat is cut, frozen, and readied for market. The bones and scrap are prepared for cattle feed. When the factory ship arrives back at port, its whale products are ready for the retail markets.

With the onset of modern whaling techniques, the populations of whale species greatly declined. Countries interested in preserving whale populations formed the International Whaling Commission (IWC) in 1946. The commission's goals were to regulate the whaling industry by setting caps on whale catches. The commission completely stopped the hunting of blue, bowhead, gray, humpback, and right whales. The member nations also exchanged observers to oversee each other's whaling operations. In 1971, the United States ended its whaling industry and refused to import whale products into the country. In 1972, the Marine Mammal Protection Act prohibited all commercial use of whales in the United States waters. In 1982, the IWC doubled its membership and voted to ban all commercial whaling operations. In 1990, the commission reviewed the ban to determine if whaling quotas should be started again. An inventory by whaling experts revealed the number of whales had not increased enough to lift the ban. Although the commission has no power to enforce its regulations, public pressure groups have been successful by threatening to boycott the products of any country that does not abide by the commission's rules. These steps have helped to slow the decline of the whale population.

Fur Seals

Seals are divided into three groups: eared, which includes fur seals and sea lions; earless; and walruses. The fur seal is highly prized because of its soft coat.

The fur seal spends the winter off the coast of California and the summer in Alaska. The male seal is called a bull and weighs between 500 and 700 pounds. The female seal, called a cow, weighs 50 to 100 pounds and bears one offspring per year, which is referred to as a pup, whelp, or calf.

The bulls arrive at the breeding grounds, called **rookeries**, in May or June to stake out their territories, which are about 40 feet in diameter. When the cows arrive in July, they immediately join a harem and bear their young. The cow mates again one to two weeks after the birth of the pup. The bull is always being challenged by other bulls for his territory and usually must be at least 10 years old or older before he develops enough strength to defend his territory.

The main predators of the seals include sharks, killer whales, and parasitic roundworms; but humans are even more dangerous. Thousands of fur seals were killed annually at their breeding grounds and on the open sea before protection and management procedures were adopted in 1911 with the formation of the North Pacific Fur Seal Convention. The Convention included Japan, Russia, Canada, and the United States.

The Fur Seal Act of 1966, as amended in 1983, 1988, 1990, 1992, and 1993, generally prohibits the taking of fur seals. Exceptions are granted for Indians, Aleuts, and Eskimos living along the North Pacific Ocean. The Act committed the United States to the Interim Convention on the Conservation of North Pacific Fur Seals signed in Washington, D.C., on February 9, 1957, as amended. That Convention provided for limited commercial harvesting and harvesting by governmental agencies in the United States and Russia. The Convention lapsed in 1984. Since the cessation of commercial operations only a subsistence harvesting of seals by Native Americans has been allowed. The subsistence hunting of seals has resulted in the taking of fewer than 2,000 animals a year since 1986. Neither commercial nor subsistence harvesting of fur seals is allowed on any other U.S. rookeries or haulouts.

Walrus

Walruses, the only tusked seals, live in the Arctic, North Atlantic, and North Pacific regions. Their bodies have developed flippers that make them excellent swimmers. During the winter and spring months, the walruses drift on pieces of floating ice; they spend their summers resting on shorelines. The main food of the walrus is clams. The Eskimo is the most common user of walruses. The meat is used for food, the hides for shelter, and the oil for lamps.

THE ESTUARINE ECOSYSTEM

Characteristics

The area where a freshwater source opens into the ocean is called an **estuary**. This transition area supports a variety of life that is found nowhere else, life that can withstand rapid changes in salinity, temperature, and density. The estuary continuously receives fresh water from the rivers and streams and salt water from the tides and currents.

The classification of the estuarine ecosystem is usually varied because so many bases are used. The common classification is based on geomorphology, which refers to the land relief areas of the earth. This method includes estuaries from drowned river valleys, fjord-type estuaries from glaciers, bar estuaries built from low tides, and tectonic estuaries from the heavy streams pouring into the ocean.

Although the estuary is a transitional zone, it has characteristics common to neither the rivers nor the ocean:

1. The estuary is usually shallow and turbulent, which results in a high amount of dissolved oxygen in the water.

2. The tides cause the area to be nutrient rich—the rivers bring nutrients to the estuary from above and the tides bring nutrients in from the ocean; the estuary acts as a nutrient trap.

3. Because of the high level of dissolved oxygen, the bacteria count is also high, and this causes a rapid decomposition of organic wastes; the breakdown of organic materials to soluble nutrients causes plant life to prosper, and this plant life attracts large numbers of plant-eating fish.

Life in the Estuary

The life in the estuary may be grouped into three areas:

1. species that travel only a limited distance into the estuary

2. species found in both the estuary and in other parts of the ocean

3. species whose entire life cycles are in the estuary

Estuaries are economically important to marine fisheries. About 90 percent of the marine fish harvested by American fishers either comes from the estuarine ecosystem or passes through the ecosystem between spawning and maturity.

The best-known marine life in the estuary, along with fish larvae, are oysters, clams, lobsters, crabs, and shrimp. A major problem existing with respect to estuaries is the conflict between land developers and fishers. Shorelines near estuaries have become one of the most sought-after types of real estate at the present time. Some estuaries have also been used as dumping grounds, which destroys the natural habitat of the area.

ARTIFICIALLY CULTIVATING THE OCEAN

Although our knowledge of the complexities of the ocean is limited, scientists have developed ways to artificially propagate marine animals. This form of agricultural production is called aquaculture, and the most commonly "farmed" marine resource "crop" is the oyster. Experiments have also been conducted with shrimp, salmon, and milkfish; but success seems some time away.

Oysters

As we saw earlier, oysters live in estuarine waters close to shore, making them one of the easiest marine animals to raise. We can start by improving their habitat. This may include providing a place for the larvae to attach, controlling parasites, and keeping their enemies, such as the starfish and oyster drills, away. French scientists have even developed special algae as an improved food for the oysters to grow more quickly.

An example of one of the most productive oyster farms is Japan's Inland Sea near Hiroshima. The water flow and mineral content are correct, making this a very desirable oyster-producing area. The only management item needed is a stopping apparatus to which the larvae can attach. The Japanese oyster farms consist of bamboo poles tied together and floating on barrels. Wires are hung from the poles to a depth of 20 feet. The wires hold clam shells, which act as the stopping place for the oyster larvae. In July and August, billions of oyster larvae attach to the artificially hung shells. The oyster larvae are thinned to the correct amount. In midseason, around

October, the wires are pulled and the young oysters are cleaned and brushed, which helps them grow. The oysters are harvested in January and February.

The standard oyster raft measures 35 to 40 feet square. An individual farmer may own over 100 rafts. The number of barrels needed to support the raft increases as the oysters grow and weigh it down. The oysters feed on native plankton; therefore, the costs are low. The average yield amounts to about 13,000 pounds of oyster flesh per acre farmed. The shells of the oyster are then cleaned and sold for poultry grit and lime.

Japan can attribute its successful oyster farming both to low labor costs and to the preservation of its natural oyster waters. Pollution of natural oyster waters has almost stripped every other nation of an economically feasible oyster industry.

OCEAN LAWS AND REGULATIONS

Who owns the ocean? How many miles offshore can a country lay claim to? Who owns the mineral resources under the ocean's floor? These questions face the leaders of many countries bordering the oceans.

What are the major difficulties involved with developing and enforcing world fishery policy? The first problem stems from the fact that fishing involves the catching of mobile animals. Fish cannot be claimed as property as land wildlife can. The second problem involves the issue of how much marine resource an individual nation should harvest. Most recent legislation is aimed toward conservation problems rather than to the political and economic questions addressed by the early regulations and treaties.

Due to the long history of international disputes regarding ocean resources, a variety of commissions and committees have been formed. Most of these bodies deal with the regulation of individual resources such as the fur seal, whale, or tuna. The major examples are the Inter-American Tropical Tuna Commission (IATTC), the International Commission for Northwest Atlantic Fisheries (ICNAF), the International North Pacific Fisheries Commission (INPFC), and the International Whaling Commission (IWC), which was discussed earlier. Most commissions are formed only after a particular resource develops serious problems such as low population numbers. In addition, the commissions are only as strong as the member nations want them to be. The most recent attempt to develop ocean regulations is the United Nations' "Law of the Sea" bargaining conferences. These conferences involve 158 nations with interests in ocean resources. The first conference was initiated in 1958 to discuss international fishing limits.

The United Nations Convention on the Law of the Sea entered into force in November 1994. Today 158 nations are parties to the Convention, and 133 have formally signed the Convention. The United States is a party to the Convention but had not formally ratified it as of May 2000.

The provisions of the Convention are as follows (*Source:* United Nations, Division of Ocean Affairs and the Law of the Sea):

■ Coastal States exercise sovereignty over their territorial sea which they have the right to establish its breadth up to a limit not to exceed 12 nautical miles. Foreign vessels are allowed "innocent passage" through those waters.

■ Ships and aircraft of all countries are allowed "transit passage" through straits used for international navigation. States boardering the straits can regulate navigational and other aspects of passage.

■ Archipelagic States, made up of a group or groups of closely related islands and interconnecting waters, have sovereignty over a sea area enclosed by straight lines drawn between the outermost points of the islands; all other States enjoy the right of archipelagic passage through such designated sea lanes.

■ Coastal States have sovereign rights in a 200-nautical mile exclusive economic zone (EEZ) with respect to natural resources and certain economic activities, and exercise jurisdiction over marine science research and environmental protection.

■ All other States have freedom of navigation and overflight in the EEZ, as well as freedom to lay submarine cables and pipelines.

■ Land-locked and geographically disadvantaged States have the right to participate on an equitable basis in exploition of an appropriate part of the surplus of the living resources of the EEZs of coastal States of the same region or subregion. Highly migratory species of fish and marine mammals are accorded special protection.

■ Coastal States have sovereign rights over the continental shelf (the national area of the seabed) for exploring and exploiting it. The shelf can extend at least 200 nautical miles from the shore, and more under specified circumstances.

■ Coastal States share with the international community part of the revenue derived from exploiting resources from any part of their shelf beyond 200 miles.

■ The Commission on the Limits of the Continental Shelf shall make recommendations to States on the shelf's outer boundaries when it extends beyond 200 miles.

■ All States enjoy the traditional freedoms of navigation, overflight, scientific research and fishing on the high seas. They are obliged to adopt, or cooperate with other States in adopting, measures to manage and conserve living resources.

■ The limits of the territorial sea, the exclusive economic zone, and continental shelf of islands are determined in accordance with rules applicable to land territory, but rocks that could not sustain human habitation or economic life of their own would have no economic zone or continental shelf.

■ States bordering enclosed or semi-enclosed seas are expected to cooperate in managing living resources, environmental and research policies and activities.

■ Land-locked States have the right of access to and from the sea and enjoy freedom of transit through the territory of transit States.

■ States are bound to prevent and control marine pollution and are liable for damage caused by violation of their international obligations to combat such pollution.

■ All marine scientific research in the EEZ and on the continental shelf is subject to the consent of the coastal State, but in most cases they are obliged to grant consent to other States when the research is to be conducted for peaceful purposes and fulfills specified criteria.

■ States are bound to promote the development and transfer of marine technology "on fair and reasonable terms and conditions," with proper regard for all legitimate interests.

■ States Parties are obliged to settle by peaceful means their disputes concerning the interpretation or application of the Convention.

■ Disputes can be submitted to the International Tribunal for the Law of the Sea established under the Convention, to the International Court of Justice, or to arbitration. Conciliation is also available and, in certain circumstances, submission to it would be compulsory. The Tribunal has exclusive jurisdiction over deep seabed mining disputes.

We once thought of the world's oceans and the ocean's harvest as boundless. Total marine harvests of all kinds may well be reaching the limits of sustainability. In fact, many scientists fear that humans are already harvesting more of the most popular marine species than the world's oceans can replace. According to a recent *Washington Post* article, a big bluefin tuna can weigh as much as 1,500 pounds and sell for as much as $20,000 at the dock. The fish is considered a delicacy in Japan and can bring as much as $350 per pound in the most exclusive restaurants in Tokyo. Less exotic catches, such as shrimp, salmon, and grouper, bring prices as high as $10 to $20 per pound in U.S. fish markets.

At the same time people have developed an increasing desire for marine products, our technology has advanced dramatically. The *Washington Post* article indicated that about

$92 billion were spent to operate the world's fishing fleet in 1989. The result of such huge expenditures is a world fishing fleet that is taxing the oceans' ability to replace the fish and other marine species that are harvested.

An encouraging trend is that aquaculture production has more than doubled in the decade from 1988 to 1998. Table 25-1 shows the total annual worldwide aquaculture and commercial fisheries output for 1988 through 1998. Aquaculture production includes all species: aquatic animals, aquatic plants, crustaceans, diadromous fishes, freshwater fishes, marine fishes, and mollusks. Commercial catch includes all marine and freshwater species: aquatic animal products, aquatic animals, aquatic plants, crustaceans, diadromous fishes, freshwater fishes, marine fishes, mollusks, whales, seals, and miscellaneous species.

TABLE 25-1 World Fisheries Production (in million metric tons)

Year	World Aquaculture Production	World Marine Fisheries Catch	Grand Total
1988	15.5	89.8	105.3
1989	16.5	90.5	107.0
1990	16.8	86.8	103.6
1991	18.3	85.6	103.9
1992	21.3	86.4	107.7
1993	24.5	87.5	112.0
1994	27.8	92.6	120.4
1995	31.3	92.8	124.1
1996	34.0	94.7	128.7
1997	36.0	94.9	130.9
1998	39.4	87.4	126.8

(Source: United Nations Food and Agriculture Organization [FAO])

SUMMARY

The ocean is a vast resource covering 70 percent of the earth's surface. Its movements and characteristics determine which marine resources prosper and which fade away. The ocean can no longer be regarded as a giant dumping ground. The potential of the ocean for providing food must be examined. The steps we take to harvest this most important resource must be economically and ecologically sound. We cannot allow any mistakes or miscalculations.

DISCUSSION QUESTIONS

1. List and explain the characteristics of the zones of the ocean.
2. How is salinity measured?
3. What are the common types of waves?
4. How do tides differ?
5. What role does plankton play in the biological ocean?
6. Explain the life cycle of the salmon.
7. What is meant by each of the following?
   - ■ bait fishing
   - ■ long-lining
   - ■ purse seining
8. For what is menhaden used?
9. What is the life cycle of shrimp? of the oyster? of the lobster?
10. How are lobsters caught?
11. What are the three major groups of baleen whales?
12. What is the life cycle of the whale?
13. Explain modern whaling techniques.
14. What are rookeries?
15. How are fur seal harvests controlled?
16. What are the major uses of walruses?
17. What is an estuary?
18. What are the characteristics of the estuary ecosystem?
19. What is the most common form of marine life found in the estuary?
20. Explain how oysters are artificially "farmed" from the ocean.

SUGGESTED ACTIVITIES

1. Prepare a report on your favorite marine resource. What is its life cycle, its economic importance, and its future potential?
2. Organize a class debate on the topic: "Resolved—that the killing of all whales should be prohibited by international law."

26 FRESHWATER FISHERY MANAGEMENT

OBJECTIVES

After reading this chapter, you should be able to

- explain the zones of the lake and the habitat of each
- discuss the uses and management of a farm pond
- list the characteristics of the common freshwater fish
- explain the main management procedures for freshwater fisheries

TERMS TO KNOW		
littoral zone	embankment pond	fish sampling
limnetic zone	excavated pond	
acre foot	watershed	

As a child, you may have dreamed of sitting on the bank of a deserted pond, fishing pole in hand, and just watching the bobber bouncing up and down with the waves. Mark Twain, in his classic *Tom Sawyer*, painted this vivid picture. Freshwater fisheries have come a long way since these early times. It has been estimated that over $1 billion are spent each year on bait, on food and lodging, and on rods, reels, and other equipment as well as on transportation for freshwater sport fishing in the United States alone. The fish of our freshwater habitats have been affected by increased population movements and continued pollution of freshwater supplies. In this chapter, we examine the lake habitat, farm pond construction techniques, common freshwater fish, freshwater fish habitats, and common management techniques.

THE LAKE HABITAT

The lake, depending on its size, can be divided into three zones: littoral, limnetic, and profundal. This zone classification is based on the depth of the lake. The **littoral zone** is the shallow zone and contains rooted vegetation, such as pond lilies and cattails. This zone starts at the shoreline and extends to where the rooted vegetation ends. Photosynthesis takes place on the floor of the lake because light penetrates the water. Some small ponds have only this zone. One of the most important organisms in this zone is the tiny greenish-brown plankton. They may be plantlike (phytoplankton) or they may be animal-like (zooplankton). These tiny organisms provide food for many varieties of fish in the lake. The littoral zone also contains a vast amount of aquatic life, including both invertebrates and vertebrates. Invertebrates are animals without spines. Invertebrates include insects, mollusks, arachnids, crustaceans, and others. Vertebrates are animals having spines. Examples include mammals, amphibians, reptiles, birds, and fish. (See Figure 26-1.)

In the **limnetic zone**, rooted vegetation is no longer present. Photosynthesis, however, reaches to the bottom of the zone. An abundance of phytoplankton is usually present, giving off oxygen in this area, and fish are very characteristic.

The bottom zone of the lake is the profundal zone. Because sunlight cannot penetrate to this zone, no photosynthesis takes place. This zone is the warmest in the winter and the coolest in the summer. Bacteria is the most common organism in this zone, and the bacteria continually decompose wastes that settle from the upper zones.

FARM PONDS

A farm pond is a good way to supply a habitat for fish. The main uses of a farm pond are livestock watering, irrigation, fish production, fire protection, recreation, and wildlife.

FIGURE 26-1 Workers harvesting catfish at the Delta Pride Catfish Farm in Mississippi.

Water is a very important part of the livestock diet. Ponds for livestock should be available in each pasture or grazing unit. The livestock should never have to travel more than one quarter mile to water. Irrigation from ponds is usually limited to home gardens or small acreages of special crops such as nurseries. The amount of water needed for spraying field and orchard crops to control insects and plant diseases usually is small, but it must be available when needed. (See Figure 26-2.)

The farm pond also can yield a dependable supply of fish and provide recreation for the family. A pond with a surface area as small as one quarter acre can be managed for fish production. The USDA indicates that the ideal family farm pond is less than two acres.

If the pond is located close to the farmstead, it also can be used for fire-fighting purposes. As little as one tenth of an **acre foot** of water will provide a good fire protection water source. (An acre foot is defined as the equivalent of 1 acre in area and 1 foot deep. An acre is a unit of measure commonly used to denote land area. It is any rectangular surface area of 43,560 square feet. It is equivalent to a square area 209 feet in length on each side.) This amount of water will supply 250 gallons/minute for more than two hours.

The farm pond can also provide the family with a place for recreation such as boating and swimming. Again, a one-half-acre pond is ideal for this purpose. If the pond is to be open to the public for a fee, it needs to be at least 5 acres in size.

A pond can also be used to attract and support many kinds of wildlife. Game birds and migratory waterfowl often use ponds as resting places. Ducks breed near ponds if there is an ample supply of good food and enough surrounding vegetation for protection.

FIGURE 26-2 A small farm pond in Tennessee.

The water in a pond can be used for more than one purpose, such as for livestock, fish production, and irrigation. Some multiple-use practices may not be compatible with each other. Using a pond for swimming might make it less usable for irrigation. The water for both uses is most important during hot, dry weather, and this might not be practical.

Management of the Farm Pond

The exact beginning management techniques will depend on whether a new pond is being built or an old existing pond is being renovated.

In new pond management, we are concerned with new pond care, pond area development, what fish to stock, and correct fish protection techniques. Old pond management usually is concerned with correcting problems in the pond—aquatic plants, muddy water, leaking, fish diseases and parasites, and animals such as turtles, muskrats, and groundhogs. It is important to note that even new ponds may have all the problems that were mentioned previously for existing ponds.

There are basically two kinds of artificial ponds, **embankment ponds** and excavated ponds. In embankment ponds, water is impounded behind an earth embankment or dam built across a watercourse. They are generally suited to areas where slopes range from gentle to steep and where there is enough water in the watercourse so that the ponded area is at least 6 feet deep. **Excavated ponds** are made by digging a pit below the surrounding ground level. They are built primarily in nearly level areas that are not suitable for embankment ponds. Excavated ponds are best suited to uses for which a small pond is adequate.

Once the kind of pond needed is selected, a site must be considered. There are several points to keep in mind when the selection process is conducted. The most important items to remember are

- watershed care
- the dam and spillway
- pond basin
- pond banks

The **watershed** is that land that drains into the pond. It should be surrounded by some sort of permanent vegetation buffer strip before the pond is filled. This allows the pond to be filled with clean water. The fish stocked will grow much better in clean water. Usually, the cover should be at least 100 feet wide around the pond and should consist of orchard grass or a grass-legume mixture. If the watershed area filling the pond is not the correct size, the pond will not fill to the correct level.

The dam and spillway should be covered as quickly as possible with a fast-growing grass. Do not plant trees or shrubs on dams because the roots can weaken the structure. The spillway will pass the overflow water when the dam is full. The most common arrangement is to establish a vegetated earth spillway around one end of the dam and install a pipe spillway through the dam. (See Figure 26-3.) A spillway 10 to 50 feet wide is generally acceptable. The general rule is usually to make the spillway 3 feet or more below the top of the dam. This way, the water will never flow over the dam.

The pond basin is the area to be flooded. It is recommended that cover crops be planted in the basin before it is flooded. In late summer or fall, rye or wheat can be planted; in spring or early summer, oats; in midsummer,

FIGURE 26-3 Spillway of Mississenewa Reservoir, Indiana.

Sudan grass. If the cover grows to 6 inches or more before flooding, it needs to be mowed. There is no need to remove these clippings from the basin; they add nutrients for fish food. Super phosphate can also be added on the pond basin.

Pond banks, like the watershed, need to be protected from washing rain. Millet may be planted on pond banks in a strip surrounding the pond. The cover may be extended even to areas that eventually will be covered with water. This prevents a wash-in once the pond is full and the water's wave action tears the soil loose.

Pond Development

Many ponds are developed for wildlife, recreation, and fish production. Wildlife will be more comfortable if there is cover present. Trees and shrubs may be planted around the pond to protect the cover. Windbreaks can help stop wave erosion and provide cover, food, and nesting areas for many species of wildlife. Care should be taken not to place the trees too close to the shoreline, which interferes with fishing and causes spillway damage by the roots.

The pond can be used for many kinds of recreation. Swimming, fishing, boating, and ice skating are a few activities farm pond owners enjoy. Safety, comfort, and beauty should be considered when planning and developing the area. Safety equipment should be available to everyone to use. An approved life ring attached to a long rope is excellent. Warning signs at danger points along the shoreline where drop-offs of 8 to 10 feet are located are also necessary. A rope barrier is even better. Mark the safe swimming areas with a buoy system. A long pole or wooden plank painted white should be kept close to the pond. Swimming beaches should be sloped gradually to the deep waters,

FIGURE 26-4 A typical farm pond.

with 6 to 12 inches of sand on the floor. The areas of heavy use should be well groomed. The addition of a picnic table, fireplace, or shelter might add to the beauty of the pond.

A farm pond should be stocked with both the right kinds as well as the right number of fish for its size. Deciding the purpose of the fish (i.e., food or sport) is also important. If the purpose is food, then the pond could be stocked with channel catfish. For sport fishing, a smallmouth bass-redear sunfish combination may be the best. The most common stocking that provides both food and sport is the largemouth bass, bluegill, channel catfish combination. All three species are harvestable and provide good fishing and eating. (Figure 26-4 shows a typical farm pond.)

Using a Farm Pond as a Production Site

As the demand for pollution-free fish increases, the economic feasibility of producing fish in a controlled environment also increases. If the farm pond is properly constructed and maintained, the landowner may consider producing fish, such as catfish, in floating cages. (See Figure 26-5.) The pond can produce an additional

FIGURE 26-5 Floating catfish cage.

income for the landowner without destroying the environment. However, the landowner must be more aware of water quality necessary for fish production, as well as keeping abreast of problems due to watershed runoff. It is critical that the landowner ensure that there is a market for the fish before investing in fish production.

COMMON FRESHWATER FISH

To produce good fishing, it is important to know something about the fish. Some common freshwater fish include

- largemouth bass
- bluegill
- channel catfish

Largemouth Bass

The largemouth bass, a common fish stocked in farm ponds, can be recognized by its large mouth and dark stripe or blotches down the side. The bass is preferred by many fishers because of its fighting abilities. They are well adapted to ponds and spawn readily in shallow water.

The diet of the bass consists of aquatic insects and fish. When available, they will eat frogs and crayfish. The adult bass feed by day or night, but their main feeding periods are early morning and late evenings.

The bass prosper when food supplies are good. It usually takes about 3 years for a bass to reach 10 to 12 inches long, but they can grow 3 to 5 inches in their first summer. The current world bass record is 22 pounds, 4 ounces, caught in Georgia. The largemouth bass reach spawning maturity in the spring of their third summer. The males fan out depressions (nests) with their body and fins, and they lure the females over the nest to deposit their eggs. Nesting begins when the water temperature reaches 60° to 65°F (16° to 18°C). Approximately 2,000 to 7,000 eggs are produced per pound of fish. The male will then stand guard for 8 to 10 days or until the eggs are hatched.

Bluegill

The bluegill is most commonly used in combination with the largemouth bass. Bluegill are a food source for the bass as well as good fishing. The bluegill is most commonly caught in the middle of the day when the bass are not biting.

The bluegill has a short head, small pugged mouth, and an irregular blackish spot located at the base of the top soft fin. Other names given to the bluegill include sunfish, perch, pond perch, bream, and brim. The young bluegill eat tiny plankton in the water. As they mature, they feed on aquatic insects, snails, small crayfish, and occasionally small fish.

The bluegill usually average less than a pound in weight, but three-pounders have been recorded. Their growth is dependent upon the amount of food available, temperature, and number of fish in the pond. On the average, it takes at least three or more growing seasons to produce an adult bluegill. Their life span averages 5 years, but bluegills can live up to 10 years.

The bluegill spawn from May to August. They are frequently sexually mature in their second year or when they are about 3 inches long. During spawning, the male constructs a saucer-shaped nest when the water reaches 70° to 75°F (21° to 24°C). The female deposits eggs in the nest. The male fertilizes the freshly deposited eggs and remains to guard the nest. The female usually deposits 12,000 to 50,000 eggs depending on her size, and the eggs hatch in two to five days depending on water temperatures. There may be more than one spawning bed in the pond.

Channel Catfish

Channel catfish are a favorite fish in the farm pond. They are usually stocked in combination with largemouth bass and bluegill. They eat a wide variety of foods ranging from fish to aquatic plants. The catfish feed off plant life, insect larvae, and frogs until they reach 12 to 14 inches in length; then they feed on small bluegill fish. The channel catfish grow to 11 or 12 inches the first year, reaching 14 to 15 inches after 2 years and 16 to 18 inches after 3 years. Channel catfish may live to be 12 to 15 years old, but average life spans are 6 to 8 years.

The channel catfish can reproduce in ponds if a place is provided for spawning. In clear ponds, the catfish eggs and young are eaten by bass and bluegill. If only channel catfish are stocked in the pond, then fish population management must be used to maintain the population level at no greater than 60 or 70 adult channel catfish per acre.

FISHERIES MANAGEMENT

Fishery management techniques are necessary if the pond or reservoir is to survive for long periods of time. It is necessary for the fishery biologist to know both the population and habitat of the area. The main management procedures include

- vegetation control
- fish sampling
- population removal and adjustment
- fertilization techniques
- fish regulations
- water-quality control

Vegetation

Vegetation is an important part of any lake, pond, or reservoir. It is important in providing food, shelter, oxygen, and spawning and nesting habitats. Vegetation can also cool surface waters, trap excessive nutrients, and stabilize bottom sediments. When the aquatic vegetation grows too much, fish experience problems.

An overabundance of aquatic vegetative growth can create problems in the multiple-use aspect of water management. In addition to harming fish, too much vegetation can greatly impair the use of the water for activities such as recreation. The desirable fish population decreases and smaller, unwanted fish are able to find refuge in the cover. The excess plants compete for nutrients in the water with desirable plant life such as phytoplankton, a principal food for the bluegill. Some plants, such as blue-green algae, produce a poison harmful to fish populations. With high amounts of plant life, the dissolved oxygen in the water also tends to be lower. Fish exposed to a low-level oxygen supply are more susceptible to disease and parasitic infections.

There are many ways to control the aquatic plant population in our waters. The most effective ways are watershed improvement, water level management, biological control, and chemical controls. If we can keep the soil and nutrients on the land and out of the water, aquatic vegetation is controlled. One of the most common practices that creates a vegetation problem is the spreading of manure on frozen land. Many nutrients are washed into the water before they are incorporated into the soil. As was discussed in the chapter on water, nutrients in water can cause excessive plant growth.

Water level management involves changing the level of the water to expose plants to adverse conditions. You may want to lower the water level in the winter and expose the weeds to freezing, drying, and bottom compaction. If the weeds freeze in the ice, water can be pumped under the ice and weeds, which will be ripped from the bottom of the lake when the ice rises, then can be removed.

Biological control involves using some other living organism, either plant or animal, to control aquatic plants. This method may be potentially dangerous. The organism introduced may become more of a problem than the original weeds. The introduced organism should destroy the weeds, breed and maintain itself, provide long-term weed control, and be legal to import across state boundaries. Examples are crayfish, turtles, ducks, geese, and snails.

Chemical controls involve using herbicides on weed-infested waters. This practice is highly dangerous and should be used only as a last resort. It should also be mentioned that chemical treatments are only temporary. If the excess growth is from an overabundance of nutrients in the water, then chemicals will only kill the existing weeds. New weeds will follow after the effects of the chemical wear off.

Fish Sampling

Fish sampling is a major procedure in keeping track of what species are in the managed water as well as how well the sport species are growing. It is virtually impossible to inventory the actual fish in a pond, lake, or reservoir. Therefore, a plan must be used to get an accurate sample of the fish life in the water. When a problem develops, it is the biologist's job to examine both its causes and its solutions. Some lakes might seem to be fished out, whereas they might contain only stunted, overcrowded varieties.

There are many methods of fish sampling, and each should be used for the specific fish you want to obtain data on. The main methods are nets, spot poisoning, boat shocking, and angling.

Various kinds of nets may be used in a fish sampling program. Gill nets are fine nets that entangle fish; they are effective in catching herring and trout species. Trammel nets are used in a current or around a school of fish. These nets trap fish that can be frightened into the net. Examples of fish sampled with this method are carp and catfish. Seines are nets held on the bottom by weights and on the surface by floats. Seines encircle an area and trap fish within them. This method is used annually to record growth increment in the pond. Seines are used to manage bluegills and sunfish.

Another method of fish sampling yielding fairly accurate results is that of spot poisoning. This method involves poisoning an inland area of a larger body of water. If possible, the bay is blocked by a seine apparatus. The poisoning is usually done at night when the fish are moving in the shallow parts of the lake. The fish are counted, measured, and weighed. The data are recorded in the yearly record log of the lake.

Using a boat shocker as a means of fish sampling involves applying alternating current into the water to stun the fish. This method is fairly selective in the fish stunned and counted. It is most effective in shallow water areas. It is not as effective in sampling catfish and bullheads, because they go to the lake bottom when stunned. Stunned fish usually revive after a minute or two. Occasionally, fish are killed to determine their eating habits by examining stomach contents.

Angling fish with a fish hook and line is an effective method of fish sampling for members of the bass family. Nets are usually avoided by large- and smallmouth bass. The use of a fly rod and artificial popper can sample more bass than a week's work with nets.

Population Removal and Adjustment

Many times a lake or pond will become completely populated with undesirable species of fish. They may be suckers, shad, or small bullheads. Many of these species travel lake bottoms in search of food, which stirs up the silt and makes it an undesirable habitat for other fish. Species such as crappie can overpopulate small ponds, crowding out other species. To gain control of the pond, an adjustment of the fish population or a complete removal of fish may be called for.

Most ponds, lakes, or reservoirs are equipped with a drain. It may be necessary to completely drain the lake or pond. This method is not used for reservoirs. An apparatus called the "wolf-type weir" is placed at the drain to catch all the fish. The desirable species are kept and used for restocking purposes, the undesirable species are destroyed. Using this method saves many good fish. Some fish populations are removed by poisoning the entire lake with a compound of rotenone or sodium sulfite. Chemical application kills all fish, both good and bad. It should be avoided if possible.

Management often may call for only an adjustment in population. Usually, adjustment rather than population destruction is recommended when there are desirable but stunted fish populations present, when destroying the population is impractical (such as in a reservoir), and when there is a low demand for a certain species of fish.

Adjustment procedures used include the use of nets, water level alteration, and poisoning. In each case, only a selected small portion of the population is changed.

Fertilization Techniques

Fertilizing ponds to increase fish production had its beginning in the southern part of the United States where the soil is less fertile. The annual fertilization program consists of applying eight to ten treatments of a commercial fertilizer. The fertilizer causes an increased production of plankton, a principal food source of fish. Nitrogen, phosphorus, potassium, and lime are the principal nutrients applied.

Applying fertilizer to a pond or lake poses many dangers. Many of the results of fertilizer programs have been unpredictable. The most common problem is the excess growth of vegetation. As was previously discussed, the vegetation causes many problems. With the increased plant growth, there is an abundance of dead, decomposing organic matter, which can tie up the oxygen in the water and release toxins. In the winter, the bacteria decomposing the waste use most of the oxygen, thus increasing the incidence of winterkill. Therefore, this management practice is not recommended in areas outside the southern United States.

Fishing Regulations

Most states have enacted regulations controlling the fish taken from their public waters. Most of the regulations are expressed in the size of fish taken. Some fish, such as coho salmon, are regulated by limiting the number caught. In both instances, the goal of the regulations is to provide the best possible environment to desirable fish.

Water Quality

Good-quality water is essential for successful fish production. The main factors associated with water quality include temperature, oxygen, acidity, and muddiness.

Fish grow best if the temperature is above 65°F (18°C). The most common way to control water temperature is to take any overflow from the cold bottom waters. This allows the sun-heated top water to remain in the pond.

All animal life in the pond requires oxygen. Oxygen is added to the water as a by-product of photosynthesis. Oxygen deficiencies commonly occur as a result of the action of bacteria. Oxygen can be added to the water by aeration or sprinkling. An outboard motor churning the water adds dissolved oxygen to the water. Water may also be pumped into the air to increase the oxygen content.

The acidity of water is measured by its pH. A pH of 4.0 is poisonous to fish. Spawning will not occur in waters with a pH of 5.0. The ideal pH appears to be between 6.5 and 9.0. To correct an acidity problem, limestone must be added to the pond. The waters can become too acidic if the fertilizer is not correctly regulated or acids, such as sulfuric acid, are washed in from the watershed.

Muddy water is a problem affecting every part of the pond. Fish are stunted if they grow in water with a bad siltation problem. A good rule of thumb is this: Insert your arm into the water up to your shoulder; if you can still see your fingers, the water is clean enough to promote fish production. Muddiness is usually dependent upon how much sedimentation enters the lake as well as how much activity, such as fish feeding, takes place at the bottom of the water. A simple test to find out where the muddiness originates is to take a sample of water in a clean half-gallon jar. If the mud does not settle out, then water chemistry is the problem. If the mud settles out within a few days, bottom activity is stirring up the water.

SUMMARY

Our freshwater fishery resources are an important asset to our environment. The farm pond can provide much enjoyment for the family. A larger lake or reservoir can provide flood control as well as recreation.

A functional pond, lake, or reservoir must be managed like any other resource. The most common management techniques include controlling the vegetation, keeping a record of growth through fish sampling, adjusting and removing unwanted species, fertilizing when necessary, obeying fishing regulations, and maintaining a high quality of water.

DISCUSSION QUESTIONS

1. What are the three zones of a lake? What are their characteristics?

2. What are the common uses of the farm pond?

3. What are the differences between embankment ponds and excavated ponds?

4. How is a pond developed for recreational activities?

5. What are the food requirements of largemouth bass, bluegill, and channel catfish?

6. How is vegetation growth controlled in a pond or lake?

7. Explain the common fish sampling techniques.

8. How are unwanted fish removed from an area?

9. Why would you want to fertilize a lake?

10. How are fishing regulations determined?

11. What temperature is the best for fish production?

12. At what pH level do fish grow best?

13. How do you determine if the water is too muddy to produce fish?

SUGGESTED ACTIVITIES

1. Find out how many farm ponds are in your area. How many total acres of farm ponds are involved? What are the common freshwater fish produced?

2. Study a catfish farming operation. How is it operated? What is its profitability?

27 CAREERS IN FISH AND WILDLIFE MANAGEMENT

OBJECTIVES

After reading this chapter, you should be able to

- compare opportunities in fish and wildlife management, comparing nature of work, salaries, and requirements

TERMS TO KNOW

conservation officer	animal control biologist	zoologist
game biologist	fishery biologist	
fish and wildlife technician	fish culture technician	

Public concern over the mismanagement of fish and wildlife resources has led to greater opportunities in the management field. Careers in fish and wildlife management fall into two large areas: fish and wildlife technicians, and fish culture technicians. Employment is by federal and state governments, universities, industry, zoos, museums, and the armed forces.

CONSERVATION OFFICER

The governmental watchdog of the game management program is the **conservation officer** or game warden. The role of the conservation officer is one of great importance. He or she is responsible for enforcing and upholding all federal and state statutes, with particular attention to those related to the conservation of natural resources. In most states, the conservation officer has attended the training academy of his or her own department as well as the law enforcement academy for police personnel. The officer must possess a knowledge of state laws; Department of Natural Resources regulations, rules, and directives; and sporting practices such as hunting, fishing, and trapping. The officer must also be able to operate such equipment as autos, trucks, snowmobiles, all-terrain vehicles (ATVs), fire-fighting trucks, and motorboats. Because the officer works with people, he or she must be able to speak publicly and to deal effectively with people. The officer must be able to use two-way radios, fingerprint kits, cameras, drug-testing kits, firearms, and sporting equipment.

The conservation officer's duties include

- enforcement of laws, regulations, and rules dealing with fire prevention and permits, and with licenses for hunting, trapping, shooting, boating, snowmobiles, and all-terrain vehicles

- enforcement of criminal and traffic laws
- aiding and instructing the general public on such matters as water safety, firearms, boating, camping, fishing, hunting, and trapping
- patrol by foot, boat, car, ATV, and plane of reservoirs, lakes, rivers, and streams on state and federal forests and lands, as well as in rural public areas
- providing information and explaining laws, rules, and regulations to the public
- aiding farmers and other citizens in planning and implementing conservation projects
- making arrests, issuing citations and warnings, and appearing and testifying in court, as well as obtaining, preparing, and submitting evidence
- performing public relations duties such as public speaking and attending meetings, fairs, sporting events, and related activities
- conducting investigations of swimming, boating, automobile and other accidents, as well as thefts, vandalism, wildlife, and pest damages
- assisting in emergencies such as floods, storms, tornadoes, and forest and grass fires

GAME BIOLOGIST

The second person responsible for working with game resources is the wildlife or **game biologist**. Unlike the conservation officer, the wildlife biologist deals with the management practices of game rather than the un-

lawful activities against game. The employment of the biologist may take him or her to national or state parks, wilderness areas, or reservoirs. The duties vary with location.

The most important duty of the wildlife biologist is the planting of food plots. The biologist must determine what crop goes where so the most possible game will benefit. Most areas issue low-cost leases for governmental parks or reservoirs to local farmers. In return, the farmers leave a portion, usually 10 percent, of the crop for wildlife purposes. The biologist must then evaluate each food plot for future years.

The wildlife biologist also keeps records of the game harvests in the area. This is accomplished by maintaining hunter check-in stations. The stations are positioned around the park or reservoir for hunters to record the game taken or seen. The biologist then looks at wildlife populations to determine the game management procedures to follow. (See Figure 27-1.)

In addition to food-plot plantings and hunter check-in stations, the wildlife biologist is responsible for pond construction, property boundary maintenance, nesting structure construction, fish surveys, equipment maintenance, duckblind checks, wildlife surveys, and the management of workers. They also must answer questions about wildlife from the public.

Wildlife biologists require education beyond the high school level and must like the outdoors. The biologist must work in all extremes of weather. The wildlife worker needs only a high school diploma and a love of the outdoors.

FISH AND WILDLIFE TECHNICIAN

A career as a **fish and wildlife technician** is for persons who like working with animals.

FIGURE 27-1 Career areas: Wildlife biologist/ manager/technician. Habitat analysis and game counts are techniques used by wildlife specialists to advise government agencies on setting hunting and fishing limits. *(Courtesy USDA Agriculture Research Service)*

FIGURE 27-2 Management of deer is an important part of the fish and wildlife technician's job. *(Courtesy USDA)*

Most of the time, technicians examine methods that could conserve and aid animals having trouble surviving in changing habitats. A fish and wildlife technician looks at the causes of disease and pollution and works to establish hunting and fishing regulations. (See Figure 27-2.)

Nature of Work

One who works as a fish and wildlife technician will often specialize in one of the diverse positions available in this broad area. A refuge manager is concerned with those species of wildlife that are considered endangered or rare. He or she spends most of the time developing habitats and food crops favorable to the protected species. The manager must be able to work with people, as most of the refuge areas are open to the public. The federal government employs most of the refuge managers through the United States Bureau of Sport Fisheries and Wildlife.

The game biologist is concerned mainly with birds and mammals. The biologist works with the problems of predators and economic loss of wildlife. He or she studies and examines past history, habitats, and distribution of the wildlife. Game biologists develop management programs to propagate wildlife species. Wildlife technicians implement those plans and work with the game biologists.

An **animal control biologist** works mainly with problems caused by rodents and predators. These animals often carry disease to other wildlife species. The animal control biologist will try to find ways of keeping these harmful populations to a minimum.

A **fishery biologist** works with habitats, spawning, and artificially grown young. He or she handles stocking procedures and policies, and may work in a federal- or state-owned hatchery.

Wildlife technicians may work in the park system as park naturalists and park rangers. They study wildlife in the park, keep records, conduct tours, preserve samples for future use, and perform many other tasks.

A fish hatchery manager supervises fish breeding operations. Catfish, trout, and salmon have been artificially propagated. The manager works closely with state and federal agencies in restocking programs.

Many people work at the university level in fish and wildlife management. About two-thirds of all doctoral degree holders in this field teach at the university level. A survey taken of the fish and wildlife graduates found that 58 percent were employed by state agencies, 28 percent by federal agencies, 8 percent by universities, and 6 percent by private employers.

Requirements

A career in fish and wildlife management requires a person who likes the outdoors. He or she must be in good health and possess the ability to walk, climb, and stand for several hours. He or she must be able to lift heavy objects and withstand extreme climate conditions.

The technician must be able to make correct decisions when given a series of circumstances about a situation and should have the ability to carefully observe situations. The technician is involved with public relations, frequently talking to sports enthusiasts, school officials, and individual citizens. Along with public speaking skills, the technician must possess good writing skills to prepare the reports the job requires.

Many jobs in this area require a college degree. High school students should take classes in geometry, trigonometry, chemistry, physics, natural resources management, and physical education. College courses depend on the position desired. A fish biologist should study aquatic biology, fishery biology, oceanography, fish culture, and limnology. A game specialist should study zoology, anatomy, entomology, parasitology, ecology, and genetics. Many state universities offer courses for the fish and wildlife technician. Such programs are offered in many community colleges.

Employment Outlook

As leisure time increases for the American public, many people increase their activities in fish and wildlife areas. Even with this increased activity, the employment opportunities offered by the state and federal governments are not expected to increase. Opportunities will increase with companies interested in fish and fur. There should also be an increase in the number of fish and game clubs.

Earnings

Game biologists employed with the federal government could expect starting salaries of about $26,100 per year in 1999. Those people employed with the federal government could expect median earnings of $48,600 a year in 1999. State salaries were somewhat lower. As federal employees, the fish and wildlife specialists fall under the Civil Service Commission.

The top salaries go to those people employed as management biologists, supervisors, animal control officers, and wildlife district supervisors.

For additional information, contact

Fish and Wildlife Service
United States Department of the
   Interior
Washington, DC 20240

National Wildlife Federation
1412 16th Street, NW
Washington, DC 20036

Wildlife Management Institute
709 Wire Building
Washington, DC 20005

FISH CULTURE TECHNICIANS

The main job of the **fish culture technician** is to produce fish for either food or game. He or she may hatch fish from eggs and raise them for stocking purposes. With the increased pollution of our rivers and streams, fish numbers have greatly decreased and people interested in artificially propagating fish are more and more in demand.

There are three main areas of fish culture: fish hatchery technology, fish wildlife conservation technology, and experimental biology technology. The employee of a fish hatchery works with hatching new fish. The technicians work with both the eggs and the parent fish. They care for the young fish, clean spawn tanks, and maintain equipment. The fish culture technician will also be responsible for keeping records of the operation.

A fish wildlife conservation technician assists in the fish conservation programs. He or she is responsible for gathering data and evalu-ating programs. The technician helps in the construction of fishways and fish ladders.

An experimental biology technician assists biologists in the field and laboratory. Much of the work consists of sampling commercial fish landings and shellfish for tags. Data obtained from this work may result in new hybrid fish as well as new food items.

The basic requirement of the fish culture technician is a high school diploma. The student should specialize in biology, chemistry, physics, and mathematics. Courses in natural resources management would be useful. With the large number of reports the technician is required to write, English should be a part of the studies. Many positions require a two-year college degree for initial entry.

The technician needs to be familiar with the classification of fish and their habitats. The technician is also required to use laboratory equipment. While in the laboratory, he or she must be able to keep accurate records of the projects being performed.

The loss of many marine areas to pollution and urban/industrial growth has caused the field of fisheries management to expand. Along with commercial needs, the need for professionals in sport fishing and home aquariums has increased.

The Sea Grant College Program, administered by the federal government, provides assistance in the marine areas.

The fish culture technician may advance to a hatchery manager or fisheries management specialist. About 95 percent of the people employed are in positions with the federal government.

For additional information, write

American Fishers Society
1319 18th Street
4th Floor
Washington, DC 20036

Sport Fishery Research Foundation
719 13th Street, NW, Suite 503
Washington, DC 20005

ZOOLOGIST

A large field connected with fish and wildlife management is zoology. A **zoologist** works with many kinds of animals, both domestic and wild. A zoologist is sometimes referred to as a life scientist.

About 66 percent of the zoologists work in research and development. Most of this research takes place at universities, much of it funded by the government. The zoologist may direct research or be a research assistant. There are also supporting teams of biotechnicians, laboratory assistants, or laboratory animal technicians.

The zoologist may work with a government agency. The most common government employers are the Food and Drug Administration, the Department of Agriculture, the Fish and Wildlife Service, the Public Health Service, and the National Park Service. The main government jobs are fisheries biologist, park ranger, park naturalist, and research biologist.

There are also great demands for zoologists in zoos and museums. A curator of a museum of natural history collects and exhibits animals. Most of these people hold advanced college degrees. A zoo is a collection of live, wild animals; zookeepers handle the feeding and diet of these animals.

The employment outlook for zoologists looks very strong for the coming years. It is estimated that 9,600 more zoologists will be needed each year to meet the demands. The strongest demand will be in the area of research. As regulations become stronger, more researchers will be needed in medicine, health, and environmental control.

Earnings of zoologists are similar to those of game biologists.

For additional information write

American Society of Zoologists
Box 2739
California Lutheran College
Thousand Oaks, CA 91360

American Institute of Biological
  Sciences
1401 Wilson Boulevard
Arlington, VA 22209

SUMMARY

The employment outlook in all areas of fish and wildlife management is good. People are becoming more and more aware of the importance of fish and wildlife. They realize fish and wildlife are parts of a very complex world and their management is equally complex.

DISCUSSION QUESTIONS

1. What does a fish and wildlife technician do?

2. Describe the major duties of a
   - wildlife biologist
   - animal control biologist
   - fishery biologist
   - park technician
   - fish hatchery manager

3. What classes should a high school student take to prepare for a career as a fish and wildlife technician?

4. What was the average starting salary of a zoologist in 1999?

5. What are the main duties of a fish culture technician?

6. What are three major areas dealt with by a fish culture technician?

7. Explain the requirements for a career as a fish culture technician.

8. What are the main duties of zoologists?

SUGGESTED ACTIVITIES

1. Have a conservation officer visit the class to discuss job opportunities and requirements for his or her field.

2. Write to your state agency in charge of fish and wildlife management. Ask for career opportunities and training requirements.

3. Invite your district wildlife biologist to visit your classes.

THE WHALING CONTROVERSY

In the beginning of this book, we discussed how the usefulness of any specific natural resource changes. Whaling typifies this change in attitudes. Whale oil once provided the world with much needed products like lighting, heating, food, margarine, soaps, and lubricants. In fact, proper young ladies often wore whalebone and baleen corsets to enhance their figures. Ambergris, a substance formed in the intestines of whales, was used in cosmetics, potions, pain remedies, and perfume. Today, many see whaling as unnatural and feel that there is no need to commercially harvest whales. Technologies such as the advent of petroleum-based products and hydrogenation of vegetable oils to make margarine have dramatically decreased the need for whale products. Others argue that whaling provides meat more efficiently than agriculture in some countries and that antiwhalers are overlooking cultural values. (See Figure V-A.)

FIGURE V-A. Humpback Whale "breaching." The humpback is on the Federal Endangered and Threatened Wildlife and Plants List. All species of whales were commercially harvested in the 1800s and early 1900s before legislation and international agreements protected them. *(Photo courtesy U.S. Department of the Interior, Fish and Wildlife Service)*

THE CASE FOR WHALING

- In the past, whales were overharvested because of high demand. Today the demand comes from a small segment of the world's population who value the meat because of the historic contribution whale meat has made to the diet and culture of these populations.

- The Norwegian quota of minke whales (425 in 1996) can be harvested from a northeast Atlantic population of over 110,000 with little or no risk to the population.

■ Whaling in Japan, Norway, and Iceland uses less fossil fuel than would raising chicken, pork, or feedlot beef. To these countries, whaling **is** the most economical and environmentally friendly method of meat production.

■ Whales and humans compete for the same marine fisheries. If whales were no longer harvested, they could become a serious competitor with the seafood industry.

■ In Japan, some coastal whaling villages use whale products for more than 30 cultural events.

THE CASE AGAINST WHALING

■ Whales are sentient beings (capable of feeling pain) that are able to communicate with humans through echolocation. We have no right to take their lives.

■ Even with modern technology, death associated with whaling is still a very painful, if shortened, experience.

> *What an outcry there would be if we hunted elephants with explosive harpoons fired from a tank and then played the wounded beasts upon a line.*
> Professor Sir Alistair Hary, Zoologist
> on the Discovery Expedition

■ Whales have been overharvested for so long that stocks are depleted. The extinction of some species or populations is at risk. Humans should not cause the extinction of any species of population.

Why should the whaling controversy concern students who do not live in countries that wish to continue whaling? Should governments prohibit or promote whaling? How many of the points for or against whaling are value-based, economic-based, or culture-based? Which points are most convincing to those who share the other viewpoint? Is it possible to reason about such a value-laden issue using pure logic? You decide.

Section VI
Outdoor Recreation Resources

OBJECTIVES

After reading this chapter, you should be able to

- discuss the recreational possibilities on public land

- explain the federal government's main natural resource and recreation programs

- explain how our public lands are misused and abused

As our leisure time becomes greater and greater, the demand for recreational activities and facilities also increases. Recreation deals with an activity, or inactivity, that one wants to do rather than being paid or forced to do. How does outdoor recreation rank with the other forms of recreation? Just as people once made their way to the hustle and bustle of the city, more and more today want to escape back to the country. City dwellers frequently visit outdoor recreation areas more so than do rural people. As vacation time and vacation pay have increased, so has the money made on outdoor recreation. As a whole, rapid means of transportation, longer vacations, earlier retirements, longer lifespans, and more money to spend have been the forces behind the surge in outdoor recreation. (See Figure 28-1.)

FIGURE 28-1 Recreation on public lands is an important use of our natural resources—and a significant marketplace for goods and services.

FEDERAL PROGRAMS

Public lands have been developed by considering the main ways people like to spend their time outdoors. The government has been developing public lands for recreational areas since the founding of our country. In 1965 Congress passed the Land and Water Conservation Fund Act to provide matching funds to states wishing to develop outdoor recreation facilities. A total of $248 million was matched by the states to purchase 615,000 acres of recreational land and to fund the development projects. The money was received from entrance fees, motor boat fuels tax, sale of surplus real estate, and mineral rights on the outer continental shelf.

The federal government's main programs that combine the conservation of natural resources and recreation are

- national park system
- national forests
- national island trust
- national wilderness preservation
- national trails system
- national system of wild and scenic rivers

These lands are maintained and managed under a wide array of federal laws. The most comprehensive law is the Federal Land Policy and Management Act of 1976 (FLPMA). That act defined U.S. policy as follows: "The public lands be retained in Federal ownership, unless as a result of the land use planning procedure provided in this Act, it is determined that disposal of a particular parcel will serve the national interest." FLPMA made it clear that the federal lands are held in public ownership and are to be managed for "multiple use." *Multiple*

use is defined as "the management of the public lands and their various resource values so that they are utilized in the combination that will best meet the present and future needs for the American people."

Federal lands are managed as follows:

- Bureau of Land Management (270 million acres)
- U.S. Forest Service (191 million acres)
- Fish and Wildlife Service (91 million acres)
- National Park Service (81 million acres)

National Park System

Congress set aside Yellowstone National Park over a century ago. This was a unique step because Yellowstone was not a location of some historical event. It did not mark a major battle. It was not a famous religious site. Instead, the Yellowstone National Park Act of 1872 established a protected area "for the benefit and enjoyment of the people." This was the first time public lands had been preserved just for public enjoyment. The land was "reserved and withdrawn from settlement, occupancy, or sale under the laws of the United States, and dedicated and set apart as a public park or pleasuring-ground."

It was almost half a century before a real system of national parks existed. On August 25, 1916, the responsibility of managing federal parks was assigned to the Department of the Interior. The National Park Service Organic Act of 1916 provided, "There is created in the Department of the Interior a service to be called the National Park Service, which shall be under the charge of a director. The service thus established shall promote and regulate the use of the Federal areas known as national parks, monuments, and reservations hereinafter specified . . . to conserve the scenery and the natural and historic objects and the wildlife therein and to provide for the enjoyment for the same in such manner and by such means as will leave them unimpaired for the enjoyment of future generations."

The General Authorities Act of 1970 placed all areas administered by the National Park Service in one National Park System and clarified the authority of the National Park System. The Redwoods Act, as amended in 1978 emphasized that the National Park Service's mission involves protection for all the lands it administers.

Today, the National Park Service is organized into seven regions. It administers lands in all 50 states, Guam, Puerto Rico, American Samoa, the U.S. Virgin Islands, and the District of Columbia. (See Figure 28-2.) It administers 379 facilities of 20 different kinds. (See Table 28-1.)

The government classifies national parks into three basic categories: historic, recreational, and natural.

Historic areas are those areas used to maintain and restore historic wholeness. There are 241 historic areas. The historic areas include military parks and battlefields, battlefield parks, battlefield sites, historic sites, **memorials**, monuments, cemeteries, and historic parks.

Recreational areas are used to promote public recreation, and the primary objective is to provide outdoor recreational opportunities. The 46 areas include national recreation areas, national seashores, national riverways, national lakeshores, and national parkways.

The park system contains 74 areas designated as natural areas. These areas contain lands minimally disturbed by humans. These areas do not allow hunting, trapping, or fishing, so they act much like wildlife sanctuaries.

TABLE 28-1 National Park System (as of 11 November 1998)	
International Historic Site	1
National Battlefield Parks	3
National Battlefield Site	1
National Battlefields	11
National Historic Sites	77
National Historical Parks	38
National Memorials	28
National Military Parks	9
National Monuments	73
National Lakeshores	4
National Parks	54
National Parkways	4
National Preserves	16
National Recreation Areas	19
National Reserve	2
National Rivers	6
National Scenic Trails	3
National Seashores	10
National Wild and Scenic Rivers	9
Parks (other)	11
Total	379

It is also illegal to gather any type of plant material. These areas provide an excellent opportunity for the study of life as it exists in the wild.

The national park system provides much enjoyment to millions of people each year. In 1991 a total of 267,840,999 people visited sites managed by the National Park Service. That had increased to 286,739,115 in 1998. As of May 2000, the forecast number of visits for 2001 was 290,976,170. Many national parks do allow fishing and some allow hunting. Many provide areas for camping, and most provide for hiking and nature viewing. Food and lodging are available on many national parks. The federal government also brings the park system to the people. In 1970, the government made plans to develop 14 recreational areas in

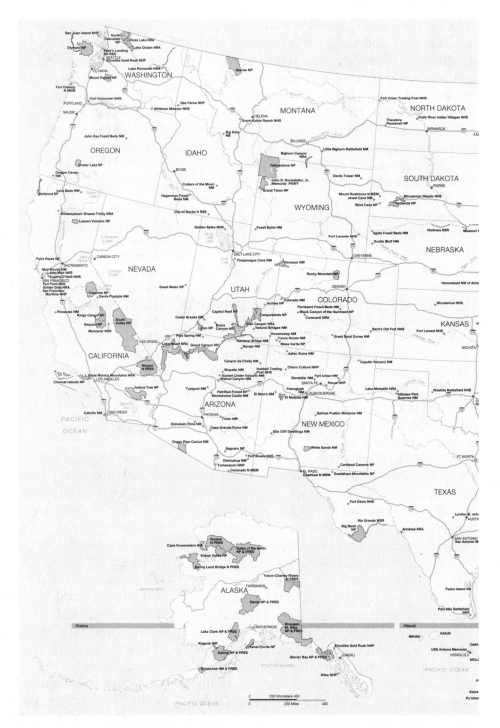

FIGURE 28-2 The national park system. (*Courtesy of the Office of Public Affairs and the Division of Publications, National Park Service*)

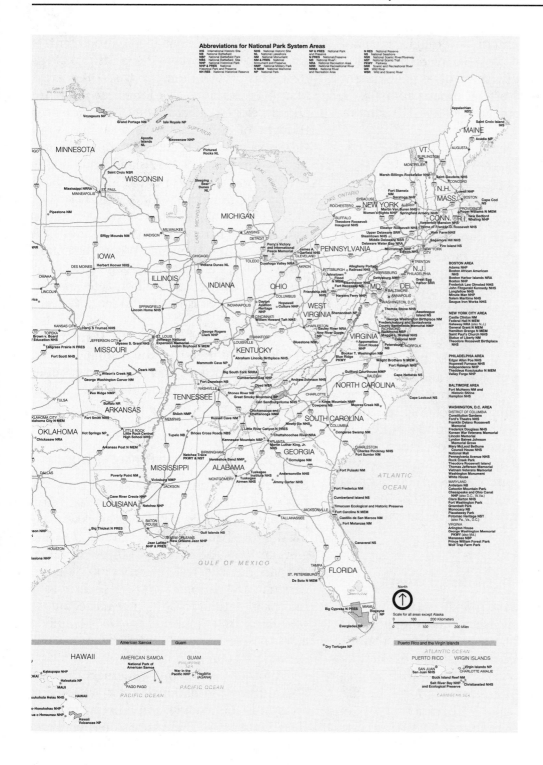

Abbreviations for National Park System Areas

IHS	International Historic Site	NHS	National Historic Site
NHP	National Battlefield	NL	National Lakeshore
NBP	National Battlefield Park	NM	National Monument
NBS	National Battlefield Site	NMEM	National Memorial
NHP	National Historical Park	NM & PRES	National Monument and Preserve
NHP & PRES	National Historical Park and Preserve	NR	National River
NH RES	National Historical Reserve	NMP	National Military Park
		N MEM	National Memorial
		NP	National Park

NP & PRES	National Park and Preserve
N PRES	National Preserve
NR	National River
NRA	National Recreation Area
NRR	National Recreational River
NRRA	National River and Recreation Areas

N RES	National Reserve
NS	National Seashore
NSR	National Scenic River/Riverway
NST	National Scenic Trail
PKWY	Parkway
SRR	Scenic and Recreational River
WR	Wild River
WSR	Wild and Scenic River

metropolitan areas. This new trend was a pilot plan for national parks in major population centers. Eventually, the federal government wants to place at least one urban national park in cities with a population of 250,000 or more.

National Forest System

Our national forest areas are very similar to our park system. Many areas are different only in name. Forest areas can provide many recreational activities to the public. The national forest system supports thousands of camping areas with safe water supplies, picnic tables, and garbage disposal systems. Many national forest areas have even developed extensive winter sports programs such as skiing and ice skating. There are over 250,000 miles of roads and trails. Along with these there are over 83,000 miles of streams and about 2 million acres of lakes to support recreational activities such as fishing and boating. The forest system allows game hunting under individual state statutes. The national forests are less rigorously regulated than the national parks system, but the amount of artificial construction is held to a minimum to ensure that the areas maintain pleasing and aesthetic views for their visitors.

National System of Island Trusts

One potential recreational area not often thought of is the islands surrounding the United States that are under federal control. A study completed by the Bureau of Outdoor Recreation found 26,325 islands with an area of 10 acres or more; this amounts to 28.6 million acres of land with various types of topography as well as climate. The major problem with developing these islands is their remoteness from major population centers. Alaska's large number of islands (5,688) reflects this problem more so than other areas. Many of the islands have long been inhabited and developed. In fact, they have not been thought of as "recreational areas" for some time. Examples of these areas are New York's Manhattan Island, Staten Island, and Long Island, as well as the Hawaiian Islands, Guam, Samoa, and the Virgin Islands. Although each of the islands has a highly developed tourist trade and recreational areas, industries have also moved into the areas. When these areas are excluded, the remaining island areas number 20,637 with over 7 million acres of land. Most of this area is located in Florida, Texas, and Michigan.

The concept of **national island trust** was developed to restore and maintain islands with historic, scenic, and recreational values. The trusts employ planning from government agencies as well as private individuals.

National Wilderness Preservation

Early conservationists realized that some national lands needed to remain as unchanged as possible, but these **wilderness** areas were being destroyed quickly as populations grew and grew. In 1964, the federal government started the national wilderness preservation system. It designated 54 national forest areas, which were to remain natural forever. Since its beginning, over 88 units, amounting to over 14 million acres, have been brought into the national wilderness program.

Another part of the wilderness program goes outside the national forest areas. These areas must be roadless and amount to more than 5,000 acres. These areas could be a part of the national park system, national wildlife refuge, or roadless islands. This added another 200,000 acres of wilderness to the program. Two of the first national parks were Craters of the Moon National Monument in Idaho and Arizona's Petrified Forest National Park. Most of the wilderness areas lie west of the Mississippi. The first eastern wilderness area was the Linville Gorge Wilderness Area in the mountains of North Carolina.

National Trails System

The purpose of the **national trails system** is to provide outdoor recreation by the use of trails. This system came into being in 1968 and includes horseback riding, motorcycling, hiking, and walking trails to name a few.

Under the National Trail Act, an area can be classified into one of two types of trails: scenic and recreational. The scenic trails do not allow motorized equipment. They are authorized by Congress and provide conservation and enjoyment to people without much traffic. Highways and power lines are prohibited in these areas. At the present time, there are two scenic trails in the United States. One is a 2,350-mile trail running along the Sierra Nevadas and Cascades to the Canadian-Washington border. The second is the Appalachian Trail, which extends 2,144 miles from Georgia to Maine.

The national recreation trails are usually located near urban cities. Their main goal is to provide outdoor recreation for the public. Land that is under the jurisdiction of the Department of the Interior or the Department of Agriculture can be used. This trail system usually allows snowmobiles, trailbikes, and horses. They may also have power lines and roads. There are about 30 such trails in the national trails system.

National System of Wild and Scenic Rivers

The last of the national areas is that of the wild and scenic river. Congress designated these areas after six years of debate and discussion. The goal of the program was to preserve those rivers that provide environments of scenic value. These rivers are set aside to preserve the natural condition of pure water that will support fish and wildlife.

This national system classifies the rivers into three categories:

1. wild river areas
2. scenic river areas
3. recreational river areas

The *wild river* areas are those not accessible by road. These rivers are basically in a primitive, unpolluted state. The *scenic river* areas can be accessed by roadways. These rivers are unpolluted and remain in a primitive, undeveloped state. The *recreational river* areas have access to roadways and developed shorelines. These rivers usually have houses and boat docks built on the shorelines. These rivers are also subject to much pollution. At the present time, portions of 80 rivers have been included in programs.

STATE RECREATION AREAS

Many states have a large variety of outdoor recreational activities for their residents. A state department of natural resources or its equivalent usually operates the recreational areas as a part of the multiple uses of the other areas. These state areas usually include

- state parks
- state recreation areas
- United States Corps of Engineer reservoirs (Recreational activities are generally given to the state.)
- state forests
- state fish hatcheries
- state memorials
- state museums
- nature preserves
- historic preservation sites
- fish and wildlife areas

What is available in each area varies. The most popular activities in state areas are camping, fishing, hunting, boating, swimming, and hiking, and photographers find them attractive. State areas are usually popular because they are close to people's homes.

ABUSE AND MISUSE OF PUBLIC RECREATION AREAS

Conservationists have worked hard for many years to develop vast recreational areas for the public. There is a public area designed to accommodate almost every form of recreation a person could desire. Therefore, it is a shame when these areas are abused and misused. It costs the government millions of dollars to repair and clean areas vandalized by inconsiderate people. Much of the vandalism is a result of an I-don't-care or everybody-else-does-it attitude. The shooting of road signs, damaging of restrooms, and scattering of garbage are evidence that this attitude exists. Along with vandalism, tourists collect for souvenirs materials that should be left in place. It seems so harmless

to take a rare blooming flower; but if this is done often enough, the flower will not be able to reseed, thus eliminating the chance for anyone else to enjoy it. How can the abuse be stopped? Will laws stop the problem? These questions are constantly arising with the conservation authorities. Strict fines for violations are definitely a way to curb some vandalism. Most experts feel that a better way would be through education. If field trips and group discussions could teach the public, especially young people, the value of the natural resources at their disposal, much would be gained. The attitude that resources will last forever, no matter how they are handled, must be changed.

SUMMARY

As we look at the outdoor recreational activities at our disposal, the number seems limitless. Between federal, state, and local governments, almost any form of outdoor recreation activity we can imagine is available. Some activities may require you to travel some distance, but most common activities can be found very close to home on public land. The American public has created this great resource, and it will be up to the public to preserve it. If we allow a few inconsiderate people to deface, mar, and destroy what has been built, all of us will lose an irreplaceable resource. Always be extra considerate of our natural recreational resources.

DISCUSSION QUESTIONS

1. Why has there been an increase in recreational activities in recent years?

2. List the six major federal programs offering recreational activities.

3. How are the national parks classified?

4. List the national parks located in your state.

5. Which government agency (Parks or Forests) employs stricter rules?

6. What is the system of island trusts? How many islands are in the trust?

7. Which type of national trail does not allow motorized vehicles? Which does?

8. What are the categories of wild and scenic rivers?

9. How do state governments provide recreation areas?

10. How are recreation areas abused?

SUGGESTED ACTIVITIES

1. Make a list of the national and state recreational areas in your state.

2. Pin a state map to a bulletin board and place a pushpin at the location of each national and state recreation site. Identify each area and describe its size.

29

OUTDOOR SAFETY

OBJECTIVES

After reading this chapter, you should be able to

- list the Ten Commandments of Gun Safety

- explain the Code of Ethics for Hunters

- list the correct safety procedures for using bows and arrows

- know the ten rules for safe snowmobile operation

- explain basic survival and first-aid techniques

- explain safe boating procedures

- know the common traffic rules for boats

- list the water skiing signals

TERMS TO KNOW

Ten Commandments of Gun Safety	bag limit	snowmobile safety code
	hunter orange	hypothermia
Code of Ethics for Hunters	nock	

In Chapter 28 we discussed the large increase in outdoor recreational activities: More and more people are enjoying the great American outdoors. As the outdoor activities increase and the areas become more crowded, the chances for accidents increase and the need for information on outdoor safety practices becomes very important. This chapter deals with safety in land-based activities, such as hunting and snowmobiling, and water-based activities, such as fishing and boating.

LAND-BASED ACTIVITIES

Land-based recreational activities examined here include hunting, archery, and snowmobiling.

Hunting

Many hunting accidents are due to human carelessness. It is estimated that about two-thirds of all gun casualties are caused by persons under the age of 21. Many states, therefore, require a person to pass a hunter safety class before he or she is allowed to purchase a license, ammunition, or a firearm.

The hunter must be knowledgeable in the correct and safe use of a firearm. He or she must obey public safety laws and be considerate to landowners and other hunters. Much work on hunter and gun safety has been done by the National Rifle Association (NRA) to curb hunting-related accidents.

When working with firearms, one should always think safety first. Guns are not toys; they must be handled properly.

NEVER handle a gun without adult supervision.

NEVER let anyone you are with mix alcohol, or drugs, and guns.

ALWAYS follow these IMPORTANT safety rules: The Sporting Arms and Ammunition Manufacturers Institute offers the **Ten Commandments of Gun Safety**:

1. Always keep the muzzle pointed in a safe direction.

2. Guns should be unloaded when not actually in use.

3. Do not rely on your gun's "safety." Treat every gun as if it can fire at any time.

4. Be sure of your target and what's beyond it.

5. Use correct ammunition.

6. If your gun fails to fire when the trigger is pulled, ***handle with care***. Keep the muzzle pointed in a safe direction and get adult help.

7. Always wear eye and ear protection when shooting.

8. Be sure that the barrel is clear of obstructions before shooting.

9. Do not alter or modify your gun, and *do* have your gun serviced regularly.

10. Learn the mechanical and handling characteristics of the firearm you are using. Do not use any firearm that you have not had adult instruction in handling.

In addition to these very important rules, as the authors of this book we would like to add one more "commandment" of our own:

11. Avoid alcohol and drugs anytime, but especially when you will be handling firearms.

Hunting remains popular today. As more and more people pursue outdoor recreation, including hunting and fishing, it is more important than ever that hunters be "good neighbors." The National Rifle Association has published the following **Code of Ethics for Hunters** that all hunters should follow:

1. I will consider myself an invited guest of the landowner, seek his permission and conduct myself so that I may be welcome in the future.

2. I will obey all the rules of safe gun handling and I will courteously but firmly insist that my companions do likewise.

3. I will obey all game laws and regulations and insist my companions do likewise.

4. I will do my best to acquire those marksmanship and hunting skills that assure clean, sportsmanlike kills.

5. I will support conservation efforts that assure good hunting for future generations of Americans.

6. I will pass along to younger hunters the attitudes and skills essential to a true outdoor sportsman.

The hunter should follow the code of ethics whether with companions or deep in the woods alone.

Many states have also installed legislation that has become necessary as hunting has increased. Most states require a license to hunt. The money collected from these licenses usually is spent on wildlife conservation measures. Many wildlife species have also had **bag limits** placed on them. A hunter may only take a certain number of animals per day or season. Many state and national areas allow only a limited number of hunters to any area. This prevents the area from being overhunted and gives the wildlife a chance to prosper. Some laws specify the method of taking game, such as bow and arrow, muzzle loader, or shotgun.

Legislation has been passed in several states prohibiting the use of alcohol or drugs while hunting. It is also unlawful to shoot from or across a public highway, or to hunt from your vehicle. Most states do not allow the carrying of loaded firearms in a vehicle. These rules, as well as any other regulation your particular state passes, must be followed if you wish to continue to have the right to hunt.

The hunter adheres to some unwritten laws also. The hunter has respect for wildlife and its habitat. He or she is not "trigger happy," shooting at anything, and has a responsibility to the property on which he or she hunts. Every year, unsuspecting hunters are mistaken for game and tragically killed. The responsible hunter never shoots at signs or chases game with vehicles and obtains permission before hunting. He or she does not litter on the property. The responsible hunter always tries to leave a camp in better condition than he or she found it.

Hunters also have a responsibility to themselves. If they are hunting in big game country, they should wear **hunter orange** so that they can be seen. This is a bright, fluorescent color that can be easily seen at dusk or dawn when visibility may be impaired. This color is not found in nature, so it cannot be confused with anything else. Along with clothing, hunters must be aware of their physical limitations. When one becomes fatigued, safety procedures become lax, and the chances of a hunting accident greatly increase.

The hunter also has a responsibility to the wildlife. The hunter should shoot only the wildlife he or she is hunting. The gun sights should be properly adjusted before going into the field. The hunter should be familiar with all the wildlife in the area and be able to quickly identify each type. Improperly identified game often results in illegal kills. The hunter should be familiar with the proper use and care of game meat. It should be kept clean, dry, and cool.

Archery

Archery has gained considerable interest in recent years. Bows and arrows are being used for target practice as well as for hunting. The safe use of this equipment will provide many hours of enjoyable recreation.

A bow and arrow can be as dangerous as a gun if it is improperly cared for. You should never nock an arrow or draw a bow if someone is in front of you. (The **nock** is the end of the arrow that fits onto the bow string.) Only draw the bow if you intend to shoot something. When shooting the bow, always know the full path of the arrow; this includes what is in front of as well as what is behind the target. Never shoot straight into the air. The equipment should also be in good condition. A frayed bowstring, loose nock, or feather can injure you or a companion.

If you are bowhunting, extra caution must be taken. Often the bowhunter must sit for long hours in a blind before seeing game. In the excitement of the hunt, safety procedures can be forgotten; this is when accidents are most likely to occur. Before going on a hunt, all hunters in a group should agree on the lines of fire each will take, and only those lines should be used. Each hunter should know where every other hunter is located. If stalking game, be aware of your footing. Remember, undrawn or carried arrows can inflict injury also. Never shoot blindly into the brush at sounds. It could be another hunter. Following these simple precautions and rules can give the hunter much accident-free enjoyment.

Snowmobiling

Since the 1960s, the sport of snowmobiling has gained much popularity.

As with other sports, when the number of people participating increases, so does the incidence of accidents. As winter sports increase, there is a need for additional land for the sports; you are responsible, as in hunting, for gaining permission before entering private property.

Much snowmobiling is done after dark. The winter days are short; therefore, by the time the snowmobiler gets home from work or school, daylight may be gone. Safety at night is of extra importance. If you are going to go snowmobiling after dark, be sure that your headlight is in good working condition. Never overdrive your headlight beam. If your speed is too fast, it is easy to hit an object before you can stop. Carry a flashlight or emergency flare. Be familiar with your trails if you are traveling at night.

The snowmobiler needs to be very careful at all times. Snows may cover dangerous objects such as rocks and fences. You should be aware of low or hidden wires, drainage culverts, fallen trees, and limbs. Always have your machine under control.

On snowmobile trails, road signs have been designed to aid and inform the rider. Heed the directions of signs and follow them as you would if you were on the highway in your automobile.

All snowmobilers should adhere to the **snowmobile safety code**. It contains basic rules for safe snowmobile operation.

- Be sure that your snowmobile is in top-notch mechanical condition at the beginning of the winter season and throughout the months of use.

- Familiarize yourself with the snowmobile you are driving by reading, in detail, the manual accompanying the snowmobile.

- Wear sensible, protective clothing designed for snowmobiling.

- Use a full-size helmet and goggles or visor to prevent injuries from twigs, stones, ice chips, and flying debris. Avoid wearing long scarves. They may get caught in the moving parts of the snowmobile.

- Know the terrain you are going to ride. If unfamiliar, ask someone who has traveled over it before.

- Know the weather forecast and, especially, the ice and snow conditions in the area.

- Always use the buddy system. Never ride off alone or unaccompanied.

■ Do not pursue domestic or wild animals. No true sportsperson would stoop to such conduct.

■ If you see a violation of this rule, re-port it to the nearest law enforcement officer.

■ If you snowmobile at night, be sure you have a properly operating lighting system on the snowmobile.

■ Drowning is one cause of snowmobile fatalities. When not familiar with the thickness of the ice or water currents, avoid the area.

Snowmobilers are responsible sportspersons who adhere to a code of ethics that give pride and credibility to the sport. They do not litter trails or camping areas. They also do not pollute lakes or streams. They do not damage living trees, shrubs, or other natural features. They will respect the rights of people and property. Snowmobilers are eager to assist search-and-rescue parties in lending a helping hand to someone in distress. A snowmobiler is enthusiastic and will not interfere with hikers, skiers, snow-shoers, ice anglers, or other winter sports participants. They also obey all federal, state, and local rules regulating the operation of snowmobiles in areas where they use the vehicle. The sportsperson will not harass wildlife and will avoid wildlife feeding areas. The sportsperson will stay on posted trails and not travel in prohibited areas. Following the above ethics, snowmobilers will always have areas to travel and use.

Survival and First Aid

Many land sports and recreation activities take place during cold or inclement weather conditions. Also, whether gun hunting, bowhunting, or snowmobiling, accidents do happen.

A person who participates in such activities should be familiar with survival and first-aid techniques.

The basic survival rules include both preparation and logical techniques if problems occur. A person should wear the proper clothing and use the proper equipment. Always take enough food for several days in case of an emergency. The sportsperson should know how to build a fire and treat common accidents.

If stranded, he or she should take care to avoid physical injury, find protection from the elements, acquire food and water, and prepare for rescue. In cold weather, frostbite will be the biggest problem. Numbness or a change in skin color to gray or yellow-white spots indicates that frostbite has taken place. In hot desert areas, heat exposure, heat cramps, heat exhaustion, and sunstroke are the problems. Staying in the shade and avoiding overexertion will keep these problems to a minimum.

A stranded or injured person must also be aware of the ground-to-air rescue signals and carry a copy of the signals in a first-aid kit or wallet. Ground-to-air rescue signaling methods are shown in Figure 29-1.

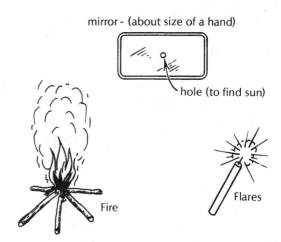

FIGURE 29-1 Ground-to-air rescue signals.

A sportsperson must be aware of basic first-aid procedures. Attending a Red Cross first-aid class is time well invested.

One of the common problems caused by improper clothing is **hypothermia**. Hypothermia occurs when the body loses heat faster than it can be produced. Hypothermia drains energy from the body. Early signs are uncontrolled shivering, memory lapses, and numbness. Internal temperature drops to a dangerously low level. To prevent hypothermia, one needs to keep dry. Wet clothes lose more heat and have a lower insulating value than dry ones. Give warm drinks to a person suffering from hypothermia. Keep the victim awake. Attempt to keep the victim warm by using a warm sleeping bag if possible.

Any survival kit should contain a first-aid kit. There are a variety of kits available on the market (or you can make one yourself). They come in large and small sizes. A typical first-aid kit contains two 2-inch compresses, four 4-inch compresses, four triangular bandages, one roll of 2-inch gauze, and one roll of 1-inch adhesive tape. Place the items in a waterproof container and make it easily accessible in case of an emergency. Having the correct first-aid equipment, and knowing the proper procedures for using it, will keep the chances of a fatality to a minimum.

WATER-BASED ACTIVITIES

The common water-based recreational activities include boating and other water sports. Activities such as water skiing and swimming are included in the safety procedures for water sports.

Boating

Safe boating procedures are a must if the skipper of the vessel wants all passengers to have an enjoyable outing. This includes operating a safe boat, knowing lifesaving techniques, knowing basic weather conditions, observing the rules of the water, and being familiar with accident procedures.

A safe boat contains a fire extinguisher. Fires can be extinguished by carbon dioxide, foam, dry chemical, freon, or water. Most boat fires are caused by gas and oil. Some are caused by electrical wiring problems. Check the extinguisher frequently to be sure that it is properly stored and undamaged. Testing an extinguisher by trying it is generally not recommended because the valve may not reseal.

A boat should also contain a whistle or horn that will blast for at least two minutes. This equipment is used to signal the boat's whereabouts or condition. Short blasts signal the boat is turning. One indicates a starboard (right) turn, two indicate a port (left) turn, three indicate reverse, and four indicate danger. Long blasts are signals in fog, snow, or sleet. A long blast is also sounded when leaving the slip, pier, or wharf.

The boat, by law, must display lights between sunset and sunrise. Light requirements under inland and international rules are illustrated in Figure 29-2.

A boat must contain lifesaving equipment. There are four kinds of personal flotation devices, commonly called life preservers. Type I devices are the typical life preservers, type II are buoyant vests, type III are marine buoyant devices, and type IV are throwable devices.

A boating enthusiast should also be familiar with weather signs. The wind and currents greatly affect a boat's course. A boat can best be handled going against the current, and

Under Power alone	Auxiliaries under Sail and Power	Auxiliaries under Sail alone

INLAND RULES

Under 26 Feet

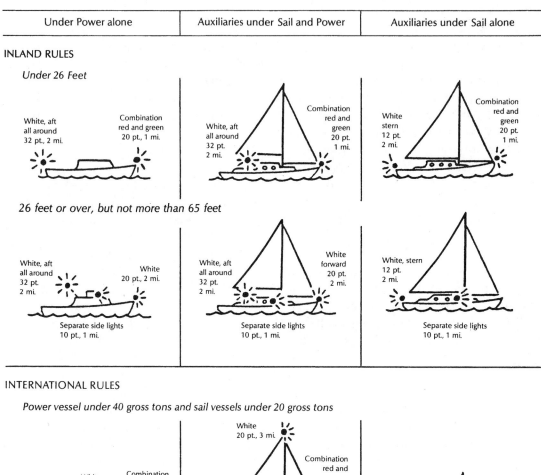

26 feet or over, but not more than 65 feet

INTERNATIONAL RULES

Power vessel under 40 gross tons and sail vessels under 20 gross tons

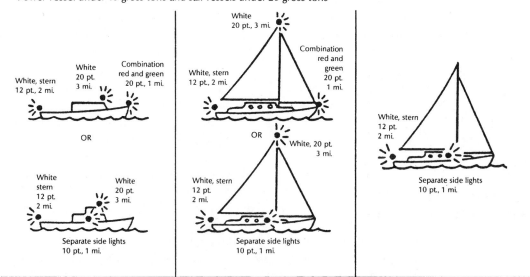

FIGURE 29-2 Inland and international rules for lights required on motorcraft underway between sunset and sunrise.

drifting in heavy currents should be avoided. If caught in foul weather, reduce speed and proceed with caution. Head for the nearest shoreline that is safe to approach. Head the bow (front) into the waves at a slight angle and watch for floating debris. Secure loose items and have emergency gear ready. Be sure that all passengers are wearing their life jackets and sitting on the bottom of the boat. If the motor fails, drag a sea anchor. Figure 29-3 depicts the signals used by the National Weather Service to advise boaters of bad weather.

Just as there are rules to be followed when traveling on the highways, there are rules to follow on the water. The common traffic rules include

- When two boats are approaching each other head on, each turns to pass to the right, or starboard.

- When two boats are approaching each other at right angles, the boat to the right has the right-of-way.

- Boats traveling in channels or rivers must bear to the right.

- A boat may overtake another vessel on either side, providing each is equally safe.

- Boats leaving docks or wharfs have the right-of-way over vessels approaching the moorage area.

- Motor boat operators must not travel closer than 200 feet from the shoreline of a lake.

Even though boat owners work for safety, accidents can still happen. Knowing how to respond to an accident will save time and

STORM SIGNALS

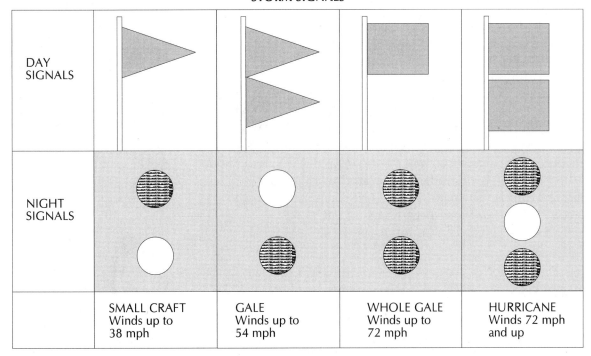

FIGURE 29-3 National Weather Service storm advisories.

FIGURE 29-4 Distress signals. In an emergency, anything can be used to improvise a distress signal. During the day, tie a shirt or jacket to an oar and wave. At night, a flashlight can be used to flash the Morse Code SOS signal (• • • ––– • • •).

maybe lives. The most common causes of accidents include an engine the wrong size for a boat, high-speed turns, overloading, riding on decks, damage from waves, no life preservers, and fires from refueling.

If an accident does occur, always be ready to help. Remember to help save lives, but do not risk lives to save equipment. One should never take unnecessary chances. Above all, *do not panic.* If a person falls overboard and a life preserver is unavailable, use anything that floats to get the person aboard the boat. If you become stranded because of an accident, distress signals can be improvised. (See Figure 29-4.)

Water Sports

The common water sports include water skiing, rowing, canoeing, fishing, and swimming.

As watershed programs have developed more reservoirs, lakes large enough for water skiing have grown in number. Safety becomes even more important when we consider the additional number of people on the water.

Water skiing is a three-person team sport. The skipper steers the boat safely, the observer passes signals between the skipper and the skier, and there is one or more on skis. There must be at least two people in the boat. The boat pulling a skier should not make sharp turns that cause the skier to fight the wake of the boat. The boat should also stay a safe distance away from docks, swimmers, people who are fishing, and danger areas. Always give the boat plenty of turning room. Skiers should never show off when skiing. Horseplay is the chief cause of accidents. Everyone involved should know and understand the water skiing signals shown in Figure 29-5.

Canoeing has also become a popular sport in recent years. Some simple safety rules will

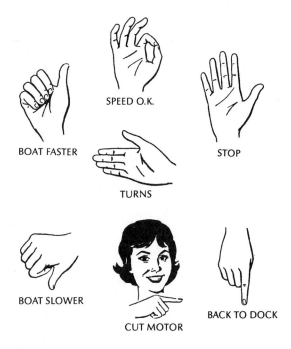

FIGURE 29-5 Water skiing signals.

make the sport safer for its participants. Stay out of a canoe if you cannot swim. Keep your weight low and centered, and sit still to eliminate the chances of upsetting the canoe. As a rowboat or canoe operator, watch for motorboats with large wakes. Wear life preservers at all times and practice paddling and steering in shallow water.

Swimming should be done in roped-off areas only. A calm water surface may have severe currents underneath. An approved Red Cross swimming class is invaluable for participating in any water sport.

Fishing as well as swimming should never be done alone. It is wise to have a friend close by in case help is ever needed. Fishhooks can also cause serious injuries. Never carry a rod with a freely dangling line or hook, and be careful when casting. Don't hook your partner instead of a fish. Also, be aware of sharp fish spines and teeth. Always pick up a fish by its gills, never by its mouth. It is also wise to wear gloves when handling fish. If you are cut or bitten by a fish, treat the wound with a disinfectant and see a doctor. Also, be careful with sharp knives. When carrying a knife, keep it closed or in the sheath. If cut, apply an antiseptic, then a sterilized bandage, and then some pressure to stop the bleeding.

SUMMARY

Recreational activities can be very beneficial to a person's attitude. Activities provide relaxation to participants. (See Figure 29-6.) An accident, however, can turn an enjoyable time into disaster. Most accidents can be prevented if proper safety procedures are followed. Whether the activity is on land or water, always know the proper safety techniques and follow them precisely. If this is done,

FIGURE 29-6 Outdoor recreation can be fun for all if safety rules and safety equipment are used. This group is safely enjoying rafting an inland river.

many safe, outdoor recreational activities can be enjoyed.

DISCUSSION QUESTIONS

1. What are the ten commandments of gun safety?

2. How should you always treat a gun?

3. What is the hunter's code of ethics?

4. What are the responsibilities of hunters to wildlife, the environment, themselves, and the habitat?

5. At what time is an archery accident most likely to occur?

6. If you snowmobile at night, what safety practices should be followed?

7. What is the snowmobiler's safety code?

8. What are the common survival procedures for winter sports?

9. What should a first-aid kit contain?

10. What equipment should a safe boat contain?

11. What are the common boat traffic rules?

12. What are the common boat distress signals?

13. List the common water skier's hand signals.

14. What safety precautions should be followed in fishing? in swimming? in canoeing?

SUGGESTED ACTIVITIES

1. Have a conservation officer discuss the fundamentals of gun safety.

2. Organize and conduct a safety lesson in a junior high or middle school dealing with guns, snowmobiling, or boating.

3. Organize a hunter safety course for the students at your school.

30 CAREERS IN OUTDOOR RECREATION

OBJECTIVES

After reading this chapter, you should be able to

- describe the opportunities in an outdoor recreation career

TERMS TO KNOW	
park ranger	recreationists

Careers in outdoor recreation are large in number as well as varied in opportunities. The field offers opportunities on water or land. They might be technical, professional, skilled, or unskilled jobs. A person might be employed by the federal, state, or local government. You might choose a position in a private industry or business. If a career in outdoor recreation is what you want, the choices are many.

GOVERNMENT EMPLOYEES

Government employees in outdoor recreation include park rangers, park aides and technicians, park guides, and park guards. There are also careers as recreation workers, snow rangers, and wilderness rangers.

Park Ranger

Several government agencies hire people for outdoor recreation careers. The National Park Service provides several of these career oppor-

tunities. The **park ranger** performs professional duties to maintain the park in good condition. The ranger works to see that his or her area is free from fire hazards and visitor vandalism. He or she conducts educational programs to inform visitors about proper safety in the park. He or she may guide tours through points of interest as well as set up displays. The ranger attempts to carry out programs that show the park's visitors the value of the resources preserved in the park and may work on planning recreational activities for visitors. He or she works closely with the other employees to manage the park. The park ranger may take his or her knowledge to areas outside the park itself and be invited to speak at local functions on the park and its activities.

The ranger should like working with people. (See Figure 30-1.) Because the ranger is in a management position, he or she must be able to manage workers, be willing to help people, and be patient with them. The ranger must enjoy the out-of-doors. With advancement, however, he or she may be moved indoors to an office with more management duties.

339

FIGURE 30-1 A Park Service Ranger provides instruction during a guided tour for visitors to the Golden Spike National Historical Site. The site is home to fully functional replicas of the locomotives Jupiter and #119 that met at the famous "Wedding of the Rails" ceremony on May 10, 1869. *(Photo courtesy of the National Park Service)*

The park ranger is required to have at least a four-year bachelor's degree when entering the career field. The ranger should have completed classes in natural science, park and recreation management, business administration, sociology, behavioral science, and archeology. As experience increases, the ranger may advance to district ranger, park manager, or park planning manager.

Park rangers must take the federal service examination and qualify under the Civil Service Commission. Beginning salaries are based on the current federal civil service pay scale.

In 1999, park rangers working for the federal government started at salaries ranging from $25,500 to $31,200. Part of the difference was based on location and part on the degree held. The average salary for all federal park rangers was $42,100 in 1999.

Other National Park Service Opportunities

With an annual budget nearing $2 billion, the National Park Service (NPS) employed almost 16,000 full-time workers in June 2000. An additional 5,500 seasonal and part-time employees work for the NPS, along with as many as 90,000 volunteers. NPS jobs range from very low paid entry-level, part-time positions all the way to the NPS Director. When we talk about the Park Service, most of us think first about the park ranger wearing a uniform with a brown "Smokey Bear" hat (refer to Figure 30-1). In reality, most people who work for the NPS have jobs very different from the ranger's. The salaries paid to these workers vary greatly based on the kinds of education and training required and on the job responsibilities. The pay scale is based on the Government Service pay scale in effect for all federal employees. The jobs' GS (annual salary) or WG (hourly wage) rating indicates where on that pay scale the worker would fall. As of June 2000, the federal pay scale for a GS 04, with no experience started at $19,100 annually. That pay scale went to $24,833 with 10 years of experience. For comparison, a GS 11 started at $39,178 with zero experience and ranged to $46,359 after 10 years. A WG 04 started at $9.15 per hour with no experience and went to $11.90 per hour after 10 years. A WG 11 started at $18.77 and went up to $24.40 after 10 years.

A recent web search (June 2000) of NPS job opportunities showed 225 openings with such job titles and GS-ratings as follows:

■ Lifeguard supervisor (GS 06)

■ Maintenance mechanic (WG 10)

■ Masonry worker (WG 07)

■ Park guide (GS 04 to 06)

■ Park ranger (Enforcement) (GS 05 to 07)

- Park ranger (Interpretation) (GS 05 to 07)
- Park ranger (Supervisor) (GS 11)
- Preservation specialist (Welder) (WG 09)
- Tractor operator (WG 06)
- Trail laborer (WG 06)
- Visitor use assistant (GS 04 to 06)
- Wildlife biologist (GS 11)
- Writer-editor (GS 07)

Fish and Wildlife Guides

The recreational programs in the national wildlife refuges have greatly increased in the past years. Most careers involve guiding visitors through their areas. Most people who visit the refuge are there to view the many types of wildlife that make their homes in the area. Many species are rare and uncommon to the typical visitor, and the guides explain various things concerning the refuge.

Vandalism has not only struck the park system. It has struck the refuges. Guides may have duties such as patrolling and watching the area. The government is attempting to stop as much damage as possible.

The employment requirements and salaries are similar to those of the park guide. A civil service examination is required.

Recreationists

Recreationists are employed in planning, coordinating, and developing recreational programs. These programs may involve a federal park, state park, or private industry. Therefore, we cover their job as a whole instead of separating it under either government or nongovernment heading.

Outdoor recreationists may be technicians, supervisors, leaders, or directors. They may be involved with the young, old, or handicapped. They may work with hiking, boating, camping, canoeing, or nature study. These activities may also change from day to day.

The recreationist may work in or supervise a camp. This includes camp counselors or camp instructors. Recreationists direct campers in nature-oriented forms of recreation, such as swimming, hiking, and horseback riding. They also provide campers with specialized instruction in drama, gymnastics, or tennis, and may also provide instruction in games, sports, and crafts. Recreationists are expected to keep records and maintain equipment.

The working conditions of the recreationist involve an average work week of 35 to 40 hours. The hours may be irregular, and some night work is necessary. The outdoor recreationist may work in sparsely populated areas and must like to work outdoors.

In 1996, there were 233,000 people employed as recreation workers. This total does not include the summer people employed at seasonal businesses. About 40 percent work for governments—federal, state, and local. About 25 percent work for civic, social, and fraternal associations such as the Boy Scouts and Girl Scouts. The remainder are found in commercial and private industries such as amusement parks, entertainment centers, wilderness and survival enterprises, tourist attractions, and vacation areas.

The recreation field also has a large number of part-time, seasonal, and volunteer positions. They may be on local park boards or commissions. Many of these part-time jobs offer experience that may lead to full-time jobs. Seasonal jobs include summer camp counselors, playground leaders, lifeguards, and weekend program leaders.

A college degree with a major in parks and recreation provides the best qualifications for the recreation field. There are many jobs available to high school graduates with specialized training. Most supervisors have a bachelor's degree, and this may be the minimum qualification for an administrator. Many people obtain training in other areas such as social work, forestry, and resource management.

If teaching recreation appeals to you, the standards are somewhat higher. About two-thirds of the junior colleges require a master's degree, one fifth a bachelor's degree, and one-tenth a Ph.D. degree.

In 1996, about 200 two-year community colleges offered an associate degree in recreation and park management. There were also about 300 colleges and universities offering bachelor's, master's, or doctoral programs. In their studies, students gain knowledge of natural and social sciences and history. They also spend time studying professional park theory, history, and philosophy; community organization; recreation and park services; leadership supervision and administration; and working with the elderly and handicapped. Students may specialize in park management, outdoor recreation, park and recreation administration, industrial recreation, and campground management.

The people in recreation careers must be able to motivate people and be sensitive to their needs. Good health and physical stamina are required. They must be willing to accept and carry out responsibility. They must be willing to work alone in some jobs.

As leisure time increases, so do the employment possibilities. Many government positions depend on funding to the agencies. The best opportunities will come to those who are best qualified for the current needs in the field.

According to the United States Bureau of Labor Statistics, the average earnings of full-time recreation workers were about $18,700 in 1996. That was well above the average for all full-time workers of $15,600 for the same year. Half the recreation workers earned between $12,900 and $28,900. The top 10 percent earned from $37,500 or more. Directors and managers of major recreation facilities earn substantially higher salaries. Workers in full-time positions in the recreation industry generally receive full benefits such as health insurance and vacations. Most recreation workers are part-time and so receive little or no employment benefits except hourly wages, which are generally below the average wages for all workers.

For additional information, write to

> National Recreation and Park
>   Association
> Division of Professional Services
> 3101 Park Center Drive
> Alexandria, VA 22302
>
> American Camping Association
> Bradford Woods
> Martinsville, IN 46151

PRIVATE BUSINESS

Recreation in the outdoors also provides career opportunities to the businessperson. The business can range from a campground to fishing or hunting for pay. The business can be a franchise such as Kampgrounds of America (KOA) or your own private development. The investments will vary with each type of business. The following is a partial list of possible business ventures:

- recreational vehicle campground
- campground franchise
- condominium campground manager
- youth camp
- fishing bait shop

- hunting preserve
- ski area developer
- riverboat owner
- pack train owner
- dude ranch operator
- fishing and hunting guide
- travel agent
- sporting goods store
- fish farmer

Each of these businesses provides opportunities to interested people. If one of the careers is of interest to you, explore the possibilities. Look at its potential and its investments. It is wise to take a job in your area of interest before investing your money.

SUMMARY

Outdoor recreation holds great opportunities for the new century. As leisure time increases, so does the demand for outdoor recreation specialists. There is money to be made in this field and at the same time conserve the beauty and resources of our valuable lands. If a career in outdoor recreation seems to be for you, examine your goals, background, and education, and choose a field you will enjoy working in for a lifetime.

DISCUSSION QUESTIONS

1. What types of government jobs are available in outdoor recreation?

2. What does a recreation worker do?

3. What are the working conditions for a person in outdoor recreation?

4. What are the educational requirements for a person wanting a career in outdoor recreation?

5. What is the expected salary in outdoor recreation?

6. List the possible occupations in outdoor recreation in private business.

SUGGESTED ACTIVITIES

1. Survey your community or county to find the number and types of outdoor recreation opportunities. Categorize them as private, state, and national opportunities. Which area yields the most possibilities? Report your findings to the class.

2. Conduct an Internet search for careers in natural resources. (*Hint:* Try http://www.bls.gov/ocohome.htm as a start.)

OF FORESTS AND FOES

Imagine having a friend whom you met on a camping trip. You are drawn towards this friend because in that person you see someone just as concerned about the world as you are. Now imagine several years later. You are both embroiled in a bitter dispute over how America's forests should be managed . . . and you are both the leaders . . . on different sides. This was the fate of two of the most influential men in Conservation History: John Muir and Gifford Pinchot.

TWO PHILOSOPHIES

Muir embraced the belief that forests should be protected for the sake of having this natural beauty available. In essence, Muir was a preservationist. It is this philosophy that stands as the underlying premise behind our National Park Service. (See Figure VI-A.)

Pinchot's utilitarianism held with conservationists. Forests were a resource to be managed by and used for mankind. (See Figure VI-B.) As a renewable resource, trees are capable of reproducing. Therefore it was a waste to him to see forested land set aside without making the harvestable timber available to the American taxpayer through logging, hunting, and the like.

These views clashed during the late 1800s and early 1900s as professionals debated on how to manage national reserves. With tracts of land available in all conditions, both parties seemed to have good arguments. So who won?

A DISPUTED DECISION

Some will argue that Pinchot won. The National Forest Service controls millions of acres of land and the multiple-use laws fit neatly into the utilitarian conservationist view. How-

FIGURE VI-A Craters of the Moon National Park, Idaho. Trees are not harvested in national parks. This site is being "preserved" in its current state. (*Photo courtesy U.S. National Park Service*)

FIGURE VI-B National forest land such as this in Colorado is "managed" rather than being "preserved." Trees may be harvested for use in constructing homes and other buildings, but the forest may also be used for recreation and other things. *(Photo courtesy U.S. Department of Agriculture, Forest Service)*

ever, Muir and the National Park Service can also claim victory. Today, National Park Service Lands are protected so that management for forest harvesting and hunting are forbidden. Only in extreme cases of wildlife damage can wildlife pests be removed from these areas. So who is right?

Should America's public forests be controlled under preservation- or conservation-based management? You decide.

Section VII
Energy, Mineral, and Metal Resources

CHAPTER

31

FOSSIL FUEL MANAGEMENT

OBJECTIVES

After reading this chapter, you should be able to

- explain the various ways coal is mined from the earth
- discuss oil exploration and drilling techniques
- explain how natural gas is obtained and distributed
- discuss oil shale, tar sands, and the petroleum potential

TERMS TO KNOW		
lignite	drift mine	jack-up rigs
bituminous coal	room and pillar	drillships
anthracite coal	longwall	semisubmersible rigs
strip mine	gravimeter	fixed platforms
cut	magnetometer	tar sands
shaft mine	seismograph	oil shale
slope mine	derrick	

Have you ever wondered why the United States developed into a leading nation of the world and, at the same time, nations of a similar size remained undeveloped? Among the reasons we could give is the vast amount of natural resources this country contains. Of the resources, fossil fuels play a dynamic role in how our country operates. Without the fuel, our country would literally stop. Fossil fuels are those minerals formed over time from compressed vegetation. They include coal, oil, natural gas, tar sands, and oil shale. The management of these nonrenewable resources is critical if our country is to remain a leading power and influence in the world.

COAL

Coal is a black or brown rock developed from plants that died between 1 million and 400 million years ago. Most of the coal was formed from swamp areas. The extreme pressure in the earth caused the concentration of carbon matter and, thus, formed the coal. Coal is ranked and classified according to the carbon it contains.

The lowest carbon coal and the first step of coal formation is **lignite**. Lignite contains about 33 percent carbon and is formed from peat deposits. As the rock and soil above the lignite increases, the high pressures cause the coal to become harder. This next step in coal formation is called subbituminous coal. With additional pressures, the subbituminous coal is transformed into **bituminous coal**. The hardest and the oldest coal is **anthracite coal**, which contains up to 90 percent carbon. Most coal was formed about 300 million years ago during a period of time called the Carboniferous period. Most of the plant life of this period were tall ferns growing in swamp areas. It takes over 7 feet of plant matter to make a 1-foot vein of bituminous coal.

Most of the coal found today occurs in veins or seams. Coal has been discovered in seams from as small as one inch up to as large as several hundred feet thick. It is estimated that there are over 2 trillion tons of known coal reserves in the world. About 479 billion tons are in the United States, and most of this coal lies in the eastern half of the country. Nearly all the anthracite deposits are found between the Appalachian Mountains and the edge of the Great Plains. The principal coal-producing states include West Virginia, Pennsylvania, Kentucky, Illinois, Ohio, and Virginia.

Coal has a great variety of uses. According to a recent estimate of the United States Department of Energy, 68 percent of the coal was used in the production of electricity, 13 percent in steel production, 9 percent each in general industries and exports, and 1 percent for home heating. Coal with a high heating value is used to drive steam turbine generators. In recent years, concern has developed over the use of coal because of the air pollution problem. As coal burns, sulfur dioxide is released. This poisonous gas irritates the body's respiratory tract. Many scientists believe it also contributes to the growing problem of acid rain. Much research has been done to design filtering systems to trap this gas. Another concern is the powder ash material, called fly ash, that is also emitted from the smokestacks. The adoption of screens has limited this problem.

Coal is mined from two types of mines: surface (or strip) mines and underground mines. Most **strip mines** (Figure 31-1) are within 50 feet of the earth's surface. Strip mines produce about 60 percent of the coal in the United States. Underground mines may extend hundreds of feet under the earth's surface and supply about 40 percent of our coal.

Strip mining of coal requires huge machinery to remove the overburden, or top earth, from the coal vein. This soil is placed in large piles, which can cause vast environmental problems. In past years, the strip mines were left as they were on the last day coal was removed. Huge pits and tall piles of soil were left behind to erode and wash away. The federal government now requires the strip-mining

FIGURE 31-1 Strip coal mine. *(Courtesy of Ashland Oil Inc., by the American Petroleum Institute)*

company to replace or reclaim the land before moving to a new location. The company can have no soil slopes steeper than 33 percent. Today, when an area is selected for a strip mine, the large earthmovers peal a strip of land away until the coal is exposed. This ditch is called a **cut**. The coal is removed from the seam by large trucks. Once the coal is removed, the earthmovers make another cut, dropping soil on the first cut. When the coal supply has been exhausted, only small ridges need to be leveled in the reclamation process.

In underground mining procedures, the work of getting coal from the earth is very hazardous. There are always chances of cave-ins, explosions, and exposure to poisonous gases. The underground mines require more human labor than do strip mines. Recent advancements have produced better machines to help the underground miners.

There are three types of main underground mines: shaft mines, slope mines, and drift mines. The **shaft mine** access passageways are vertical to the coal seam. These passageways are equipped with an elevator to bring machinery and personnel to and from the coal. In a **slope mine**, the access tunnel is on a slope from the surface to the coal seam. These mines are not as deep as those with shafts. A **drift mine** is used when a passageway is bored into a hill or mountain. These tunnels may be parallel to the surface.

Once the access tunnels are constructed, mine experts must decide how the coal is to be removed. There are two common methods: (1) the room and pillar system, and (2) the longwall system. The **room and pillar** system employs the idea of leaving pillars of coal to help support the ceiling. Seams of coal 40 to 80 feet wide are cut. Once the parallels are cut, then the blocks of coal are left to support the ceiling. The underground mine may then be expanded into rooms. The first rooms serve as main entries and coal transport rooms.

The **longwall** system uses the method of mining the coal from one face about 300 to 700 feet long. As the coal is cut from the seam, it falls into a conveyor belt and is transported to the surface. This method is more common in Europe than in the United States. Mine safety laws require longwall mines to have both subentries as well as main entries.

Because mining is so dangerous, mine safety is of the utmost importance. Most mining accidents come from roof and wall failures, accumulation of gases, concentration of coal dust, and accidents with machinery. In early mining history, about 3 in every 1,000 miners were killed annually. Today, that rate has dropped to about 1 in every 2,000. The federal government, under the United States Bureau of Mines, has the responsibility of setting standards. The U.S. Congress enacted the Federal Coal Mine Health and Safety Act in 1969 to strengthen the standards for ventilation, coal dust concentration, and mining structure. The act also gave financial benefits to miners disabled by black lung disease. The standards and safety measures are enforced through the Mine Safety and Health Administration, the U.S. Department of the Interior, and the Environmental Protection Agency.

The newest research in the coal industry is involved in finding new uses of coal and developing more economical methods of converting coal into a liquid or gaseous fuel.

Coal gasification converts coal into fuel. The fuel can then be used as raw product of chemicals and fertilizer. Large-scale production is still in the experimental stage but shows promise in providing uses of high-sulfur grade of coal.

OIL OR PETROLEUM

Many people consider petroleum a black gold, especially since the price of its products has skyrocketed. Like coal, it is a vital backbone of our industries. It keeps America moving. Today's research has found a large variety of uses for oil.

Oil is formed in a way very similar to coal. As plants and animals living in the water died, they settled to the bottom of the ocean, pond, or swamp. Although coal requires several million years to form, oil usually requires only about 1 million years. As pressures were exerted on the materials, oil was formed and squeezed into the rock openings or into specialized rocks called reservoir rocks. These reservoir rocks are porous, allowing oil to fill them. Many times the shifting of the earth trapped oil deposits in the rock layers. It is the job of petroleum engineers to find those oil deposits.

The most widely used machines today for finding oil are the gravimeter, the magnetometer, and the seismograph. The **gravimeter** uses the principle that the gravitational pull of oil-filled rocks differs from rocks containing no oil. The **magnetometer**, on the other hand, measures differences in the magnetic pull of the earth to find oil-bearing rocks. This machine enables the geophysicist to locate rock layers that might contain oil. The **seismograph** uses sound waves to identify various layers and formations under the earth's surface. This method is useful on offshore drilling rigs.

Once oil is thought to be present, then the oil company must lease the land before drilling. The owner of the land is usually paid royalties if oil is found on the property. After obtaining the lease, the drilling site is prepared. Roads may be built, land leveled and cleared, and, possibly, site reinforcements made. A power plant must be readied as well as a water supply system. After much preparation, the drilling rig is ready for construction. A hoisting apparatus called a **derrick** is constructed first. The drills, tanks, and pumps will then follow.

The actual drilling can be completed in one of three ways: cable-tool, rotary, or directional. The early oil crews used the first method, cable-tool drilling. The drill contains a tool such as a chisel that is lifted and dropped to loosen the soil and rock. Fresh water is then poured into the hole to loosen the soil. This process was very slow. Rotary drilling uses a bit similar to a carpenter's wood bit. As the drill is lowered, breaking loose rock fragments, water is pumped down the pipe to carry the sediments to the surface. Several drilling teams work 8- to 10-hour shifts to keep the rig working 24 hours a day. Directional drilling involves drilling the shaft at an angle. Special drills called turbodrills and electrodrills rotate and bend, directing the cutting bit. This method is used when the well cannot be drilled directly over the deposit. Figure 31-2 shows an oil drilling rig.

In recent years, offshore drilling rigs have been constructed. They are more expensive and more dangerous than land rigs. The offshore operations include **jack-up rigs**, **drillships**, **semisubmersible rigs**, and **fixed platforms**. Most exploratory rigs are one of the first three mentioned. Once oil is discovered, the well is completed with a fixed platform. When the oil is brought to the surface, it is piped to supertankers and taken to a refinery.

At the oil refinery, the crude oil is distilled into various products. These may be fuels, lubricants, or petrochemicals. Fuels include aviation gasoline, diesel fuel, gasoline, jet fuel,

FIGURE 31-2 Oil drilling rig. *(Courtesy of the U.S. Department of Energy)*

kerosene, home heating oil, and distillate oils. The lubricants include greases, road oils, technical oils, and medicine oil. The petrochemicals include alcohol, ammonia, ink, paint, plastic, synthetic rubber, and food additives, to name a few.

Oil companies and geologists have been estimating how much oil is left to be recovered. As technology has changed over the years, it has become possible to recover crude oil that could not have been recovered earlier. Also, as prices of crude oil rise, it becomes economically feasible to recover oil that would be left in the ground at lower prices. Finally, as oil explorations techniques have improved, new sources of oil have been found that were unknown previously. The key estimate is what

is called Estimated Ultimately Recoverable (EUR) oil. Over the last 50 years, the world EUR oil reserve has actually increased. Estimates in the 1950s averaged around 1.2 trillion barrels of EUR oil. Today, the estimates generally range between 2.0 and 2.4 trillion barrels of EUR oil. The countries in the former Soviet Union are estimated by the USGS to have 0.3 TGG and China is estimated to have 0.23 TGG of recoverable oil. Much of the oil has not even been found yet—these are USGS estimates.

Of the 2.0 to 2.4 trillion barrels, only 1 trillion barrels are actually proven to be available. Of that, two thirds are in the Persian Gulf. About half of the Persian Gulf proven reserve is in Saudi Arabia. Most of the rest is about equally divided among Kuwait, Iraq, and Iran.

Using these estimates of EUR oil, we can expect world production of crude oil to peak between the years 2007 and 2019. Nobody believes that we will "run out" of oil in the foreseeable future. When crude oil production capacity falls, as it must some day, prices will rise. When oil prices rise, it will become economically feasible to extract oil from coal, **tar sands**, heavy oil, and **oil shale**. Huge reserves of oil are stored in those resources. The only limit on oil extraction from those sources is the immense cost of building the extraction facilities. If oil prices get high enough, oil can be removed from coal, tar sands, heavy oil, and oil shale for centuries.

How long will the petroleum last? We know it cannot last forever. It is time to start a conservation program to reduce the overuse of oil. The petroleum industry is working on methods to recover more oil from the wells we now have as well as to develop more precise location methods. The consumer can save oil as well. This can be done by driving the automobile less and lowering thermostats in the winter and raising them in the summer.

Lower speed limits, greater use of mass transportation systems, and purchasing fuel-efficient automobiles are other ways to conserve oil. Each person must play a part in a conservation program.

NATURAL GAS

Natural gas is one of the most perfect and demanded fuels in the United States. We use it to heat our homes as well as to cook our food. We use it in the production of plastics, detergents, and drugs.

The gas industry can be divided into three areas: production, transportation, and distribution. In the exploration of gas, most gas and oil companies work together. The gas is usually located above an oil deposit. One of the deepest wells ever drilled was in Oklahoma, in 1972, to a depth of 30,000 feet at a cost of about $5 million.

Gas taken from a well must be cleaned in an extraction unit. This unit removes impurities such as water, sulfur, and dust. At the processing plants, butane, propane, and gasoline are removed from the gas. The finished gas is then fed into a transportation pipeline where it is compressed to about 1,000 pounds per square inch. It travels through the pipeline at about 15 mph. When it reaches a city, distribution lines carry it to the consumers. The consumers receive the gas from individual service lines connected from the main lines. The United States has about 1 million miles of gas lines in operation.

One of the main problems with natural gas is supplying the gas when the consumer needs it. The peak usage time for natural gas is during the winter months. The amount used on a cold winter day can be more than five times the summer usage. The oil and gas companies are presently storing the gas in huge underground reservoirs. These reservoirs are actually old oil wells that are no longer in use. When the demand for gas is low, the excess is pumped into the well and sealed to be used later when demand for the gas increases. In the early days, this excess gas would be burned off and wasted. Some countries in the Middle East continue this practice.

Gas can also be stored in a liquid state. When the temperature of the gas is lowered to –260°F (–168°C), it changes to a liquid. This fuel is termed LNG or liquid natural gas. It takes up less space (about 600 times less); therefore, it is a good way to ship gas overseas. Simply by raising the temperature, the LNG changes back into a gaseous form.

In the early 2000s, natural gas is the most rapidly growing source of fossil fuel energy. Improving exploration technology has been resulting in rapid increases in natural gas reserves. In spite of that, the known reserves of natural gas are considerably less than known reserves of crude oil. Russia has the largest natural gas reserve at 48.2 trillion cubic yards (1993 estimates). That represents about one third of all known world reserves (141 trillion cubic yards) and over ten times U.S. reserves (4.6 trillion cubic yards).

Most of the natural gas is found in Texas and Louisiana. The increased use of natural gas has caused concern over our future supplies. With the passage of the Natural Gas Act of 1938, the Federal Power Commission regulated the industry to prevent exploitation of the consumer. This act consisted of establishing a cost rate base system of controlling operating and capital costs of the oil companies. Recently, deregulation of natural gas has produced rapid increases in cost to the consumer. Gas producer profits have climbed in proportion.

OIL SHALE AND TAR SANDS

When oil, natural gas, and coal were considered inexhaustible, oil shale and tar sands were of little interest to us. It was not economically feasible to mine and develop this resource. In recent years, however, interest has increased and more research is being applied to this area.

The U.S. Geological Survey has published estimates of the large oil shale deposits in Colorado, Utah, and Wyoming. The survey estimates the reserve to contain 80 billion barrels of oil. The deposit was estimated to be 25 feet thick yielding 25 gallons of oil per ton of shale. Upon further examination and extension of the surveyed region, additional deposits estimated at 3 trillion barrels of oil were found. In Canada, tar sand was discovered with a reserve of 300 billion barrels of oil.

If there is so much oil in these reserves, why all the publicity on the energy shortage? There are problems in getting the oil from the oil shale and tar sands. The main problems deal with the vast cost differences in recovery of oil from shale and sand rather than from natural crude. The shale and sand are solids and require huge mining equipment. Most shale and tar sands contain only 13 to 16 percent oil, so about four fifths of the work is wasted or useless. The technology at our disposal is also very limited when it comes to extracting the oil economically. The shale must be heated to 800° to 1,000°F (427° to 538°C) to release the oil. For a plant to be commercially important, a capacity of 50,000 barrels per day is required. This means the plant would have to process about 75,000 tons of shale per day, of which 60,000 tons would be waste. On a yearly basis, the plant would have to dispose of about 15 million tons of waste product. Because water is also required in the process, careful consideration must be given to this resource. Rock disposal and water pollution are only two of the critical environmental factors that need to be considered.

SUMMARY

For many years, fossil fuels such as coal, oil, and natural gas were believed to be our only source of energy. The United States has reserves of each, but we must remember that they are nonrenewable resources. Some day they will run out.

We must constantly be on the lookout for advanced technology that will help us conserve these valuable and useful resources. These include improved discovery and recovery methods and better processing methods as well as consumer conservation measures. As with most good things, there are some concerns. When coal is removed from strip mines, tons of soil are moved. Therefore, the balance of the ecosystem is interrupted. When oil is piped over frozen ground, or carried in huge tanker ships, there is a chance for a spill or slick. Procedures have been instituted and are continually monitored to ensure that other segments of the environment do not suffer because of our mining operations. The next chapter discusses possible alternative methods of relieving the pressures placed on the fossil fuels.

DISCUSSION QUESTIONS

1. What is coal?
2. How is coal mined?
3. What are the coal reserves of the United States?

4. What are the differences between shaft mines, slope mines, and drift mines?

5. What is the most dangerous coal mine?

6. Who governs coal mine safety standards?

7. How is oil formed?

8. What are the three devices used to find oil deposits? What do they tell the geophysicist?

9. What are the three oil drilling methods? Briefly explain each.

10. How do offshore drilling methods differ from land drilling methods?

11. List at least ten uses of oil.

12. Where is natural gas found?

13. What unit of measure do we use for natural gas?

14. List the steps natural gas takes from the well to the home.

15. What are the problems associated with oil shale and tar sands?

SUGGESTED ACTIVITIES

1. Prepare a field trip to a strip mine site or an oil field if available. If those are not available, perhaps a petroleum refinery, pumping station, or storage facility is nearby.

2. Prepare a report on the safety procedures in mining. How have mining procedures changed over the years?

32 ALTERNATIVE ENERGY SOURCES MANAGEMENT

OBJECTIVES

After reading this chapter, you should be able to

- explain the use of solar energy as an alternative energy source

- discuss the operation of a nuclear power plant

- explain the value of geothermal energy, alcohol, methane, hydropower, tidal power, wind, and wood as alternative energy sources

TERMS TO KNOW

solar energy	meltdown	hydropower
nuclear power	geothermal energy	tidal power
atom	methane	
chain reaction	methane digester	

As the prices of fossil fuels increase, many Americans are searching for other ways to supply their energy needs. When gasoline was 25 cents per gallon, few people were concerned with examining other fuel sources. In this chapter, we review the types of fuels that can be used as an alternative to the fossil fuels discussed in Chapter 31.

The main types of alternative energy sources include solar power, nuclear power, alcohol, geothermal energy, methane, wind, hydropower, tidal power, and wood. Much research has been conducted in developing each source into a reasonable alternative to fossil fuels, and individuals can adapt one or all of them to fill part of their energy needs.

SOLAR ENERGY

Solar energy is more abundant, less exhaustible, and more pollution free than any other energy source. The sun floods the earth daily with energy 100,000 times greater than the electric power capacity of the entire world. The sun as a potential energy source has interested people for a long time, but it was not until the extreme rise in cost of conventional energy sources that a closer, more detailed look at solar energy came about. More and more houses are being built that employ solar energy as a primary energy source.

Solar energy systems can be divided into two major types: active and passive. The active

FIGURE 32-1 A solar collector. *(Courtesy Photo-Disc)*

systems are those that capture, store, and distribute the energy from the sun, whereas passive systems provide avenues for the sun to enter but rely on natural airflow to provide distribution.

An active solar system includes a collector (Figure 32-1), a storage mechanism, and a distribution device. The active solar systems overcome the age-old problem of the entire solar concept: The energy is most needed when the sun is not shining—for instance, at night. This system collects the sun's energy in the form of heat and stores it for future use. The storage area can be water, such as an indoor swimming pool, or stones and bricks located in the basement. Once the heat is in the storage area, it can be distributed as the need arises. The heat is extracted by means of pumps or fans, depending on the storage method used.

A passive solar system (Figure 32-2) is less expensive to construct than an active system because it has only a collection device. This device consists of a south-facing solar panel that collects the sun's rays. This method works well in greenhouses and homes equipped with another backup heat source. Construction techniques must be carefully

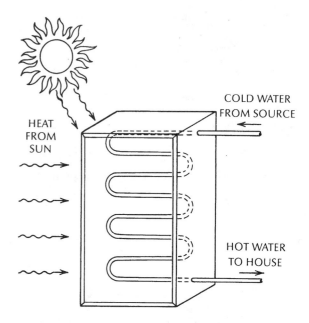

FIGURE 32-2 Passive solar system.

followed to ensure that the heat is held once it has been captured. Passive solar systems are used to heat water, dry materials such as grain, distill water, and in cooking.

The newest research conducted with solar energy concerns converting the sun's rays (radiant energy) into electricity (electrical energy). The second law of thermodynamics tells us that when we change the form of energy, heat is lost. Research indicates that about 99 percent of the radiant energy from the sun is lost in the form of heat leaving the earth's surface. The main goal of using solar energy to produce electrical energy is to more efficiently use the escaped heat of the sun. This method can be considered a rather indirect way of converting solar energy to electrical power. Research is also being conducted on the direct conversion of sunlight to electricity by the use of solar cells. Solar cells need further development before they are readily available to the public.

NUCLEAR POWER

Nuclear power plants have had much news coverage in recent years, mostly because of fears of a nuclear disaster. Here, we examine what nuclear power is, where the power comes from, and what some of its problems are.

All matter is composed of small, submicroscopic particles called **atoms**. The atom is made up of a central nucleus of positively charged protons and neutral (or uncharged) neutrons. Outside the nucleus are negatively charged electrons traveling at very fast speeds.

In 1896, Henri Becquerel, a scientist working with photographic plates, discovered that when the plates were subjected to radium they became exposed even though they had been covered. It was later discovered that the exposure of the plates was due to radiation emitted from the radium atom (thus the term *radioactive*). Radium emits positively charged alpha rays, negatively charged beta rays, and neutral X rays. When rays interact with other compounds, they split atomic nuclei. This split is called fission. During the fission process, heat is given off, which is the important component in a nuclear power plant.

One important characteristic of radioactive material is that once a fission process is started, it can continue on its own. This process is called a **chain reaction**. Uranium is a good example. Uranium oxide is the important material found in the earth's crust that allowed the development of the atomic bomb. It is also the one material most used in nuclear power plants. The nucleus of the uranium atom contains 146 neutrons. When the uranium nucleus is flooded with neutrons, it accepts one, making a total of 147 neutrons. This is a very unstable condition and causes the uranium to disintegrate. The disintegration breaks the ura-

nium into two different elements, krypton, with 47 neutrons, and barium, with 82 neutrons. Between krypton and barium, we have used 129 neutrons, leaving 18 unattached neutrons to flood other uranium particles, which causes additional reactions. Each time a split occurs, heat is produced. Fissioning one pound of uranium quickly yields an explosive force equivalent to 10,000 tons of TNT; released slowly, it can produce 12 million kilowatt hours of power.

The heart of the nuclear power plant (Figure 32-3) is the reactor. The reactor uses a mixture of U^{235} and U^{238}, in the form of uranium pellets, as a fuel source. This makes up the reactor core. As discussed earlier, the chain reaction is triggered by free neutrons in the fuel mixture. To control or stop the reaction, the reactor uses cadmium rods to absorb neutrons. The rods can be inserted or withdrawn around the core. As the reaction progresses, heat is produced. Water in tubes surrounding the core turns to steam, which is sent through turbines to turn electrical generators. An auxiliary water system is available to maintain the core at about 1000°F (538°C).

FIGURE 32-3 Nuclear power plant. (*Courtesy of the U.S. Department of Energy*)

The main concern with the increased use of nuclear power is the fear of an explosion or an uncontrolled heat buildup causing a **meltdown**. High public pressure has resulted in strict controls designed to prevent a nuclear disaster. Nuclear power plants also generate great amounts of hot water, and this water must be cooled before reentering the reactor. If the cooling is completed in a nearby stream or river, there can be environmental damage through thermal pollution.

Another main concern with nuclear power plants is the disposal of radioactive wastes from the reactor. The uranium in the reactor core will last up to two years before it must be replaced. The wastes are first sent to a processing plant that recovers any unused uranium. Once the recovery process is complete, the wastes are packaged in stainless-steel containers and buried. The wastes are still somewhat active and, therefore, continue to generate heat. The containers last only about 100 years; then they must be repaired or replaced. A better technique may be to place the containers in above-ground concrete bunkers. Wherever they are placed, extreme care must be used to ensure that the containers are not broken open.

GEOTHERMAL ENERGY

Geothermal energy involves tapping the underground reservoirs in volcanically active areas. The steam is piped through the ground to turbines that turn electric generators. The largest geothermal plant in the United States, which uses steam from geysers, is located in northern California. Operated by Pacific Gas and Electric Company, this geothermal plant supplies electricity to the city of San Francisco.

The two main disadvantages of geothermal energy are (1) the energy is not uniformly located around the country and (2) the minerals in the steam are very hard on machinery. The development of machinery to withstand the abrasives is needed to make this energy source economically feasible.

In areas where geothermal energy is available, it is cheap and clean. It is nonpolluting, and setting up a plant requires only a small investment.

ALCOHOL

Alcohol is another possible new energy source. We have heard numerous stories of backwoods moonshiners who made their own alcohol. In the 1930s and 1940s, several cornbelt states passed legislation giving farmers tax rebates on alcohol blends. With the end of World War II, however, petroleum prices dropped and interest in alcohol production faded. With the increased petroleum prices in the 1970s came a renewed interest in alcohol as a fuel.

Alcohol is produced by growing yeasts in a grain solution. The yeasts take in sugar, proteins, vitamins, and minerals, giving off carbon dioxide and ethanol. The grain solution can be anything from moldy corn to cheese waste. Supplies will never run dry like an oil well and need no fancy refineries.

Alcohol can be used in a blend with gasoline, called gasohol, as a substitute for diesel fuel, kerosene, and heating fuel oil. It is less explosive and more stable than gasoline, also less polluting, more efficient, and more economical.

Alcohol production also produces some useful by-products. The carbon dioxide can be used for the carbonation of beverages, for drying grain, in fertilizer production, in fire extinguishers, in refrigeration systems, and in dry-ice manufacturing. The residue from the alcohol can be used as feed for livestock. Research has shown there is an increase of 15 to 30 percent in feeding efficiency when using the alcohol residue.

METHANE

As cities continue to grow and grow, the problem of waste management also increases. One method of putting the waste to good use is by using anaerobic bacteria to decompose the wastes. Anaerobic bacteria grow and prosper in the absence of oxygen. **Methane**, sometimes referred to as "bio-gas," is a by-product of decomposition by anaerobic bacteria. The decomposing wastes can be from human, animal, or plant material.

Methane is an odorless gas with a heating rate of 600 to 700 British thermal units (BTUs) per cubic foot of gas. (1 BTU is, approximately, the amount of heat needed to raise 1 pound of water 1°F.) Under natural conditions, decomposing wastes produce methane and a compound called hydrogen sulfide. The hydrogen sulfide gives the gas a very unpleasant odor and is sometimes referred to as "sewer gas." Through devices called scrubbers, the odor-bearing compound can be removed from the methane.

Methane as can be produced artificially in a device called a **methane digester**. (See Figure 32-4.) The digester is an airtight container holding the wastes to be decomposed. Anaerobic bacteria break down the wastes giving off bio-gas as a by-product. Bio-gas contains about 60 percent methane, 35 percent carbon dioxide, 3 percent nitrogen, 0.1 percent oxygen, and a trace of hydrogen sulfide. For optimum gas production, the digester should be gently agitated and maintained at a temperature of 100°F (38°C). The digester will produce about 2 to 5 cubic feet of gas per pound of waste. Although over 50 percent of the gas is extracted from the waste in the first two weeks of production, gas can be obtained for up to six weeks.

After the wastes have been digested, solids, called sludge, remain. Sludge is an important by-product of methane production because of its use as a fertilizer. Analysis of methane sludge shows that it contains nitrogen, phosphorus, potassium, boron, calcium, copper, iron, magnesium, sulfur, and zinc, all of which are essential for plant growth.

One problem with the digester is the gas storage technique. Because of its characteristics, methane cannot be highly compressed like other gases. Storage structures at the present time must be large and bulky.

The heat from burning methane can be used many ways. Home methane production can be done at very low cost, and the wastes can be used on home gardens, thus lessening fertilizer costs. Some farmers are constructing methane digesters to be used as a heat source for the production of alcohol from grain. The methane digester also gives the farmer a way of getting something extra from livestock wastes. Methane is one method of lessening the amount of fossil fuels used in this country.

HYDROPOWER

Hydropower is water power. In its simplest form, the power from moving water is used to do work, whether it be grinding wheat or producing electricity. Water power has been used for many centuries. To gain more control over

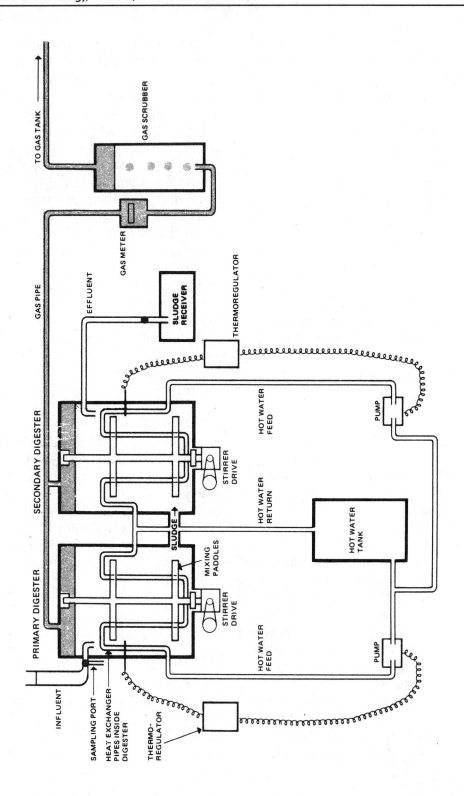

FIGURE 32-4 Diagram of an experimental livestock waste digester.

FIGURE 32-5 Power dam. *(Courtesy of the Bureau of Reclamation, U.S. Department of the Interior)*

water, dams, like the one shown in Figure 32-5, are constructed in water pathways. Water, released through openings in the dam, is used to drive turbines connected to electrical generators.

The continued development of hydropower as an alternative energy source has been opposed by some environmental groups. When a dam is constructed, the waters are slowed and backed up, and many wilderness areas are destroyed by the artificial reservoirs. Legislators have already stopped future hydropower projects on the Colorado, Columbia, and Snake Rivers. People will have to decide between environmental beauty and electricity. This conflict is discussed in some detail in Chapter 3.

TIDAL POWER

An alternative energy source being researched at the present time is **tidal power**. The tides work like clockwork and carry huge amounts of force behind them. Finding a way to harness this energy source has been intriguing to researchers.

The method most often used is the building of small basins that collect the water during high tide. When the tide waters recede, the water is released through openings containing turbines that drive electric generators. The potential of tidal power is estimated at 2.9 million megawatts of power.

The principal drawback of tidal power is the massive dams and levees that must be constructed initially. Once the construction is complete, the operation is relatively inexpensive and power is always available.

WIND

Wind has been used as an energy source for many years. The early settlers relied on the wind to bring water up from deep wells. Many parts of the country continue to use wind power to provide water for livestock where electricity is not available. The incredible power of the wind can be examined just by watching a tornado at work.

In recent years, scientists have been experimenting with new and more efficient windmills. The new structures appear to be more like tall "windtowers" equipped with propeller blades. The main use of the new towers is to turn generators that produce electricity. The rotating generator may produce either alternating current (AC) or direct current (DC), depending on the needs of the individual.

As with most energy sources, this alternative source has its limitations. The wind does not blow all the time, and the speeds at which it blows vary. This creates a need for some sort of storage mechanism. The most widely used storage device is the DC battery. Another problem is in the construction of the high towers. With airplane traffic increasing daily, the builder must be careful where he or she places the tower. Building codes need to be explored and followed to avoid unnecessary accidents.

WOOD

Wood is one of the oldest energy sources known to humans. During the 1800s, wood was the only supply of fuel available for fac-

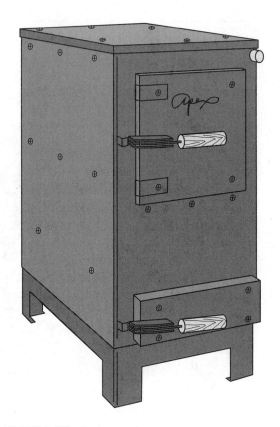

FIGURE 32-6 A wood stove.

tories and home heat. As with other sources discussed, wood has become more attractive with the rise of traditional energy prices. (See Figure 32-6.)

Wood as a heating source has several advantages. Wood is widely available, so transportation costs can be low. It comes from a renewable natural resource. Much research has gone into developing faster-growing trees. With correct forest management practices, the supply of wood should be abundant. Wood also gives us an energy edge in case of a disaster. Water pipes can be prevented from freezing, cooking can continue, and lives can be saved if this alternative energy source is available.

Wood energy does have its disadvantages. It is less convenient to burn than other energy sources and is bulkier and less efficient than oil and gas. Most people who enjoy heating with wood usually like the independence from other expensive fuels.

SUMMARY

Energy costs are high. People are always in search of alternate sources of energy, whether for their home, office, or automobile. Each alternate source examined in this chapter can be used. Each individual must look carefully at each source to be sure the right selection is made. Not all alternative sources are for everyone. If each uses only one, then the dependence on high-priced fuels will be lessened.

DISCUSSION QUESTIONS

1. What is the potential of solar energy?
2. Explain the difference between an active and passive solar energy system.
3. How can solar energy be stored?
4. How was nuclear energy discovered?
5. What is fission?
6. Why is the chain reaction important in a nuclear power plant?
7. How is the chain reaction started? How is a chain reaction controlled?
8. What is the fuel used in a nuclear power plant?
9. What are the major concerns of nuclear energy?
10. Where is geothermal energy obtained? What are the problems related to geothermal energy?
11. How is alcohol produced?
12. Where is alcohol used as a fuel? For what are alcohol by-products used?
13. What is bio-gas?
14. How is methane produced?
15. How are the by-products of methane production used?
16. What are the problems associated with methane production?
17. What is hydropower? What are the environmental concerns of hydropower?
18. What is the potential of tidal power?
19. How has modern technology changed wind power?
20. What are the advantages and disadvantages of using wood as an alternative energy source?

SUGGESTED ACTIVITIES

1. Choose one alternative energy source. What changes would you have to make in your home to implement the new energy source?
2. Invite a representative from your local electric company or gas company to speak to your class about energy sources for the future.
3. Design and build a passive solar energy collector. There should be plans available in your school or public library.

33

METALS AND MINERALS

OBJECTIVES

After reading this chapter, you should be able to

- explain the principal metal and mineral resources

- list and explain the various metals and minerals, including ferrous, nonferrous, scarce, and plant minerals

- discuss mining principles, resources available, and uses of minerals

TERMS TO KNOW		
ferrous metals	pig iron	Bayer process
taconite	ferroalloys	red mud
Bessemer process	nonferrous metals	fertilizers
benefication	bauxite ore	

Correct management procedures are extremely important when working with our mineral and metal resources. There is no second crop of minerals—they are nonrenewable resources. Are the mineral resources of our earth abundant or are they scarce? It is estimated that the earth contains about 2 billion tons of metal resources per square mile of land surface. The problem stems from the fact that the resources are at such low concentrations and at such depths that it is not usually economically feasible to remove them. The shallow concentrations could actually be placed in a 1,000-square-mile area.

In this chapter, we examine the principal metal and mineral resources of importance. They are classified as ferrous, nonferrous, scarce, and plant minerals. We examine mining principles, the resources available, and uses of the minerals.

FERROUS METALS

Of the **ferrous** (iron-containing) **metals**, iron ore is by far the most important. It is the basis of the steel industry—in fact, it is the foundation of all modern industry. According to the U.S. Bureau of Mines, 185.9 million tons of iron ore were produced in 1968. Huge iron ore discoveries in the Great Lakes region spurred the tremendous growth of the steel industry in Indiana, Illinois, New York, and Minnesota.

Data from the U.S. Geological Survey, reported by the World Resources Institute, provides more current information. As of 1995, those estimates were 554.8 million metric tons of iron produced worldwide. Of that, the United States produced 39.6 million metric tons, down substantially from the 1965 total. Only Brazil (120.9 million metric tons) and Columbia (75 million metric tons) produced more iron ore than the United States.

There are many kinds and varieties of iron ore. The most valuable are hematite (Fe_2O_3), magnetite (Fe_3O_4), limonite ($2Fe_2O_3 \cdot H_2O$), and siderite ($FeCO_3$). Low-grade ores are called **taconite**. A high-quality ore contains 60 percent metallic iron; a low grade, 40 percent or less. When iron ore is processed, steel is made. The development of the **Bessemer process** led to the cheap, practical production of steel. The process consists of making steel by forcing a blast of air through molten cast iron to remove impurities.

Once the iron ore deposits are discovered, normal mining procedures, as discussed in this chapter, are used. If the deposit is close enough to the surface, strip-mining techniques are used. Otherwise, tunnels and shafts are constructed. The ore can be obtained by hand or by using machinery. The ore is then sent to a mill to be separated from the rock. This separation process is called concentrating, milling, ore dressing, or **benefication**. This process is done near the mine to reduce the amount of material that must be transported for further processing.

The next step is smelting (melting) the iron ore to remove impurities from the concentrate. It takes about two tons of coal to process one ton of steel. The smelting plants are usually located near coal sources.

Once the molten iron is processed, it is poured into molds to form bars called **pig iron**. These bars are then shipped to manufacturing plants to be processed into various forms of steel, such as sheets, rods, tubes, or wire. (See Figure 33-1.)

FIGURE 33-1 Iron ore smelting plant. (*Courtesy of Bethlehem Steel Corporation*)

Other ferrous metals are combined with iron to add various properties to steel. The compounds are called **ferroalloys**. These include chromium, manganese, molybdenum, nickel, tungsten, and vanadium.

NONFERROUS METALS

Nonferrous metals have gained importance in recent years. These include copper, aluminum, lead, zinc, tin, and mercury.

Copper

Copper is one of the most beautiful metals known, probably second only to gold. It is very versatile, with a characteristic red color. It is ductile (it can be drawn out and hammered thin) as well as malleable. When it is heated and placed in water, it becomes soft. If it is heated to a higher temperature it becomes brittle.

Copper is commonly used for tools and weapons because of its natural characteristics. It is used in kitchen utensils, coins, screens, and piping. It is noted for its noncorrosive quality. One half of all copper is used in electrical products. It is used in radio communication equipment and electrical power transmission lines. The military uses copper in weaponry, and medical scientists use copper as an antiseptic as well as an emetic to induce vomiting.

Copper is commonly combined with other metals to add additional qualities. Copper and tin make bronze. When it is combined with zinc, we call it brass, which is widely used in musical instruments and for other devices requiring a hard, noncorrosive alloy.

The world copper reserves amount to about 212 million tons, located mostly in South America. The United States places its reserves at 32 million tons, located mostly in the Lake Superior region. Worldwide, copper ore mining totaled 10 million metric tons in 1995. Of that, the United States mined just over 1.8 million metric tons. Only Chile mined more copper ore that year, at 2.5 million metric tons. Much of the copper used is recycled; if we want to continue to use copper, even more will have to be recycled.

Aluminum

Aluminum is one of the most common metals known on earth. Because of its availability, light weight, heat reflectivity, corrosion resistance, and electrical conductivity, aluminum is often used in place of steel. Aluminum accounts for about one seventh of the earth's core and is found in many forms. The most common ore in which aluminum is found is bauxite. Although other ores such as kaolin and corundum contain aluminum, bauxite is the easiest to find and mine.

Bauxite ore is found in the earth as sheets or layers anywhere from 1 foot to 75 feet deep. It contains large amounts of aluminum hydroxide, a combination of aluminum oxide and water. (The aluminum oxide is the part from which aluminum comes.) Bauxite contains about 50 percent aluminum oxide. It can be various colors and can be rock-hard or as soft as clay. Most bauxite (85 percent) comes from Arkansas. The remainder comes from the states of Alabama and Georgia.

The method of mining bauxite ore is very similar to that of strip-mining coal. The soil and rocks are removed from the ore by huge bulldozers and earthmovers. The ore is blasted loose with explosives and then loaded into trucks and railroad cars to be shipped to a processing plant. Once the bauxite ore reaches the processing plant, the aluminum oxide, or aluminum as it is commonly called, is extracted from the bauxite. This is called the **Bayer process**, after the Austrian chemist who developed the technique. It involves mixing the crushed bauxite with caustic acid. When the solution is heated and put under pressure, the aluminum is released from the ore and combines with the sodium, creating a compound of sodium aluminate in liquid form. The remaining compounds, in solid form, are called **red mud** because of a characteristic red color.

The sodium aluminate solution is sent into a tank or precipitator to cool. This tank is called a precipitator. While in the tank, aluminum hydroxide crystals are added. When the solution is agitated, the aluminum breaks away from the sodium and collects on the crystals. The crystals are then heated to 2000°F (1094°C) to drive off the water, and all that remains is the white powdery alumina.

The second major step in the making of aluminum is the separation or smelting of the alumina. Alumina contains aluminum and oxygen. To separate the two, the alumina must be placed in a smelting pot with a carbon lining. The alumina is placed in solution with cryolite and heated to 1740°F (950°C). Another carbon block is then placed in the center of the pot. When electricity is applied to the center block, it flows down the block, through the solution, and on to the carbon lining. The current causes the alumina to spread apart. The oxygen combines with the carbon and is released as carbon dioxide gas. The pure aluminum collects at the bottom of the pot. An aluminum plant may have as many as 200 pots producing aluminum all the time. One pot can process about 2 tons of aluminum daily. The molten aluminum is then removed and placed in blocks or ingots. These ingots are shipped to foundries where aluminum products are constructed.

Worldwide, bauxite mining totaled 27.3 million metric tons in 1995, with the United States mining just 50,000 metric tons. The largest bauxite producing countries in that year were Guinea (3.6 million metric tons), Jamaica (2.7 million metric tons, and Brazil (2.2 million metric tons). The industry supplies jobs to over 150,000 people. The recent growth of the aluminum industry was due mostly to advanced processing techniques.

The demand for aluminum has steadily increased since World War II. One of the largest users in recent years has been the automobile industry. To increase efficiency and decrease weight, aluminum has been substituted for steel for various automobile parts. Aluminum is also used in solar units.

Because of the large amount of electricity needed to produce aluminum from bauxite, recycling aluminum is important. Recycling involves melting scrap aluminum back into ingots. It only takes about 5 percent as much energy to produce recycled aluminum as it takes to produce it from ore.

The estimated reserves of bauxite in the world are expected to last 200 to 300 years. In 1974, an eleven-nation association was formed to establish bauxite ore prices. The International Bauxite Association (IBA) has been able to raise revenues of the companies mining bauxite in member countries. The United States is not a member of the IBA; therefore, we should try to recycle and conserve as much as possible.

Lead

Lead ore is found mainly west of the Mississippi River. Nearly half our lead comes from Missouri, with most of the remaining resource being mined in Idaho, Colorado, and Utah. Lead is used in batteries and metal building construction, and it once was used in gasoline additives and paint. About 60 percent of the lead used annually comes from recycled sources. Lead is used because of its heavy weight, softness, corrosion resistance, and malleability. Unlike copper, it is not ductile and cannot be drawn into a wire. Lead is also known for its ability to alloy or mix with other metals.

Much of the lead used in the United States is imported. This country's annual use exceeds 1 million tons. The reusability of lead helps keep imported lead to about one third of annual use. The U.S. reserves are estimated at 18 million tons. At the present time, one third of the lead is newly mined and two thirds is recycled.

Zinc

The most common use of zinc is for galvanizing (coating) steel, thus making it rust resistant. Zinc is also instrumental in making brass.

Many parts of the automobile are made in zinc castings. About 30 percent of brass is zinc. Zinc is also used for kitchen utensils, vacuum cleaners, and washing machines.

The zinc reserves of the United States are estimated at 12 million tons. Much of the annual zinc supply is imported to cover our 1.5 ton yearly consumption rate. If we want less dependence on foreign sources, price incentives must be given for exploration purposes. Lower-grade ores have been discovered but, at the present time, are not economically feasible to mine.

Tin

Tin is usually combined with copper to make bronze. In nature, it occurs in veins in granite rocks. The production of tin has been small in past years. Of the 59,000 tons consumed, about 40,000 tons are imported. Most domestic supplies are found in Alaska, Colorado, and Texas.

Mercury

Mercury is the only metal stable in a liquid state at ordinary temperatures. Mercury in large deposits is rare. Most deposits are small and scattered in the states of California and Nevada. The main uses of mercury are for electrical equipment, special paints, and industrial chemicals. Cinnabar is the ore containing mercury; it is found in rocks such as limestone and volcanic rock.

PLANT MINERALS

It has long been known that plant minerals are an important part of producing food. These plant minerals are called **fertilizers** and include nitrogen, phosphorus, and potassium along with many others in smaller quantities.

Nitrogen

Most of the air we breathe is composed of nitrogen. As part of the nitrogen cycle (the continuous process in which nitrogen passes from air to soil to organisms), plants remove nitrogen from the air at a rate of about 400 million tons per year worldwide.

There are nitrogen deposits in the earth's surface. Most of the deposits are located in South America. Today, most nitrogen used is produced by a synthetic process from air. Nitrogen is used in the production of explosives as well as plant fertilizers. As of 1995, production of nitrogen was measured in terms of metric tons of ammonia (NH_4) or its equivalent. Worldwide, 91.6 million metric tons of ammonia was produced. China produced 19.5 million metric tons of ammonia. The United States produced 13.3 million metric tons. India produced 7.7 million metric tons, and the Russian Federation produced 7.5 million metric tons.

Phosphorus

Phosphorus is found in nature in the form of phosphate rock. Most of the phosphate rock mined is channeled toward agriculture fertilizers. About 20 percent of U.S. production is used for industrial production of phosphoric acid and elemental phosphorus, and another 20 percent is exported.

Natural phosphate rock is called phosphorite; it is found in veins created by marine sedimentary processes. Many countries contain large supplies of phosphate rock, but the United States is considered the leader. Worldwide, production of phosphorus (P_2O_5) was 40 million metric tons in 1995. The United States was the largest producer at 12.8 million metric tons, followed by China at 6.4 million metric tons and Moldavia at just under 6.4 million metric tons. At this rate, it is estimated that our reserves may last over 2,500 years. The major producing states are Florida, Tennessee, North Carolina, and South Carolina.

Potassium

Potassium is considered a basic crop nutrient and is applied frequently on America's cropland. Potassium is very prevalent in nature, making up over 2 percent of the earth's crust.

The most common name for potassium is potash. The term *potash* stems from an early accidental use of potassium. When wood was the most common heating fuel, wood ashes were quite abundant. It was noticed by the early settlers that plants grew very well in places where wood ashes were dumped. Some settlers started boiling the wood ashes in water and applying the water to their soil. If they were not careful and allowed the water to boil completely away, a white powdery residue was left behind. When this residue was applied to the soil, the plants seemed to grow even better than when only wood ashes or water from wood ashes were applied; because it came from boiled wood ashes in a pot, the residue was called potash. The important chemical in potash was later discovered to be potassium.

In nature, potassium occurs in the form of salts in beds similar to coal. The United States has about 330 million tons of potassium reserves. As of 1995, worldwide mining of potassium, in the form of potash (K_2O), totaled 24.7 million metric tons. Canada produced 9.0 million metric tons of potash. Germany was second it 3.3 million metric tons and Belarus at 3.2 million metric tons and the United States at 1.5 million metric tons. Along with the production of potash from mines, potassium can be obtained from desalination processing of seawater.

SUMMARY

Minerals and metals are essential to this country. They go into our cars as well as thousands of other products we use everyday. They are responsible for producing abundant yields on our croplands. The main goals of mineral management for the future should be exploration for new sources, the development of technologies to reclaim what we now use, and making more efficient use of what we have. Minerals and metals are a one-time resource. Most cannot be renewed, and those synthetically produced usually require more energy, thus depleting other resources.

DISCUSSION QUESTIONS

1. How is iron produced?
2. List four common uses of iron.
3. What are the common uses of copper?
4. What is brass? bronze?
5. How is aluminum made?
6. From what ore does aluminum come?
7. What are the common uses of aluminum?
8. What are the uses of lead? zinc? tin? mercury?
9. What are the three most important plant minerals?
10. Why is potassium called *potash*?

SUGGESTED ACTIVITIES

1. Plan a recycling station. Collect all metals that are recyclable. Sell the metal and use the funds for a school-sponsored project.
2. Tour a metal salvage yard. How are the different types of metals separated? How are they processed for shipping?

34 CAREERS IN ENERGY, MINERAL, AND METAL RESOURCES

OBJECTIVES

After reading this chapter, you should be able to

- explain opportunities in the oil fields

- discuss the possibilities and requirements of a career in the coal industry

- understand the opportunities available in the field of nuclear energy

TERMS TO KNOW		
petroleum geologist	petroleum engineer	stopping builder
geophysical prospector	gas treater	

Careers in energy, minerals, and metals cover a very broad field. Careers in mining, electricity, nuclear power, and alternative fuels are all part of this vast, changing area. For the purpose of this book, we consider the career opportunities in two parts: oil and natural gas; and the coal, nuclear energy, and steel industries.

OIL AND NATURAL GAS CAREERS

With the public pressure to lower American dependence on foreign oil, many opportunities have opened up in the energy career areas, including the areas of natural gas and nuclear energy.

Nature of Work

A person employed in the oil or natural gas field can work with exploration, drilling, refining, pumping, or transporting teams.

Exploration is the first step in petroleum production. This process is usually accomplished by a small crew of workers who study geological formations that might contain oil.

The exploration team is led by a **petroleum geologist** who analyzes and interprets the information gathered by the team. Other geological specialists include paleontologists (who study fossil remains in the earth to locate oil-bearing layers of rock), mineralogists (who study the physical and chemical properties of mineral and rock samples), stratigraphers (who determine the rock layers most likely to contain oil and natural gas), photogeologists (who examine and interpret aerial photographs of land surfaces), and petrologists (who investigate the history of the formation of the earth's crust). Other specialists who might be a part of the exploration team are drafters and surveyors assisting in surveying and mapping operations.

A **geophysical prospector** usually leads a crew of gravity-prospecting observers and seismic-prospecting observers who operate and maintain electronic seismic equipment. Scouts investigate the drilling, exploration, and leasing activities of other companies to identify areas to explore and lease. The business arrangements are then handled by the lease buyers.

Once oil is found or a likely area is located, the drilling team is assembled. The overall planning of the drilling operation is the responsibility of a **petroleum engineer**. The typical drilling crew consists of a driller, derrick operator, engine operator, and one or two helpers. The drilling operation runs 24 hours a day, 7 days a week, making it necessary to have several crews for each rig.

The driller supervises the crew, operates machinery that controls drilling speed and pressure, and records operations. The engine operator is in charge of the engine that provides the power for drilling and hoisting. The derrick operator is second in charge and helps see that the pipe is correctly run into the well hole. The helpers, commonly referred to as roughnecks, guide the lower end of the pipe to and from the well opening and connect and disconnect pipe joints and drill bits. Associated with the drilling team are general laborers, commonly referred to as roustabouts, who do general oil-field maintenance and construction work, such as cleaning tanks and building roads.

Once the oil is found, the producing well is supervised by the petroleum engineer. Some companies hire aides to make tests, keep records, post maps, and assist engineers. The pumps and motors are cared for by the oil pumpers and their aides. One pumper may operate several wells. Pumpers open and close valves to regulate the oil flow from wells to tanks or pipelines. The gaugers measure and record the flow and take samples to check

quality. Treaters test the oil for water and sediment and remove these impurities when they are found. The skilled positions of welders, pipefitters, electricians, and machinists all support the maintenance work of the oil well.

Careers in natural gas are mostly those of operators. The **gas treater** operates a unit that removes water from the natural gas. The gas-pumping station operator tends compressors that raise the pressure of the gas for transmission through the pipeline. The gas-compressor operator assists the other employees.

In the petroleum industry, there are also a variety of supportive jobs. Many times the oil companies contract these jobs out to other companies. The skilled workers in this area include the oil well cementer, who mixes and pumps cement into the space between the steel casing and the well walls, thus preventing cave-ins. An acidizer pumps acid down the well to increase the flow of oil. The perforator operator uses a subsurface gun to pierce holes in the casing to make openings so that oil flows into the well bore. Also assisting in the oil production is the sample taker operator (who takes samples for the geologist to test) and the well pullers (who clean, repair, and salvage the well parts).

Working Conditions

Most jobs in the petroleum fields are rugged, outdoor jobs that require work in all kinds of weather. Physical strength and stamina are important because the work requires much standing and the lifting of heavy objects. A person in the oil field may be expected to move to another area as a well nears its end. The exploration field personnel may be expected to move more frequently than those in other careers. Working in the oil and natural gas fields is also more hazardous than other kinds of work. The injury rate in the petroleum industry is about four times as high as the average in all other private industrial careers.

Training

Workers in nonprofessional positions with an exploration crew begin as helpers and advance into one of the specialized jobs. They are usually trained in the field, which may require several months to complete. In some cases, the training may last for years. Workers are usually hired by a crew chief at the job site. Many earth-science college students gain on-the-job experience through summer employment with exploration or production crews.

Members of drilling crews usually begin as roughnecks. These workers require mechanical ability and, for the long working hours, physical stamina. With experience, they may advance to more skilled jobs.

Well operations and maintenance companies generally hire people who are familiar with well operations and who live close to the wells. They prefer people with mechanical ability. This type of work is usually less strenuous than drilling and offers the advantage of a fixed locale. Members of drilling crews or exploration parties who prefer not to travel often transfer to well operation and maintenance jobs. New workers may start as roustabouts and advance to jobs as switchers, gaugers, or pumpers. Training usually is acquired on the job; at least two years of experience are needed to become a pumper.

For scientists such as geologists and geophysicists, college training with at least a bachelor's degree is required. For a petroleum engineer, a degree in engineering with specialized courses in the petroleum industry is required. There are also career opportunities for college graduates in chemical, mining, civil, or mechanical engineering as well as geology,

geophysics, or related sciences. Petroleum engineering aides include people with two-year technical degrees as well as former roustabouts or pumpers who have been promoted.

Scientists and engineers usually start at junior levels; after several years of experience, they can advance to managerial or administrative jobs. Scientists and engineers who have research ability, particularly those with advanced degrees, may transfer to research or consulting work.

Employment Outlook

In 1998, 339,000 workers were employed in the U.S. oil and natural gas extraction industry. That represents a decline of over 50 percent between 1982 and 1998. Most of the decline was a result of decreased exploration and extraction resulting from low crude oil process during the 1980s and 1990s. With the rapid increases in oil prices in the early 2000s, increased employment in the U.S. oil and gas extraction industry might result. Nevertheless, official forecasts regard the outlook for employment in this sector as uncertain and highly dependent on crude oil prices that can fluctuate wildly as the balance between international supply and demand changes. Overall, the worldwide employment in this industry is predicted to decline by 17 percent from 1998 to 2008. Although the total number of workers sounds very large, you must remember that it includes a wide range of workers: secretaries, truck drivers, oil rig workers, managers, data entry clerks, janitors, petroleum engineers, and all kinds of other workers needed for any large industry. Only about 148,000 workers are actually employed in the extraction process itself. The remainder are support workers of all sorts. As you might expect, the actual oil and gas extraction workers earn more than the support

TABLE 34-1 Median Earnings of Nonsupervisory Workers, Oil, and Gas Extraction versus All U.S. Industry

	Weekly Earnings	Hourly Earnings
All U.S. Industry	$442	$12.77
Oil and Gas Extraction (All)	$719	$16.83
Crude Oil and Natural Gas	$945	$22.66
Oil and Gas Field Services	$600	$13.91

service workers. Earnings in this industry are relatively high compared to all U.S. industry. (See Table 34-1.)

Employment in petroleum and natural gas production is expected to increase faster than the average for all industries. Increased prices for crude oil and natural gas and a national policy to move toward energy self-sufficiency are expected to provide an incentive for the rapid industry expansion. Growth will be concentrated in occupations connected to explorations and drilling. Opportunities should be good for offshore drilling.

Earnings

In 1998 the average annual salary for all petroleum engineers was $74,260. The middle 50 percent of petroleum engineers earned between $56,020 and $93,280. Earnings of the highest paid 10 percent started at $115,820 and went even higher. New college graduates with bachelor's degrees in petroleum engineering earned an average starting salary of $50,400. As of 1998, there were 12,000 jobs for petroleum engineers in the United States. Though

that number is predicted to decline over the next decade, the small number of new graduates is forcing starting salaries to remain relatively high.

On-land drilling crews work 7 days per week, 8 hours per day, and then have a few days off. In offshore operations, they may work 7 days, 12 hours per day, and then have 7 days off. If the well is far from the coast, they live on the structure containing the drilling rig or on a ship anchored nearby. Most workers in well operations and maintenance and natural gas-processing work 8 hours a day, 5 days a week.

Additional information can be obtained by writing to

American Association of Petroleum
  Geologists
P.O. Box 979
Tulsa, OK 74101
http://www.geobyte.com

Society of Petroleum Engineers of
  AIME
6200 N. Central Expressway
Dallas, TX 75206
http://www.aimeny.org

American Geological Institute
5205 Leesburg Pike
Falls Church, VA 22041
http://www.agiweb.org

COAL INDUSTRY

Nature of Work

In 1978, nearly 203,000 miners were employed in U.S. coal mining. By 1985, that number was down to 187,000. The number fell still more to an estimated 80,000 in 1998. Of that number,

51,000 work in underground mining and the remainder worked in surface coal mining operations. The news for the coal mining states is still darker, as the number of coal miners is expected to decline by another 23 percent from 1998 to 2008. About 85 percent of the people who work in this industry are production workers who mine and process coal. Coal mining increased until 1981 in this country, when an economic recession caused a crash in the coal industry. During the 1980s, employment in the U.S. coal industry fell dramatically.

Mining jobs range from apprentice miners, who usually act as helpers, to highly skilled miners, who operate expensive equipment. The jobs can be divided into three areas: mining operations, preparation plant operations, and administrative occupations.

Mining operations use two methods of obtaining underground coal: conventional and continuous. In conventional mining, the cutting-machine operator uses a huge electric chain saw, with a cutter of 6 to 15 feet, to cut a strip underneath the coal seam. This controls the direction the coal will fall when it is blasted. The drilling-machine operator drills holes into the coal where the shot firer places the explosives. After the blast, the loading machine operator scoops up and dumps the coal into cars run by the shuttle car operator. The coal is hauled to a conveyor where it is loaded for shipment. The continuous mining method eliminates the drilling and blasting operations of conventional mining. The continuous mining machine operator sits in a cab and operates levers to cut or rip out the coal and load it directly onto shuttle cars. (See Figure 34-1.)

Before miners are allowed underground, a mine inspector checks the work area for loose roofs, dangerous gases, inadequate ventilation, and other hazards. The rock-dust machine operator sprays limestone on the mine walls and

FIGURE 34-1 In mining operations, particularly surface mining, all blasting must be carefully planned and carried out by qualified blasters. Before passage of the Surface Mining Law, blasting was often done by untrained personnel and frequently caused damage to nearby structures. At this Tennessee coal mine, a federal inspector is inspecting a blast plan to be sure it is safely done. Both the inspector and the mining company employees are part of the coal mining industry. The left truck drills the holes into the ground, and the truck in the back holds explosives that the certified blaster is using. *(Photo courtesy of the U.S. Department of Interior, Office of Surface Mining)*

ground to hold down the explosive, breath-impairing coal dust. The roof bolter operates a machine that installs roof-support bolts to prevent cave-ins. The **stopping builder** constructs doors, walls, or partitions in the passageways to force air through the tunnels and into work areas. The shift boss is the supervisor in charge of all operations at the work site where coal is actually mined.

In surface-mining operations, earth and stone are removed from above the coal seam.

This operation is called *strip mining*. In many strip mines, the overburden is first drilled or blasted. The overburden stripping operator, or dragline operator, scoops the earth away to expose the coal. Sometimes the huge dragline machines are run by a team of people. Once the overburden is removed, the coal-loading-machine operator rips the coal from the seam and loads the coal into trucks to be driven to the preparation plant. Tractor operators drive bulldozers to move materials or pull out embedded boulders.

At the preparation plant, rocks and other impurities must be removed before the coal is crushed, sized, or blended to meet consumer needs. The plant supervisor oversees all operations. The number of workers needed depends on the degree of mechanization at the plant. The main jobs require workers to oversee washing, separating, and crushing operations.

The professional personnel of the coal industry include mining engineers, who examine the coal seams. They determine the depth and purity of the coal and decide what type of mine will be needed. Mechanical engineers oversee the installation of equipment, and safety engineers are in charge of health and safety programs. Technical personnel include surveyors and engineering technicians.

Working Conditions

Miners are subject to harsh conditions. The underground mines are damp, dark, noisy, and cold. Workers in mines with very low roofs have to work on their hands and knees, backs, or stomachs in cramped areas. The miners must constantly be on guard for hazards. There is also the risk of developing pneumoconiosis, also known as "black lung" disease, by repeatedly inhaling coal dust. Surface mines and preparation plants are usually less hazardous than underground mines.

Training

Most miners start out as helpers and advance to more skilled jobs with experience. Mine technology programs are available in a few colleges in coal mining areas. The programs lead to either a certificate in mine technology (after one year) or an associate degree (after two years).

As miners gain more experience, they can move to higher-paying jobs. When a vacancy occurs, miners bid for a job, and the job is filled according to seniority and ability. Miners must be 18 years old and in good physical condition.

Requirements for scientific, engineering, administrative, and clerical jobs are similar to those in other industries.

Employment Outlook

Coal is expected to play an increasingly important role in the future. The major coal users are the electric companies, and electricity is more and more in demand. The demand, however, will also depend on the price of other fuels such as oil and natural gas.

Employment is expected to increase in the long run, but for now the job outlook in this industry looks weak. Growth will depend on the level of production, the types of mines opened, and the mining methods used.

Earnings

Production workers averaged $9.53 per hour in 1978. That increased to $17.53 per hour in 1985 and to $19.16 in 1998. The pay is usually higher for underground mining. Those in surface occupations work 7.25 hours per day, while underground miners work an 8-hour day. Union miners receive 10 holidays and 14 days paid vacation per year. As length of service increases, they gain extra vacation days to a maximum of 18 days.

For additional information, write to

United Mine Workers of America
900 15th Street, NW
Washington, DC 20005
http://www.umwa.org

National Mining Association
1130 17th Street, NW
Washington, DC 20036
http://www.nma.org

Mine Safety and Health Administration
4015 Wilson Boulevard
Arlington, VA 22203

NUCLEAR ENERGY

Nature of Work

In 1990, about 400,000 people worked in nuclear energy activities. Most were employed in facility design, as facility engineers, and in weapons development. This field employs primarily scientists, engineers, technicians, and craft workers. Large numbers of engineering technicians, drafters, and radiation monitors assist engineers in conducting research and in designing and testing equipment.

Working Conditions

Except for radiation safety precautions, the working conditions are the same as in other industries. Extensive safeguards and operating practices protect the health and safety of workers. The Nuclear Regulatory Commission regulates the possession and use of radioactive material, and constant efforts are being made to provide better safety standards and regulations.

Training

Specialized knowledge of nuclear energy is essential for most workers. Many employers require college degrees. Skill requirements for craft workers are higher than in most industries, and most obtain these skills from apprenticeship programs. Many individuals must handle classified data; therefore, they require security clearances. Government contractors often provide employees with training at their own plants.

Employment Outlook

Employment is expected to be fairly stable or even decline in the United States for the nuclear energy field. There is much concern over the safety of nuclear power plants. A few years ago, the number of nuclear power plants was expected to increase rapidly. Today, the opposite appears true. Future growth in this industry is uncertain. Most new workers hired are replacements for previous workers who leave or retire. Few new jobs are expected to be created.

For additional information, write to

U.S. Department of Energy
Office of the Assistant Secretary of
    Energy Technology
100 Independence Avenue, NW
Washington, DC 20545
http://www.doe.gov

STEEL INDUSTRY

Nature of Work

There are about 2,000 different types of jobs in the steel industry. About 80 percent of all workers in a typical plant are engaged in the moving of raw materials and steel products processing the iron and maintaining machinery. The work varies with the type of furnace used and includes everything from unloading raw iron ore to rolling the steel for shipment.

Employment in the steel industry has declined steadily in the United States for at least 20 years. As of 1998, about 232,000 U.S. workers were employed in the steel industry. That is less than half the number employed in 1980. Traditionally steel production was concentrated in a few locations where iron ore, coal, and railroads for transportation are co-located. To this day, about 50 percent of all U.S. steel workers are employed in Pennsylvania, Ohio, and Indiana. In recent years, the development of a new technology has allowed for the movement of small-scale steel production into many locations in smaller plants. The technology is the electric arc furnace (EAF). This is a sort of "mini steel mill" that melts scrap metal to make steel. The expansion of EAFs means that steel can be manufactured anywhere that scrap metal is available. EAFs make about half of all U.S. steel today, and that percentage is expected to increase in the coming years. The expansion of EAF mills means that the traditional jobs in large steel mills are being replaced by a whole new kind of job requiring a different set of skills.

Working Conditions

The conditions vary with different jobs. Workers near the furnaces are subject to extreme heat and noise. The conditions have improved in recent years due to the increased use of remote controls and cabs on equipment. Most plants operate 24 hours a day, 7 days a week. Steel work is hazardous, with an injury and illness rate almost twice as high as for all other industries.

Training

New workers are usually hired as unskilled laborers. Length of service with the company plays an important part for promotions. Training for processing occupations is done on the job, and workers commonly move from one skilled position to other more difficult positions.

The minimum requirement for administrative jobs is a bachelor's degree. Salespeople generally need business administration classes.

Employment Outlook

Employment in the steel industry is expected to remain steady or decline in the coming years. As more steel-making processes are controlled by computers, there will be demands for computer systems analysts. Unless government policies change, foreign steel imports will continue to force a decline in the U.S. steel industry. A 24 percent decline is expected in steel jobs in the U.S. between 1998 and 2008.

Earnings

Earnings in the steel industry have been relatively high. Average weekly earnings of nonsupervisory steel workers were $822 in the United States in 1998. That is well above the earnings of workers in all U.S. manufacturing jobs at $563, and even further above the average for all U.S. workers, which was $442 per week. Within the steel industry, earnings differ from one sector to another. Workers in blast furnace plants and steel mills earned an average of $907 per week compared to $626 per week for workers in plants making steel pipes and tubes.

In recent years, foreign competition has forced much of the steel production in the United States out of business. Most workers are members of the United Steelworkers of

America. Workers receive 1 to 5 weeks paid vacation, and those in the top 50 percent of the seniority list receive a 13-week vacation every 5 years. Employees can retire on company-paid pensions after 30 years of service.

For additional information, write

American Iron and Steel Institute
100 16th Street, NW
Washington, DC 20036
http://www.steel.org

United Steelworkers of America
Five Gateway Center
Pittsburgh, PA 15222
http://www.uswa.org

SUMMARY

Energy is a large and varied occupational area. Coal, oil, natural gas, and nuclear energy are the main fuel sources for industry, transportation, and electricity. Exploration, production, and manufacturing open many career opportunities, with good opportunities for advancement.

Careers in metal and minerals primarily involve the steel industry, but the manufacture of all metals offers career opportunities. For those who are hardworking, excellent advancement opportunities exist here as well.

DISCUSSION QUESTIONS

1. What positions are available with exploratory teams? with drilling teams? with refinery teams?

2. What are the working conditions of people employed in the petroleum industry?

3. What training is needed to work in the petroleum industry?

4. What are the employment opportunities in petroleum?

5. What earnings can be expected by a person employed in the petroleum industry?

6. List the various jobs connected with the coal industry.

7. What working conditions are associated with the coal industry?

8. What is the employment outlook for the coal industry?

9. What are the earnings for coal industry employees?

10. What jobs are available in nuclear energy?

11. What training is required for those seeking careers in nuclear power?

12. What employment opportunities are available in the steel industry?

SUGGESTED ACTIVITIES

1. Prepare a report on career opportunities in the various fields of energy resources. Examine each field and determine the characteristics of each.

2. Invite the manager of a local electrical, natural gas, or other energy supplier to speak to the class on careers in the energy field.

TO REGULATE OR NOT TO REGULATE; THAT IS THE QUESTION.

Throughout the world, electricity improves the quality of life for everyone privileged enough to be able to use this resource. Unfortunately, many people view the methods used to produce this technological marvel as bad news for the environment. (See Figure VII-A.) Historically, regulatory commissions were more concerned with controlling the economic monopolies that naturally developed in the energy industry instead of how the production methods affected the environment.

However, in the 1970s, American policy began to heavily consider environmental impacts. Strategies varied for reducing harmful wastes from energy production. They included

- moratoriums (halts) on construction of nuclear and coal powered plants
- requirements on companies to purchase set amounts of energy produced by alternate methods such as wind
- required submission of environmental impact statements
- new bureaucracies (such as the EPA) whose standards needed to be met

Throughout the 1980s and 1990s, building trends turned toward the technological advances of turbines and cogenerators. Independent companies could use these new technologies and natural gas to fire relatively small and cost-efficient turbines modeled on jet aircraft engines. As the utilities have lost their technological and cost advantage over nonutility companies, many have considered deregulation. Following the telecommunications, commercial aviation, securities, banking, petroleum, natural gas, and trucking industries examples, many hope to see deregulation result in customers being able to obtain innovative new services at declining prices.

The Gulf War of 1991 sent energy prices soaring, and policymakers feared another energy crisis. The Bush administration pushed for market forces that would encourage technological innovations and greater energy efficiency. (See Figure VII-B.) What did this push mean? As a result, the Energy Policy Act of

FIGURE VII-A This traditional power plant in Grand Haven, Michigan, uses fossil fuel to generate electricity. *(Photo courtesy of the U.S. Environmental Protection Agency)*

FIGURE VII-B Photoelectric cell panel generates electricity at Dangling Rope Marina in Utah. *(Photo courtesy of the U.S. Environmental Protection Agency; photo by Roger Hill, Sandia National Laboratory)*

1992 was enacted into law. A potential hazard to the environment resulting from this law is that it permits state governments to deregulate power utility companies and to create free markets in electricity.

The potential environmental hazard comes from the belief that instead of opening the way for more environmentally friendly methods, companies will turn to natural gas only. Thus, the hope for a zero-emissions energy source would be destroyed. Though much cleaner than traditional electrical plants, natural gas is seen as a less desirable alternative because it will still emit some pollution, even though it will be much cleaner than coal or oil fired generators.

When gas replaces coal, it wins in the race for being environmentally sound. However, when it replaces nuclear facilities or meets new demand it is less environmentally sound. Proponents for deregulation look at natural gas as the saving grace of the energy industry. They hope it will open the way for cleaner renewable sources of energy. Opponents feel that gas will become so popular that it will bar the way for dirtier and cleaner sources alike because of its cost effectiveness.

Green power utilities are those that seek to generate electricity with minimal impact on the environment. The Knoxville, Tennessee Utility Board (KUB) and the Tennessee Valley Authority (TVA) announced an agreement in 2000. On February 8, 2000, the *Knoxville News-Sentinel* reported, "KUB customers will be able to volunteer this spring to purchase 'green power'—energy from alternative sources—by paying a surcharge on their monthly electric bills." The news article went on to say, "Their energy needs will partly be met using solar-cell arrays producing 0.25 megawatts. Three wind turbines will produce 660 kilowatts of power. Six more megawatts will come from methane gas collected from landfills."

In tests, one third of customers chose green power options. However, these tests focused on residential customers. Residential customers make up only one third of electricity users in the United States. To date, no real world green-power market controls more than 3 percent of the potential residential market.

So, which analysts are correct? Is deregulation the way to go or should the federal government take more control over the energy sector? Will natural gas open the path for green markets as a result of deregulation or will deregulation result in market barriers that will keep green markets in the minority? You decide.

Section VIII
Advanced Concepts

35 ADVANCED CONCEPTS IN NATURAL RESOURCES MANAGEMENT

OBJECTIVES

After reading this chapter, you should be able to

- explain why natural resources management decisions are so controversial

- discuss the viewpoints of the groups on the Natural Resources Management Continuum

- explain the concept of sustained yield

- explain the considerations affecting decisions on natural resources management policy issues

TERMS TO KNOW

exploitationist	preservationist	anthropocentric
developer	TINSTAAFL paradox	sustained yield
conservationist	opportunity cost	common properties

Up to this point, you have learned a number of basic concepts in natural resources management. If you have read all the preceding chapters, you have learned about the history and current status of resource management efforts in this country—soil, water, wildlife, forest, outdoor recreation, metals, minerals, and energy. You have read about various natural resource management careers.

This chapter gives you a deeper understanding of the decision-making process in natural resources management. You may find that some of the concepts challenge your beliefs about the environment and about humankind's place in nature. If it has been a long time since you read Chapter 3, perhaps it would be worth your time to review it.

PERSPECTIVES ON MANAGEMENT

What Is Management?

Management implies *doing* something. Some would argue that setting lands aside and leaving them alone is not a form of management but rather of apathy or no management at all. That idea is clearly wrong! Preserving an area in its natural state is an intentional management practice. It is important to understand that apathy is different from management designed to preserve an area in a natural state.

If management is active, the first step in management is decision-making. A specific management practice that is to be implemented may be selected from a number of alternatives. Sometimes, selecting the best management practice from among alternative solutions is a long, complex process, often requiring public participation. Making a decision about the best management practice consumes a great deal of time and effort on the part of local, state, and national agencies involved in natural resources management.

Management By and For Whom?

Our society believes (and our laws establish) that many of our natural resources (wildlife, national parks, preserves, etc.) are "held in trust" by the government for the benefit of the people. An important change in natural resources management in recent years has been the extent to which the public is allowed, even encouraged, to participate in decision-making processes. Public hearings are commonplace.

Land management plans are required of many agencies today, just as environmental impact statements were required by the National Environmental Policy Act. These plans must be readable by the public (i.e., nonprofessionals), and agencies are required to solicit public comment. Land-use planning and land-use law are rapidly becoming important specialties within the areas of natural resources management and the legal profession.

A great deal of controversy surrounds whether the public is capable of making valid, well-informed decisions regarding management of the public's natural resources. For instance, if a public hearing were conducted to determine the public's feelings about a new bridge to be built, few citizens would comment on the engineering aspects of the bridge. Few would feel competent enough to second-guess the expertise of a professional engineer. More general concerns, such as where the bridge is to be located, would dominate the comments.

Consider a situation in the area of natural resources management. Assume that the population of wolves is high. One of the animals that wolves eat is elk. Hunters who like to hunt elk would like to have fewer wolves and more elk. They would want to foster a higher elk population by controlling the wolf population. Others would advocate leaving both animal populations alone and letting the forces of nature prevail. A public hearing held on the management of wolves would represent a classic natural resources allocation issue between elk hunters and people who would like to see the wolf left alone. An emotional debate could result. (See Figure 35-1.)

The best management decision in this case is not clear-cut. There are many external concerns affecting the situation. For instance, individuals have different perspectives regarding the relative values of wolves and elk. Because this is an emotional issue, not directly answerable by an applied science such as engineering, the public is more likely to assume that it is capable of providing answers.

Sometimes natural resources managers get frustrated when the public thinks it can manage resources better than they (the professionals) can. On the other hand, public input keeps public resource agencies from becoming stagnant. We should not allow public land managers to become too comfortable with one way of thinking. The public should always be encouraged to provide new insights.

Obviously, debates regarding public participation in natural resources management abound. Public participation is legislatively mandated; it is probably here to stay. New disciplines in natural resources management will be created. These will be in areas such as public relations, communications, mass media, and extension.

FIGURE 35-1 Wolves are natural predators of elk in many of the Northwestern states and in Canada. The dispute between elk, moose, and deer hunters and preservationists who want to promote healthy wolf populations has often been very heated. *(Photos courtesy U.S. Department of Agriculture)*

CONSERVATION PHILOSOPHY CONTINUUM

A continuum, or scale, labeled exploitation on one end and preservation on the other represents the spectrum of conservation philosophies. At the most extreme, on one end is the person who believes people have no right to impose their will on nature—trees or animals should not be harvested, except as a minimum requirement for survival. On the other end is one who would take a mineral from the earth or kill an animal without regard to the impact of that action on the environment. (See Figure 35-2.)

This section may seem judgmental and condemning of some viewpoints, but there are no rights or wrongs. Without **exploitationists**, you couldn't drive your car because there would be no oil. Without **developers** you would not be able to go shopping in a mall or drive on a paved highway. Without **conservationists**, our soil and water resources would have long ago been ruined. Without

preservationists, there would be no Yellowstone Park and no Appalachian Trail. There is a need for people all along the continuum. With that in mind, let us examine the Natural Resources Management Continuum, starting on the preservationist end.

The extreme preservationist would halt the construction of a needed highway because a swamp would have to be drained. He or she would then set the swamp aside as a preserve and allow no trails or facilities for visitors to be built. There are many people who are preservationists but who would not go to that extreme. The Greenpeace movement is an

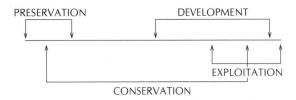

FIGURE 35-2 The Natural Resouces Management Continuum.

example of a preservationist group. One may be a preservationist with regard to only certain species. For instance, there are wide-scale movements to preserve the whale and the elephant. Both movements attract many people who are not preservationists in other areas such as anti-hunting activists.

Generally, conservationists advocate the management of nature to maintain it in a good state while using the resources to benefit people. Farmers who practice no-till crop production and crop rotation to improve their soil are acting as conservationists. People who follow rules about bag limits and creel limits in hunting and fishing are being conservationists. There are many people whose full-time professions are in the area of conservation.

Developers change some aspect of nature to make it more valuable and profitable. Some developers may be very environmentally sensitive, and others may ignore the impact of their developments on the environment, but in general their motivation is profit for themselves and not the long-term benefit of humanity or nature. For instance, the company that drains a swamp and builds a shopping center is in the business of development.

Exploiters take something directly from nature and move it somewhere else for a profit. In doing so, they may or may not abuse the environment. An example of exploitation is a coal-mining operation. The coal miner may leave the area in a restored state, but the purpose of the mining operation is removal of coal for use elsewhere to make a profit. An extreme example of an exploiter is one who would kill the last elephant to get its ivory.

Where people place themselves on the continuum depends on personal perspectives (attitude, cultural background, education, income, etc.), the resource topic at hand, and the source (or perspective of the source) from which one received the basic information. Having students place themselves on the continuum makes them realize how difficult it is to make resource decisions and, especially, to be consistent in those decisions. There are always changes over time and topic. Likewise, management goals change over time. Therefore, we cannot place ourselves on the continuum and definitely say that is how we make decisions about all resources all the time. (Refer to Figure 35-2.)

You may tend to think it is inherently better to be a preservationist rather than an exploiter, but in reality that may not necessarily work. For instance, locking up a piece of land and preserving it may not be a good idea, especially if human life is threatened by not having access to resources that may be available on that land. Everything must be done in moderation. One thing is not inherently better than the other, but one is more appropriate for a given situation than another.

The continuum helps explain a definition of conservation. Conservation can occur anywhere on the continuum except at the extreme ends. When a conservationist advocates setting aside a place to be completely undisturbed by humans, he or she is advocating preservation and not conservation. Remember, the definition of conservation includes "wise use." At the same time, if a conservationist advocates the unrestricted use of a nonrenewable resource, he or she is advocating exploitation, not conservation.

As an example of when exploitation might be good, say a certain rare tropical rain forest plant could be used to make medicine for treatment of AIDS or some other important human disease. Because burning of the tropical rainforests is happening, the plant may soon be lost forever if we don't try to remove some germplasm so that we could reproduce it in ways other than in the natural habitat. Naturally, we would need to collect all the plants we could find if it were very rare. However, if we remove the plant from the forest, that would be exploitation. We might even endanger the plant's survival in nature by removing it to produce it commercially.

Ninety percent of our medicines come from tropical species. Even though we would be exploiting that plant by removing it from its natural habitat, we would be using it to save human lives, and we would be providing for its preservation because it would have been destroyed already by the decision to burn the rain forest. Thus, exploitation in this case may be the wisest use (conservation) of the resource at the time. So, management practices aren't necessarily good or bad in general; rather, they are good or bad for given circumstances.

BASIS FOR NATURAL RESOURCES MANAGEMENT

It is a basic law of human nature that people have unlimited needs and wants. Unfortunately, natural resources are scarce—they are limited. Our land, water, mineral, energy, and other resources are not without limit, and they are subject to misuse or damage. Therefore, we must make choices. The choices we make as a society must consider our wants and needs. What else is a society for if not to fulfill the wants and needs of its members? The choices must also consider the fact that our resources need protection, maintenance, and enhancement. They must also take into account the fact that our wants and needs will continue into the future, and future generations will have wants and needs, too.

That gives rise to an interesting phenomenon. For each alternative there is an "up" side and a "down" side. If you choose to eliminate wolves, as the earlier discussion considered, some problems arise. Yes, there will be more elk—maybe too many. Without wolves to help control the elk population, the large animals would soon exceed the carrying capacity of the environment and they would become unhealthy, even face famine. Also, are elk better or do they have more right to live than wolves? Eliminating wolves means only one thing—killing them.

This problem has been called the **TINSTAAFL paradox**—*There Is No Such Thing As A Free Lunch.* For every decision we make in terms of natural resources management, we as a society should gain something. Unfortunately to gain something, we also always give up something.

The thing we gain or give up is not always money. Having an oil well to exploit a cheap energy resource is a source of money—income for the producer, refiner, shipper, retailer, and investor. However, it is also money in the form of savings for us; savings in lower costs for gasoline, for instance. Unfortunately, if that oil well is in Alaska and you live in New Hampshire, the oil must be shipped. Whenever vast quantities of oil are shipped over long distances, there will always be chances for accidents. The 1989 oil spill in Valdez, Alaska, was part of the "down" side of oil exploitation in Alaska.

Value does not necessarily mean money. There is very real nonmonetary value in having a nice park in which to go for a walk or have a picnic. If that is the case, why not build lots of new parks? One "down" side of an additional park is that we cannot use the money that the park would cost for something else. Another "down" side is that the land for the park cannot be used for a shopping mall or a movie theater. The value, either monetary or nonmonetary, of what we give up when we choose one alternative over another is called **opportunity cost**.

We must manage our natural resources to provide for the present and future needs and wants of our people. We must do so within the context of our environment—our ecosystem. We must also consider the opportunity cost of every decision.

RESOURCE CLASSIFICATION

Resource classification is **anthropocentric** (human-centered). We only classify resources because it creates an easy system through which we can understand relationships between and among different resources. However, we must get away from believing that a particular resource is firmly classified.

Classification is best assigned according to the use of a resource. Water, soil, and air (our basic resources) are usually considered renewable resources. That actually depends on use.

Soil used in agriculture for the most part is renewable if appropriate soil conservation practices are attended to. However, if topsoil erodes away by wind and water from a particular location, that soil is no more easily replaced than is oil after a well is pumped dry. Of course, the exception would be if we had 50,000 years or so to "sit around and wait" for the soil to be regenerated in that area.

Water is another case in point. If water is taken out of an aquifer faster than it is replenished until the aquifer is pumped dry, isn't that the same as mining water? For all practical purposes, once the water has been pumped out of that aquifer, it is gone! We do not have several centuries or millennia to wait for water to be replenished in that aquifer. Pollution

can be used as an example of making a renewable resource nonrenewable. For instance, consider radioactive waste, often affecting water and soil. Is that soil or water renewable for us? Not really.

We usually consider wildlife to be a renewable resource because wildlife reproduce, but are wildlife species always renewable? If they are, how do we define extinction? If we can't help the condor to survive and restore a viable population in captivity, soon it will become extinct. Endangered species may be considered nonrenewable. Once they are extinct, they are extinct forever.

It is important for you to be able to analyze the use of the resource to determine whether the resource is renewable or nonrenewable. Once that is determined, the management strategy can be determined. In general, sustained yield is used to manage renewable resources and protection (preservation) is used to manage nonrenewable resources.

SUSTAINED YIELD

Sustained yield is a very important management concept. It is the scheme by which many renewable resources are managed. The goal of sustained yield management is to protect the quantity and quality of the resource while still allowing the former or other user of natural resources to make a profit.

Management for a sustained yield means the use of renewable resources in such a way as to allow a constant rate of harvest indefinitely. Another way of defining sustained yield is harvesting a renewable resource at a rate that permits repeated harvests at similar yields indefinitely. In a sustained yield approach, we harvest a resource at a rate equal to its rate of reproduction or maturation. The resource is neither overharvested and depleted nor harvested before it is mature. In the case of certain resources, it also means that a minimum level of use is necessary to keep the resource in a healthy state. Certain rangeland and wildlife species are examples of this.

Forestry provides the best examples of sustained yield management. Federal legislation requires management of national forestland via sustained yield. Several states have followed suit and established requirements for sustained yield management in state-owned forestland. By sustained yield management, we harvest the wood we need and still have more wood each year than the year before.

Wildlife management has long applied the concept of sustained yield, whether or not it was called that in earlier years. That is, bag limits, efficiency of hunting techniques (bow versus gun versus trap), game seasons, and protection of the females in most game species are examples of techniques that protect the future resource. Wildlife managers spend a great deal of time assessing the population of a given species, its reproduction rate, and the health of its habitat. They do that so that they can decide how many birds, fish, or animals may be taken as surplus (harvested by hunters, etc.) without damaging the overall population. That way, people will be able to harvest more of that resource in future years.

Fisheries management is similar. Fisheries managers develop optimum sustained yields and maximum sustained yields. The interesting thing about fisheries management is that we can't readily see fish in their natural habitat, so counting is very difficult. Also, there is a somewhat natural, ecological, and economic limiting factor—the longer commercial fishers overfish (harvest more fish than are produced) an area, the lower the density of the pool of fish remaining to be caught. It takes a longer time to find them, which costs more. So there is a natural economic incentive not to overfish. Normally, fishing in a certain area or for a certain species is stopped after it becomes unprofitable.

Sustained yield management is not a perfect solution to problems and concerns associated with renewable resources. For instance, many would agree that managing for sustained yield does not promote improvement or increased yield in a resource. Instead, the goal is maintenance, which will not necessarily suffice when demands outstrip the ability of the resource to produce as population increases.

A second concern is that sustained yield may become a static figure for a given resource. This does not allow for market fluctuations of commercial products. Therefore, we must realize that sustained yield will not be practical for every piece of land or other resource. Additionally, on a regional basis (large scale) and over an extended period of time (long term), sustained yield is expected to produce positive rather than static results.

THE THREE *E*s OF RESOURCE MANAGEMENT

Ecology, economics, and emotion come to bear on every resource management decision. One cannot make a reasonable management decision without asking these questions:

■ Is this ecologically feasible (*Ecology*)?

■ Is this economically feasible (*Economics*)?

■ Is this socially, culturally, and politically feasible (*Emotion*)?

A revealing example is that there are many starving individuals in the United States even though we consider ourselves to be an affluent country. It is also a fact that numerous dogs are disposed of each day as a result of over-population.

Ecologically, especially from the standpoint of biology, dogs are a feasible source of food for man. Economically, it is feasible to look for ways to use that wasted, abandoned, or destroyed resource. It is not feasible; in fact, it is totally unacceptable from an emotional standpoint (politically, culturally, socially) for Americans to consider feeding the "hungry people" with dog meat. In our culture, dogs are pets, not food.

THE TYRANNY OF GEOGRAPHY/ RESOURCE DISTRIBUTION

In very broad terms, there are only two primary industries, and both are resource dependent—agriculture and mining. In this sense, agriculture includes farming, ranching, fishing, forestry, and other "growing" operations. Mining includes all forms of resource exploitation such as strip mines, oil wells, mineral mines, and other "removing" operations. In terms of generating basic materials, what is not grown is mined. Essentially all other industries are dependent on those primary industries for their raw materials.

Resources are not equally distributed around the world. There seems to be no particular pattern of distribution, at least in terms of our anthropocentric geographical regions. Soil and water, oil, diamonds, air, and other natural resources are not distributed by national or state boundaries.

Two notable examples of the uneven distribution of resources is productive soil and fresh water. If we consider these examples, it is easy to see how much of the world's poverty came about.

Practically all of the most productive soils in the world are between about 20° and 40° North latitude. The Southern hemisphere has relatively little land and much of that is mountainous. The tropical soils are very old soils with little moisture-holding capacity. Even more, the tropics tend to be warm year-round, so organic matter decomposes very rapidly, leaving the soils low in organic content. Further north, short growing seasons and permafrost make the soils less productive. The tropical and mountainous regions of the planet contain most of the world's poverty, partly because people in those regions have always had to work with unproductive soil. Thus, they have never been able to establish a productive agriculture.

The amount of fresh water on the planet far exceeds the amount people use. In fact, the total amount of rainfall on the land surface of the earth is about 27,000 cubic miles of water per year. Of that, about 9,600 cubic miles of water returns as runoff or seepage from the land back into the oceans. Yet, the annual amount of water withdrawn for human use is only about 777 cubic miles per year. The amount withdrawn is only about 3 percent of the amount of rainfall and only about 8 percent of the runoff. Of the 777 cubic miles of water withdrawn for human use, practically all is immediately returned to the water cycle for reuse. For instance, agriculture uses about 75 percent of the water, but almost all of that either soaks into the ground to add to the groundwater supply or evaporates into the atmosphere and returns again elsewhere as rain.

Clearly, there is not a shortage of fresh water worldwide. There is, however, a serious distribution problem in terms of the water supply. In southern California, the large population of the Los Angeles area must import huge quantities of fresh water each year, because there is too little water locally available. At the same time, the Mississippi River discharges about 133 cubic miles of water to the Gulf of Mexico each year. In many parts of Africa, terrible water shortages place strains on the human and wildlife populations, yet the Congo River alone discharges 301 cubic miles of fresh water per year. The problem with fresh water is not one of shortage. Rather the problem with water is one of distribution of fresh water supply.

There are several implied concepts here. First, there are "haves" and "have nots" in terms of countries with and without a good base or resources upon which to build a stable economy. Also, we are fortunate in that we were born in the United States, which is endowed with a multitude of economically important natural resources. The United States is able to maintain a strong economy because it has an abundance of basic resources and many different kinds of resources upon which to base the economy. A third implication is that our geographic divisions are human-centered. They are not rational boundaries if we were to only consider resources.

COMMON PROPERTIES

Common properties (resources usually in the public domain—but not always) are, in theory, owned by everyone, but, in reality, are owned by no one. Management of these resources is extremely complicated. The concept of the commons is important, and its implications for rational decision-making are enormous.

An example of a common property is a public lake. It belongs to a local, state, or federal government. In that sense, it belongs to everybody who is a citizen of that government. At the same time, it doesn't belong to anyone. For instance, there is probably a public highway or road that goes by your school. What part of it belongs to you? Can you make any of the normal decisions that an owner can make with regard to the road?

Actually, the only decision you alone can make regarding the commons is how you will personally use it. Normally, it costs you the same whether you use it or not. The most logical thing for you to do with a part of the commons is to use it as much as you want or as long as it is profitable to do so. Even further, it benefits the individual to use the commons without particular regard for its maintenance.

For instance, the owner of a large tractor-trailer may pay over $5,000 per year in road taxes regardless of how many miles the truck is driven. If you consider only the cost of the road tax, it makes sense to drive the truck as many miles as possible to reduce the cost per mile of the road taxes. True, driving more miles would wear the road out faster. The owner of the truck would have to pay more taxes later to repair the road; but so would everybody else, and the truck owner's share of the added cost would be very small compared to the added income from using the road so heavily.

As another example, imagine a food store that charges a set fee per year for each person who shops there, regardless of the amount of food taken. If you shopped there, would you be frugal and take a "fair" selection of foods? Would you take only the foods that you like most and large amounts of them? What about everybody else? Even if you were fair, would everybody be? Beyond that, what does *fair* mean when it comes to use and care of common properties?

That is the problem with common properties. Nobody owns them but everybody can use them. Everybody pays for them, but the amount that any one individual pays is not much related to how he or she takes care of them.

Marine fisheries and public ranges are prime examples. Fishing has an important international dimension. There is little one fisher can do to affect the population of fish in an area. If the cost of making a fishing trip is held constant, it pays to take all the fish you can on a given fishing trip. Also, as long as a fishing trip is profitable, it pays to make as many trips as possible. That is true for everyone, so it pays each one to take as many fish as possible. If each fisher independently makes the decision to maximize or limit his or her catch, there is only one logical business decision. That decision would lead to destruction of the fish population in the fishing ground, causing everyone to suffer.

In terms of public ranges, if a rancher can make a profit by putting another cow onto a section of land, he or she can be expected to do so. If all the ranchers with access to a public range add additional cows as long as each added cow continues to bring a profit, the range will soon be overgrazed. That is exactly what happened after the Civil War when the cattle barons were using open rangeland for free grazing.

RESOURCE OWNERSHIP/DOMAIN

Whether resources are owned privately or not has major implications for management. These conflicts are usually referred to in the literature as public versus private domain resources. For example, a farmer has jurisdiction over his or her soil, but does not own the wildlife that comes onto the property. The wildlife is "owned" by all Americans and "held in trust" for them by the appropriate government (state

for most species, federal for some, such as marine mammals). This distinction has obvious impacts on management options. Our system could be contrasted to the European system in which the landowner owns the wildlife on his or her property.

ANTHROPOCENTRICITY OF NATURAL RESOURCES MANAGEMENT

Everything we do with natural resources is done with the interests of humans at heart. Of course, we all have different perspectives of what our best interests are. This extends from the original definition of what a natural resource is—a non-manmade (natural) component of the earth's environment that has utility to one or more persons. Having utility means the resource has human use. For example, wolves are not howling to be managed. Elk hunters want wolves managed so that they can have more moose available to hunt. Anthropocentricity is simply a human-centered viewpoint we use when dealing with natural resources and the environment.

Consider that science is amoral. That means scientists do not consider human values in the search for scientific principles and truth. They only consider evidence. Both economics and ecology are sciences. Therefore, ecology and economics are amoral. Scientists try to be objective, at least as much as is humanly possible. Thus, an economist or ecologist operates from an objective point of view and recommends a particular opinion based on scientific evidence.

Natural resources management is not amoral or completely objective. It is a discipline separate from these sciences because resources management decisions must be done with the best interest of human beings in mind. Decisions cannot be made in a vacuum. Remember the three *Es* of resources management.

A good example is the job of an environmental engineer. Environmental engineers are in the business of protecting human health and welfare, not the environment's health and welfare. It seems astounding that a member of such a lofty-sounding profession isn't concerned primarily about the environment. Yet, ultimately we can protect ourselves and serve our best interests only by protecting the environment.

GLOBAL MODELING

The future of resource management is in modeling, especially computer modeling. Global models are complex systems for visualizing our resource base, especially the complex and dynamic interactions occurring between and among the variables and components. Global computer models will be the primary tool of natural resources managers in the future. This is true not only in terms of the future of natural resources management, but also in terms of career selection and preparation.

HUMAN POPULATION

Increased population is clearly the number one environmental problem. Population is at the root of all our resource management concerns. We humans seem to believe that we can be an exception to the concept of carrying capacity. Certainly, agriculture and other technologies have increased the earth's carrying capacity for humans. However, at some point a maximum will be reached. What will be the quality of life when that maximum is reached? Will science,

as opposed to emotionalism and humanitarian efforts, win out?

For the purpose of this book, we will not overemphasize the human population problem. Regardless of the population, we still must make intelligent natural resources management decisions.

IN THE FINAL ANALYSIS

We live in a closed ecosystem—our planet. At least in the foreseeable future, we cannot use natural resources from outside that system.

As a society, we make decisions—including natural resources management decisions—based on the benefit to the members of the society. Those decisions should take into account an understanding of our history, ecology, economics, and social values. They must also take into account world, national, state, and local politics. There are many conflicting pressures and influences on our decision-making processes. Political candidates are evaluated partially on their environmental positions.

In the final analysis, there are no "best" or "right" answers to natural resources management questions; but there are lots of "wrong" answers. Wrong answers are those that do not provide for the *long-term* benefit of humanity.

DISCUSSION QUESTIONS

1. Place yourself on the Natural Resources Management Continuum. Try to place yourself on the continuum with special consideration for cockroaches, oak trees, whales, or deer. Can you see why there is so much controversy in natural resources management decision-making?

2. If resources were the basis for dividing the world, where would the boundaries be? How would they be determined? By biomes? By major continent? By agricultural productivity? By geographic obstacles, such as mountain regions or rivers?

3. What is the commons? Why does the concept of the commons mean that governments instead of private individuals or companies must take action to manage our natural resources?

4. What do we mean by the "tyranny of geography"? What parts do politics and history play in this tyranny?

SUGGESTED ACTIVITIES

1. Organize a class debate on the following resolution: There should be no open hunting season allowed for deer in (your state).

2. Organize a class discussion on where the following people would fit on the Natural Resources Management Continuum:

    ■ Gifford Pinchot
    ■ Thomas Jefferson
    ■ Teddy Roosevelt
    ■ Our current president
    ■ The leader of Brazil
    ■ A local community leader
    ■ The editor of your local newspaper

3. Inventory the commons in your community. Try to determine who is responsible for maintaining each, who pays for the maintenance, and where the authority for collecting the money resides. Find out what kinds of abuses they have been subject to in the last year.

4. Prepare a presentation for class on "Environmental Issues Facing Our Community."

MY EXOTIC GYPSY MOTHS

What makes an animal a pest? Usually, being a pest refers to the ability of an animal to cause economic hardship to human beings. One such pest is the gypsy moth. (See Figure VIII-A.) Every year, acres upon acres of forests are destroyed due to gypsy moth infestations. A nonnative to the United States, this insect was introduced in an experiment involving silk production. Unfortunately, the moth escaped and was never able to produce silk for commerce. The gypsy moth has no predators in this part of the world. There is nothing to keep this pest in check aside from human intervention.

FIGURE VIII-A Adult gypsy moth females resting on a tree. *(Photo courtesy of Entomology Department, Virginia Tech, Blacksburg, Virginia)*

The gypsy moth starts life as an egg and then becomes a caterpillar. In the caterpillar stage, these insects cause stress on trees. All caterpillars prefer oaks that are low in tannins, apple, birch, sweet gum, and willow. Caterpillars have been known to feed on maple, chestnut, pine, beech, hickory, and elm. Caterpillars do not normally feed on tulip poplar, red cedar, black and honey locust, ash, or dogwood.

So, why do trees die? The eating of leaves or defoliation is not an immediate cause of death. Trees become stressed after defoliation attacks and then become more susceptible to other stressors such as drought, injury, and disease. After continuous cycles of stress the tree simply gives up and dies. (See Figure VIII-B.)

The following options are available to those who wish to fight the affects of gypsy moth infestations:

- Bt (*Bacillus thuringiensis*)
  - Lasts less than one week.
  - As living organisms, bacteria must be kept alive, so they are more difficult to apply than chemicals.
  - Affects all caterpillars but no other insect families.
  - Costs $20 per acre.

- NPV—Virus (GypCheck)
  - Lasts less than a week.
  - Difficult to apply.
  - Affects only the gypsy moth.
  - Very costly because it is difficult to produce.

FIGURE VIII-B A forest defoliated by gypsy moths. This is during the SUMMER! *(Photo courtesy of Entomology Department, Virginia Tech, Blacksburg, Virginia)*

■ Dimilin–diflubenzuron

— Chitin (exoskeleton) synthesis inhibitor.

— Lasts for two months.

— Affects many invertebrates.

— Cannot apply over water.

— Costs $10 per acre.

■ Sevin–carbaryl

— Lasts for one week.

— Affects most insects.

— Honey bees are very sensitive.

— Costs $10 per acre.

■ Hand Picking

— Directly kills insects.

— Not efficient for large tracks of land.

— Highly labor intensive = High costs.

■ Biological control—Introduce a predatory wasp

— High cost.

— Only effective at low population levels.

Assume that gypsy moths are beginning to infest your community. What do you do? You decide.

NATURAL RESOURCES-RELATED INTERNET WEB SITES

Current as of August 11, 2000. Please note that URLs change frequently. Any of these Web sites may have moved.

Agricultural Research Service, Image Gallery	http://www.ars.usda.gov/is/graphics/photos/index.html
American Forest and Paper Association	http://205.197.9.134/index.html
American Forest Foundation	http://www.affoundation.org/html/about_atfs.html
American Petroleum Institute, home page	http://www.api.org/
Bureau of Land Management, home page	http://www.blm.gov/nhp/index.htm
Department of Interior, home page	http://www.doi.gov/indexj.html
Environmental Protection Agency, home page	http://www.epa.gov/
Environmental Protection Agency, Environmental Education Center	http://www.epa.gov/teachers/
Environmental Statistics Programs of U.S. Government	http://www.epa.gov/ceisweb1/ceishome/digitallib/estatgov.html
Fish and Wildlife Service, Legislation Summaries	http://www.fws.gov/laws/federal/summaries/index.html
Fisheries Statistics; Economics Division	http://www.st.nmfs.gov/st1/index.html
Forest Biology and Dendrology at Virginia Tech	http://www.fw.vt.edu/dendro/
Forest History Society, home page	http://www.lib.duke.edu/forest/index.html
National Marine Fisheries Service, Alaska Regional Office	http://www.fakr.noaa.gov/
National Mining Association	http://www.nma.org/
National Oceanographic and Atmospheric Administration, home page	http://www.noaa.gov/
National Oceanographic and Atmospheric Administration, National Weather Service	http://www.nws.noaa.gov/
National Park Service, home page—Experience Your America	http://www.nps.gov/
North Carolina State University, Aquatic Botany Lab (*Pfisteria*)	http://www.pfiesteria.org/

Oregon State University, Experimental Forest photos	http://www.fsl.orst.edu/lter/navigafr.htm
Pace University, U.S. Environmental Law	http://joshua.law.pace.edu/env/fullusa.html #ESA
Society of American Foresters	http://www.safnet.org/
Stream Quality Indicator Page	http://dnr.state.il.us/orep/inrin/ctap/bugs/ default1.htm
U.S. Fish and Wildlife Service, Endangered Species site	http://www.fws.gov/r9endspp/endspp.html
United Nations, World Population statistics	http://www.unfpa.org/modules/6billion/ numbers.htm
United States Energy Administration	http://www.eia.doe.gov/emeu/cabs/usa.html
University of Minnesota, Northern Forest Photo Gallery	http://willow.ncfes.umn.edu/imagegall/ imagegall.htm
U.S. Department of Agriculture, Natural Resources Conservation Service—NRCS	http://www.nrcs.usda.gov/
U.S. Department of Agriculture, Photo library	http://www.usda.gov/oc/photo/opcphsea.htm
U.S. Department of Agriculture, U.S. Forest Service	http://www.fs.fed.us/
U.S. Department of Interior, National Park Service	http://www.nps.gov/
U.S. Environmental Protection Agency, Visualizing the Great Lakes	http://www.epa.gov/grtlakes/image/
U.S. Fisheries and Wildlife Service, Educators Links	http://wetlands.fws.gov/educator.htm
U.S. Fisheries and Wildlife Service, home page	http://www.fws.gov/
U.S. Fisheries and Wildlife Service, Photo Center	http://www.fws-nctc.org/nctc1.htm
U.S. Fisheries and Wildlife, National Image Library	http://www.fws-nctc.org/
U.S. Government Departments Links Page	http://terra.geo.orst.edu/users/raaba/gov.htm
U.S. Office of Surface Mining, photo library	http://www.osmre.gov//ocphoto.htm
Virginia Cooperative Extension	http://www.ext.vt.edu/
Virginia Department of Agriculture and Consumer Services	http://www.state.va.us/~vdacs/vdacs.htm
Weather Channel, home page	http://www.weather.com/homepage.html
World Resources Institute, 1998–99 Report	http://www.wri.org/wr-98-99/index.html
World Resources Institute, Atmosphere and Climate page	http://www.igc.org/wri/enved/trends/ atm-home.html
Worldwatch, home page	http://www.worldwatch.org/

A

abyssal zone: deep water zone of the ocean.

absorption field: a land area composing the final part of a septic system for waste treatment. The absorption field receives the liquefied, largely decomposed waste material. In the soil, the partially decomposed waste is further attenuated by bacteria, fungi, and other organisms and by dilution and adsorption.

accelerated erosion: soil loss occurring at an accelerated rate because of the loss of plant cover or a change of the natural plant cover.

acre foot: a unit of volume measure corresponding to 1 acre in area and 1 foot in depth; 43,560 cubic feet; 325,852 gallons.

activated sludge: decomposed wastes containing high levels of bacteria.

aerobic bacteria: bacteria that require oxygen to live.

alidade: the instrument used to measure the direction to a fire from a lookout tower.

all aged: a forest with trees of varying ages and heights.

alluvial deposit: soil deposited by moving water.

amoral: not considering moral questions; science is amoral—neither moral nor immoral; morality does not apply.

anaerobic bacteria: bacteria that live in the absence of oxygen.

animal control biologist: a person who works with problems caused by rodents and predators.

animal equivalent unit (AEU): the amount of forage (grazing) that is required to feed a 1,000-pound animal (such as a steer) for a given period of time.

annual ring: a visible circle in the cross section of a tree trunk produced by the rapid, soft spring growth followed by slower, denser summer growth.

anthracite coal: the oldest and hardest coal.

anthropocentric: human-centered.

artificial stocking: placing fish in a pond.

asbestos body: a thick covering of tissue over an asbestos fiber in the lungs.

asbestosis: a disease caused by inhaling asbestos.

ASCS: Agricultural Stabilization and Conservation Service of the United States Department of Agriculture.

atom: a single unit of matter containing a central nucleus and surrounded by electrons.

attenuation: the process of lessening the impact of an agent, such as attenuation of water contamination by any number of natural processes or human intervention. Purification or neutralization of impurities.

B

backcut: a cut on the side opposite and just above the undercut made when felling a tree.

backfire: use of planned fires in the path of a wildfire to burn up fuel before the wildfire reaches it.

background pollution: the introduction of containments into a water supply from natural sources. Background pollution occurs in nature and not from human activities.

bag limit: the amount of game an individual can take.

bait fishing: catching fish through the use of a line and bait.

balance: as used in this text, "balance of nature." In reality, nature is never balanced; it is always in a dynamic state and undergoing continuous change. When nature is in "balance," that change is relatively slow.

balance of nature: the tendency of ecosystems to reach a general state of equilibrium.

baleen whale: a whale that feeds on plankton and has whalebone instead of teeth.

bark borers: any of a family of beetles, the larvae and adults of which burrow under the bark of trees, eating the inner bark.

bathyal zone: ocean zone containing the continental slope and rise.

bauxite ore: ore from which aluminum is obtained.

Bayer process: the process commonly used for extracting aluminum oxide, or aluminum, from bauxite ore; named for the chemist who developed the process.

benefication: the separating of iron ore from rock.

berm: a mound of soil built in a field across the slope, designed to conduct surface runoff water safely to a desired location.

Bessemer process: the process used to make steel from iron ore.

biogeography: the study of the distribution of biological organisms throughout the world.

biological impurities: biological materials in solution or suspension in another substance. In water treatment, biological impurities range from mammals to viruses and from aquatic trees to algae. In fresh water that will be used for human consumption, those organisms must be removed or neutralized.

biological insect control: use of nonchemical methods to control insects.

biological synthesis: any change in the composition, shape, size, or structure of the plants or animals in an ecosystem, resulting from the putting together of food, energy, and water to form new combinations.

biome: a major land area characterized by dominant plant life forms, such as tundra and grasslands.

bituminous coal: the most common type of coal, softer than anthracite.

board foot: a standard unit of measure for lumber, a piece of lumber 1 inch × 1 foot × 1 foot (or the equivalent) before surfacing.

BOD test: biological oxygen demand test, used to determine the amount of organic waste in water.

bog: land area that is very damp, usually with evergreens present, and covered with moss and peat. Surface provides a spongy walk or sticky mud.

bucked: the process of cutting a tree into logs or poles.

C

canopy: the upper part of the forest, consisting of the crowns of trees that receive direct sunlight.

carrying capacity: the number of wildlife an area can support with food, shelter, and water; or the ability of a given area to provide food, water, and shelter for the population of a given animal.

catastrophic waves: waves caused by storms, hurricanes, and landslides.

CCC: Civilian Conservation Corps, a federal program in the 1930s.

certified tree farm: a farm registered with the American Tree Farm System of the American Forest Institute.

chain reaction: a continuing fission process, found in nuclear fission reactors.

chemical impurities: substances dissolved (in solution) or suspended (in suspension) in other substances. In water treatment, this refers to dissolved or suspended substances in water that cause undesirable effects in the water for its use or consumption.

cleaning: as used in this text, the selective harvesting of trees and removal of undergrowth for the purpose of assisting young seedlings or saplings to grow in a managed forest.

clear cutting: removal of all the trees in an area.

climax species: the species of vegetation that would eventually dominate an ecosystem if the ecosystem ever reached equilibrium.

coagulation: physical process of smaller particles clumping together to form larger particles. In wastewater treatment, the process of clumping smaller particles together so that they can later be allowed to settle out of the water.

Code of Ethics for Hunters: a set of sensible rules that hunters should follow to ensure that other present and future hunters can enjoy the sport.

commercial forest: a forest being managed for wood production.

common properties: natural resources that are owned by the group but really not owned by anyone—public lakes, the oceans, national forests, parks, the atmosphere, and the like.

composting: a technique of placing organic matter in a favorable environment for its partial decomposition. The finished product is referred to as compost and is used as a soil additive to promote plant growth.

conservation: use of a natural resource in such a way as to minimize waste and maintain the resource in as good a condition as is practical.

conservation farm planning: the process of planning a farming operation considering the conservation of soil and water resources as an important goal.

conservation officer: person that enforces the game laws.

conservation tillage: any one of numerous methods of cultivating the soil to control soil loss.

conservationist: a person or group who works to manage our natural resources for their long-term usefulness.

containment landfill: a landfill designed in such a way as to minimize the percolation of water through the buried materials.

contour farming: cultivation of a field with rows produced across the slope.

converted wood: wood that has been mechanically or chemically changed.

cooling lagoon: a pond used to cool heated water before returning it to the environment.

cooling tower: a device composed of coils used to cool heated water before returning it to the environment.

cord: a stack of wood 4 feet × 4 feet × 8 feet or equivalent.

cover crop: a crop grown primarily to prevent erosion and to improve the soil structure and tilth.

crop rotation: the alternation of one crop with one or more other crops to promote soil conservation, to improve soil structure, and for numerous other benefits.

crown: the branches, limbs, twigs, and leaves making up the top portion of the tree.

crown fire: a fire burning the treetops, the most dangerous kind of wildfire.

cruising: the process of estimating the pulpwood or lumber that a standing parcel of trees can produce.

cubic foot: a quantity of wood required to fill a volume 1 foot × 1 foot × 1 foot.

currents: water movement usually due to prevailing winds.

cut: a seam of coal that has been removed.

D

dam: a device used to hold water, it can be constructed of various materials such as earth, concrete, or steel.

dbh: diameter breast height, the diameter of a tree trunk, in inches, measured 4.5 feet above the ground.

decomposer: an organism that breaks down other organisms into their organic compounds, such as fungi and bacteria. A potato or orange that is rotting is being affected by a decomposer.

decomposition: the process by which complex organic compounds are broken down into simpler compounds with the release of energy and usually carbon dioxide and water as by-products.

decreaser: vegetation in a grassland area that tends to be easily depleted by moderate grazing.

defoliating insects: insects that remove the leaves from a tree by eating the leaf itself or the stem of the leaf.

dendrometer: any device used to measure the diameter of a tree trunk.

derrick: a hoisting apparatus used to hold an oil drilling machine.

desalination: the removal of salt from water.

developer: a person or group who changes the form of some natural resource, in place, to improve its usefulness to people.

diffuse source pollution: also known as non-point pollution. The introduction of contaminants into a water supply across a wide area.

distillation process: the removal of salt from water by turning the water into a vapor and then cooling, whereby the salt is left behind.

diversion ditch: a channel dug into the soil and designed to conduct surface runoff water safely to a desired location.

drift mine: a coal mine where passageway is bored into a hill or mountain.

drillships: offshore drilling rigs used to look for oil.

drip irrigation: a system of water distribution by means of progressively smaller pipes. At the point where the water is released from the pipe, it flows out very slowly (drips) and is placed directly at the base of the plant.

"duck" stamp: a stamp purchased by duck hunters, the proceeds from which are used to expand waterfowl populations.

dump: an uncovered area used to dispose of refuse.

E

ecological succession: the process by which a series of life-forms, primarily plants, replace each other as the dominant life-form in a given geographic area.

ecological wetland: as defined by the United States Fish and Wildlife Service—lands transitional between terrestrial and aquatic systems where the water table is usually at or near the surface or the land is covered by shallow water. According to this definition, wetlands must have one or more of the following three attributes: (1) at least periodically, the land supports predominantly hydrophytes; (2) the substrate is predominantly undrained hydric soil; and (3) the substrate is nonsoil and is saturated with water or covered by shallow water at some time during the growing season each year.

ecology: the study of the relationships among living things.

ecosystem: a community of interrelated plants and animals in a given environment.

electrodialysis: removing salt from water by passing an electric current through the water.

embankment pond: pond where water is impounded behind an earth embankment or dam built across a water course.

endangered: in danger of becoming extinct.

energy sources: natural resources that store energy from the sun in forms such as oil or coal, that can be used by people.

environment: all the physical surroundings of an organism.

environmentalism: social or political activities undertaken for the purpose of affecting some aspect of the environment, normally in a way that the individual perceives as beneficial or positive.

environmentalist: a person who undertakes some political or social activity with the intention of affecting some aspect of the environment, normally in a way that he or she perceives as beneficial or positive. In a very real sense, everyone is an environmentalist.

erosion: the process of the wearing away or removal of the surface layer of anything. Especially in natural resources management, erosion refers to the loss of surface soil to water and wind.

estuary: the area where fresh water meets salt water.

euphotic zone: "twilight" ocean zone where light can penetrate.

euryphagous: an animal that consumes great varieties of food.

eutrophication: process by which water becomes rich in dissolved nutrients.

evaporation: the dispersion of water into the atmosphere in the form of water vapor.

even aged: a forest with all trees being of the same approximate age and height.

excavated pond: constructing a pond by digging a pit below the surrounding ground level.

exhaustible: a natural resource that exists in a limited quantity and cannot be replaced when used.

exploitation: use of natural resource without replacement.

exploitationist: a person or group who moves some natural resource from its naturally occurring location for use elsewhere.

extinct: no longer exists.

F

farming for today: farming with only immediate economic considerations.

farmland-use planning: the process of determining the crop and livestock uses to which the various parts of a farm are to be put.

ferroalloys: ferrous metals combined with iron to add various properties to steel.

ferrous metals: metals containing iron.

fertilizers: a general term for a category of natural and synthetic soil additives designed to provide nutrients for plant growth.

field line: a porous drainage line through which largely decomposed and liquefied sewerage is carried by gravity. The liquefied sewerage flows out the openings into the surrounding soil of the absorption field.

finger system: a drainage system carrying the excess water from a septic system.

fire barrier: any obstacle, such as a road, creek, or ditch, that can help prevent the spread of a fire.

fire suppression: the process of putting out a forest fire.

fire triangle: the three things required for any forest fire: fuel, oxygen, and heat.

fish and wildlife technician: a person who examines methods that conserve and aid animals that have trouble surviving.

fish culture technician: a person who produces fish for either food or game.

fish sampling: management procedure used to keep track of what fish species are in a pond or lake.

fishery biologist: a person who works with habitats, spawning, and artificially stocking fish.

fixed platform: a permanent oil rig for off-shore drilling purposes.

flash distillation: the process of heating water and placing it under pressure, which "flashes" it to boiling; the water is collected as fresh water.

flocculation: the physical mixing of two substances. In water treatment, flocculation is the process of stirring together influent with coagulants or other additives that assist in the water purification process.

flood hazard: the probability that a flood of a given degree of severity will occur in a known period of time.

floodplain: land areas that border rivers, lakes, and streams and that are flooded periodically.

food chain: a group of plants and/or animals related to each other by the fact that one feeds or depends for food on the next.

food web: a group of organisms that depend on each other for food in a given ecosystem. A food web typically consists of a series of interconnected food chains.

forb: broad-leafed flowering plant with a fibrous root system.

forest enemies: any insects, diseases, wildlife, wildfire, or other agents that can damage a forest.

forest pathology: the study of diseases in forest trees.

forest regions: geographic areas having somewhat similar climates and forest tree species.

forester: a professional worker who plans, manages, and/or supervises a forest.

forestry technician: a paraprofessional who assists foresters.

fuel: materials, mainly organic matter, available to burn in a fire. In a forest fire, the initial fuels are dried leaves, broken and dried branches, and dry undergrowth. After a fire is started, entire trees become fuel for the fire.

G

game biologist: a person who deals with the management practices of game.

gas treater: a person who operates a unit that removes water from natural gas.

geological erosion: natural erosion that occurs over a long period of time, given a full and natural plant cover of the soil.

geologist: a person who studies the structure, composition, and history of the earth's surface.

geophysical prospector: a person who locates oil and natural gas sources.

geophysicist: a person who studies the composition and physical aspects of the earth.

geothermal energy: heated groundwater used as an energy source.

grassed waterway: a ditch, designed to conduct surface runoff to a desired location, planted to a heavy grass sod to prevent erosion in the ditch.

grassland: rangeland that has grasses as its primary naturally occurring vegetation.

gravimeter: a device using changes in the gravitational pull to find oil.

grazing capacity: the number of animal equivalent units that can be supported by a given grazing area without unacceptable damage to the grass cover.

Great Chicago Fire: a major urban wildfire occurring from October 8 to 10, 1871, which destroyed most of the city of Chicago.

green manure crop: a crop grown for the specific purpose of plowing it back into the topsoil to improve the soil structure and fertility.

gross weight: the total weight, in pounds, hundredweight, or tons, of a quantity of wood.

ground fire: a smoldering fire in the organic matter built up on a forest floor or bog.

groundwater: water absorbed into the earth's surface.

gully erosion: loss of soil in larger, often impassible trenches or ditches, resulting from run-off.

H

habitat: the place or site an animal lives.

habitat development: the development and improvement of food, water, and shelter for wildlife.

hard water: a very common chemical problem in water, usually a result of excess calcium or magnesium. Both of these form compounds that precipitate out to form scale deposits that clog up water pipes and plumbing fixtures. They react with soap making it less effective, and they generally taste bad.

harvest cutting: removal of any or all the trees in an area for sale.

hazardous waste: wastes that, in sufficient quantities and concentrations, pose a threat to human life, human health, or the environment when improperly stored, transported, treated, or disposed.

home range: the area over which animals repeatedly travel.

human water cycle: the hydrologic cycle as modified for human use of the water. Human activities remove water from the natural water cycle then return it at some other point in the natural cycle. The movement from and then back into the natural water cycle is the human water cycle.

humus: black or brown substance resulting from the decay of plant and animal matter.

hunter orange: the color required by most states, that should be worn by a hunter.

hunting: the killing of game for food or pleasure.

hydric soil: soil that is frequently or constantly saturated with water; characterized by gray, white, or light yellow discoloration resulting from lack of adequate air to allow for normal oxidation of hydrocarbons and metals such as iron.

hydroelectric power: producing electricity from water-driven turbine generators.

hydrologic cycle: the water cycle.

hydrologist: a person who studies rainfall, its rate of filtration into the soil, and its return to the ocean.

hydrology: water characteristics of the soil including presence of free water, frequency of saturation, and movement of water by percolation.

hydrophyte: water-loving/water-tolerant plant; plant that survives in soil that is underwater or frequently saturated.

hydropower: water power.

hypothermia: a condition that results when the body loses heat faster than it produces it.

hypsometer: any device used to measure usable tree height.

I

ignition temperature: the temperature at which a given fuel will begin to burn, in the presence of oxygen.

improvement cutting: an intermediate harvesting of selected older trees in a managed forest to allow remaining trees to grow better.

incendiary fire: a forest wildfire started intentionally or accidentally by a person.

incineration: the burning of waste material.

increaser: vegetation that tends to prosper under moderate grazing pressure.

industrial pollution: pollution from factories.

industrial solid waste: waste materials consisting primarily of spoilage from mining, logging, and other industrial processes that are not disposed of in landfills.

infectious diseases: diseases that are caused by parasites.

insect resistance: the ability of a tree to repel or avoid insect damage.

intermediate cutting: harvests taken from a stand of trees before the trees reach planned maturity.

internal wave: an underwater wave caused by temperature changes in the depths of the ocean.

intertidal: the area of the ocean between high and low tide.

invader: plant species that tend to take over a grassland area under heavy grazing pressure or with undergrazing.

irreversibile change: practices, such as construction of a concrete highway, that cannot be changed under normal circumstances.

J

jack-up rig: a temporary offshore drilling rig.

jurisdictional wetland: as defined by the U.S. Army Corps of Engineers, an area that has frequent flooding or saturation, is covered by hydrophytes, and includes hydric soils. According to this definition, wetlands exhibit wet hydrology, hydrophytes, and hydric soils.

L

land capability class: description of the ability of a land parcel to produce crops, given the danger of erosion and other limiting factors.

landfill: an open area into which garbage is placed to be covered by a layer of some other material, typically soil.

land-use planning: the political process of determining the uses to which a given land area can be put.

leachate: when connected with landfills, leachate is any liquid—typically water—containing contaminants, percolating from a landfill, and moving into and through the underlying or surrounding soil.

liberation: removal of dominant trees to allow younger or smaller trees to grow.

lignite: a low-carbon soft coal.

limnetic zone: the zone of a lake that extends from the end of rooted vegetation to the point where sunlight no longer penetrates the water.

liner: when used in conjunction with landfills, a liner is any layer of material, either synthetic or natural, that is relatively impermeable and intended to minimize the seepage of leachate from the landfill into the surrounding soil.

littoral zone: shallow water zone containing rooted vegetation.

loess deposit: windblown soil deposits.

logger: a laborer who harvests trees for lumber or pulpwood.

long-lining: fishing with a long line, up to 75 miles long with 2,000 hooks.

long-tube distillation: distillation process where salt water is sent through tubes heated by steam, then cooled to collect salt-free water.

longwall: a method of mining, removing coal from one face and then transporting it by a conveyor.

lumber grades: a system for determining the potential uses for lumber.

M

magnetometer: an oil-finding device employing the earth's magnetic pull.

marine deposit: as used in this text, a marine deposit is a layer of parent material deposited over time at the floor of an ocean or sea, and that forms the basis for the subsequent development of a soil.

market hunter: a person who earns his or her livelihood from hunting wild animals to sell the meat, hides, fur, or body parts.

marsh: wetland continuously or frequently covered by freshwater, tidal water, or standing salt water. Marshes do not rely on rainfall for their water supply, and soft-stemmed plants are the dominant plant type.

mature forest: a forest in which the trees dominating the canopy have reached full height.

meltdown: an uncontrolled chain reaction in a nuclear reactor.

membrane process: the filtering of salt from water through a series of filters.

memorial: an area set in remembrance of an individual or event.

meteoric water: water returned to the earth in any form, such as rain, sleet, or snow.

meteorologist: a person who studies the air surrounding the earth and forecasts the weather.

methane: a gas released from the decomposition of wastes.

methane digester: a device used to make methane gas from decomposed organic wastes.

migratory waterfowl: game birds, such as ducks, that move from winter nesting grounds in warmer climates to summer grounds in cooler climates.

minerals: inorganic compounds occurring naturally in the earth and having a distinctive structure.

Miramichi fire: the second largest forest fire in U.S. history, occurred in Maine and New Brunswick, Canada.

mulching: the use of dead plant materials to cover an exposed soil surface.

multiple use: the concept that a natural resource can be used for more than one function, such as a water reservoir being used as a water supply and for boating.

municipal solid waste (MSW): all those waste materials produced by households, businesses, and industry and disposed of in landfills, incinerators, and composting systems.

N

national island trust: the development of the islands surrounding the United States for recreational purposes.

national trails system: authorized by the National Trails Act of 1968, a network of both interconnected and nonconnected trails for scenic or recreational purposes, to include trails designed primarily for horseback riding, motorcycling, hiking, walking, and other uses.

natural attenuation landfill: a landfill designed to allow for normal percolation of water through the buried materials and into the surrounding soil, with the intention that the leachate will be diluted or otherwise neutralized with minimal impact on the environment.

natural resources: all those things with which we come in contact that can be used to perform any useful function.

natural wetland: wetlands that have not resulted from human activities.

neritic zone: the high biological zone that extends to the continental shelf.

New York Sporting Club: an organization formed in 1844 to promote wildlife management.

nock: the end of an arrow that fits in the bowstring.

noncommercial forest: an area of forest not being managed to produce wood for harvest.

nonexhaustible: a natural resource that, for all practical purposes, never runs out, such as sunlight.

nonferrous metals: metals not containing iron.

nonhazardous waste: all waste materials that are not otherwise classified as hazardous or radioactive.

noninfectious diseases: diseases caused by environmental problems.

NRCS: Natural Resources Conservation Service of the U.S. Department of Agriculture, formerly the Soil Conservation Service (SCS).

nuclear power: using energy from a fission process for an energy source.

O

ocean zonation: the classification of the ocean in zones according to depth, temperature, pressure, or light.

oceanographer: a person who studies all facets of the earth's oceans.

oil shale: stone (shale) containing crude oil.

opportunity cost: the value, either monetary or nonmonetary, of resource in one way rather than in the next-best alternative way.

organic matter: dead plant and animal material in various stages of decay.

organic waste: organic material released by plants or animals as nonusable by-products, such as animal manure.

original tissue: as used in this text, partially decomposed organic matter that still retains recognizable characteristics of the plant or animal from which it came.

overgrazing: a condition that occurs in a range area when the grazing capacity is surpassed.

oxygen: a colorless, odorless gas that occurs naturally in the atmosphere and that combines readily with many other elements and compounds. As used in this text, oxygen plays an essential part of the processes of photosynthesis, respiration, and decomposition. In forest fire management, oxygen makes up one leg of the fire triangle, allowing fuel to react with heat to produce fire.

P

parent material: those materials underlying the soil from which the soil was formed.

park ranger: a person who directs and supervises the operation of a park.

pathogen: a disease-causing agent.

percolate: to seep. Water percolates through the soil. Leachate seeps through the soil.

percolation rate: the speed at which a given quantity of water can be absorbed by the soil.

permeability: the ability of the soil to allow water to move through it.

Peshtigo fire: a huge forest wildfire near Peshtigo, Wisconsin, which burned over 3.8 million acres and killed over 1,600 people.

petroleum engineer: a person who supervises a producing oil well.

petroleum geologist: a person who interprets and analyzes information on possible oil sites.

photosynthesis: the chemical process of sugar production from water and carbon dioxide in the presence of chlorophyll and sunlight.

pig iron: raw bars of steel.

plankton: microscopic life, the principal food of several fish species.

point source pollution: the direct introduction of contaminants into a water supply from an identifiable source.

pond: a small body of water.

population density: the number of wildlife per unit of land area.

population level: the number of a given animal in a specific geographic area.

postsecondary school: a school, beyond high school but lower level than college, usually offering courses in technical or trade subjects.

potable water: water that is chemically and biologically appropriate for human drinking without further purification or boiling. Potable water almost always has some chemical and biological impurities remaining.

prairie pothole: a wetland that relies on periodic rainfall for its water supply. Prairie potholes are usually full in the spring and early summer before water levels start to drop off and the potholes start to disappear for the rest of the year. Prairie potholes are found mainly in North Dakota, South Dakota, Minnesota, and Nebraska.

prawn: a large shrimp.

predator control: management procedures used to control predators of wildlife.

prescribed fire: a planned fire used as a part of the management plan for a forest.

preservation: an attempt to prevent the use of some natural resource or the modification of an environment simply for the sake of keeping it intact.

preservationist: a person or group who seeks to maintain some resource in its current or natural state.

primary treatment: in municipal wastewater treatment, a mechanical system that involves collecting the wastewater and removing the items that settle from it. The primary system typically includes a screening device, grit chamber, settling tank, sludge digester, and drying beds.

producer: an organism that extracts energy, water, minerals, and other compounds from the environment and combines them into organic tissue. We generally think of green plants that manufacture food by photosynthesis as producers.

purse seining: the process of using a large net for catching fish. In fishing for tuna, the purse seine is drawn around the school of tuna and then closed at the end and bottom, trapping the fish.

R

radioactive material: material that spontaneously emits nuclear particles into the environment.

radioactive waste: waste materials that are characterized by the active and measurable emission of subatomic particles.

radioactivity: the property of some elements, such as uranium, or releasing alpha and beta rays by the disintegration of the atomic nuclei.

range: as used in this text: 1. the geographic area in which an organism lives and moves about; for example, the range of a specific animal. 2. a term used to describe an open, usually grassed area over which livestock move about to feed.

range manager: a person who is responsible for maintaining an open, usually grassed area in which livestock are allowed to move about to feed.

rangeland: land that is used by animals for grazing or browsing.

rare: populations are very small.

reclamation plan: a plan, required of mining operations, laying out the steps that will be taken to restore an area to usefulness after exploitation.

recreational resources: natural resources useful primarily for recreation and enjoyment, such as hiking or boating areas.

recreationists: people who plan, coordinate, and develop recreational programs.

recycling: the process of reusing materials that would otherwise be disposed of as waste.

red mud: the red compounds remaining after aluminum is extracted from bauxite ore.

refuge: an area set aside for the protection of wildlife.

regional land-use planning: the political process of determining the uses to which the lands within a given geographic region can be put, normally a major area such as a group of counties or townships, or a large part of a state.

renewable: a natural resource that can be replaced as it is used, such as trees and wildlife.

reservoir: a device used to control flood waters.

respiration: as used in this text, the process of breaking down higher-energy organic compounds into lower-energy compounds with a release of energy. We often think of respiration as "breathing."

rill erosion: soil loss in small but visible tracks or traces, like small gullies, resulting from runoff.

rookeries: breeding territories for seals.

room and pillar: a way of removing coal that employs the idea of leaving pillars of coal to help support the ceiling.

root hairs: tiny projections, near the growing tip of a root, through which most of the water and nutrients are absorbed.

root system: all the roots of a given plant.

rorqual: a type of baleen whale sometimes referred to as a finback whale.

runoff erosion: soil loss resulting from the cutting action of running surface water.

runoff water: water that runs on the earth's surface.

S

salinity: the salt concentration of water.

salvage cutting: the harvesting of damaged trees.

sandblow: sand moved by wind resulting in either removal of sand particles, as in wind erosion, or accumulation of sand particles, as in a sand dune at a beach.

sanitation cutting: removal of injured, diseased, or insect-infested trees.

SCS: the Soil Conservation Service of the U.S. Department of Agriculture; now the Natural Resources Conservation Service (NRCS).

secondary treatment: the biological processing of sewage. Over one half of all city treatment plants contain this system, which removes about 85 percent of the organic wastes in the water. Wastewater is piped through an aeration tank to increase the oxygen content in the sewage water; aerobic (oxygen-using) bacteria assist in the decomposition of the solid waste. The water is then pumped into another sedimentation tank for additional settling. The remaining water is discharged after the addition of chlorine. The settled sludge, called activated sludge, is sent through the aeration tank again to allow further bacterial decomposition. This system removes about 90 percent of the organic wastes.

sediment basin: a pondlike structure built for the specific purpose of collecting surface runoff and allowing sediment to settle out.

seed tree: a large, healthy tree left standing to produce seeds for reforestation of an area.

seismograph: a device using sound waves to find oil.

semiportable sprinkler: an irrigation system that supplies water to a lateral, which pivots around the supply, and can be moved between central water supplies.

semisubmersible rigs: rigs, partly under water, used in offshore oil drilling.

septic system: a sewerage treatment system in which waste is allowed to decompose biologically in a holding tank before it is distributed to the soil in an absorption field through field lines; the basic form of human sewage disposal in rural areas.

septic tank: a large receptacle forming a part of a septic sewerage treatment system, in which the sewerage is held while it decomposes.

shade tolerant: a tree that can grow in relatively indirect sunlight.

shaft mine: a coal mine with its access passageways vertical to the coal seam.

sheet erosion: loss of a fairly uniform layer of the soil surface, often without noticeable gullying.

shelterbelt: windbreaks designed to protect crops and livestock from the wind.

skidding: dragging logs to the loading area.

slope: the change in elevation for a given horizontal distance of the earth's surface, often expressed as a percentage.

slope mine: a coal mine with its access tunnel on a slope from the surface to the coal seam.

sludge digester: a device that uses bacteria to decompose solid wastes.

snowmobile safety code: a set of sensible rules that snowmobilers should follow to ensure their own safety and the safety of other snowmobilers and bystanders.

sodding: the use of patches or strips of mature grass to establish a ready-made lawn or other grassed area.

soil: the layer of natural materials on the earth's surface, containing both organic and inorganic materials, that is capable of supporting plant life.

soil classification: a system to describe the characteristics of a given soil in terms of its derivation and physical makeup.

soil conservation district: a voluntary organization of landowners, primarily farmers, used to help manage soil conservation in an area.

soil conservation technician: a paraprofessional employee who assists a soil conservationist.

soil conservationist: a professional person whose primary responsibilities involve the planning, implementation, and supervision of conservation practices.

soil drainage: the ability of soil to allow water to move through and out of the soil structure. Soil drainage is closely related to the amount and size of spaces between soil particles as well as to the absence or presence of impermeable layers under the soil surface.

soil horizons: layers in a mature soil.

soil scientist: a person who conducts research on soils and/or studies soil types and capabilities.

soil survey: a document classifying all the soils in a given geographic area, for the act of examining the soils to prepare such a document.

soil survey report: a document prepared as a result of the scientific examination and classification of a land area, used by farmers, planners, and soil conservationists to develop land-use plans.

soils engineer: a professional whose work centers around the use of soil in the building process or who works with soil as a building medium.

solar energy: using the sun as a heat energy source.

solid waste: waste materials that are normally disposed of in landfills, incinerators, or composting facilities.

solution: in a solution, the molecules of one substance are dissipated among the molecules of another substance; for example, table salt or sugar dissolved in water produces a salt or sugar solution.

spat: young oyster.

splash erosion: the effect of raindrops dislodging soil particles.

sprinkler irrigation: the application of irrigation waters over the top of crops.

stationary irrigation sprinkler: involves burying permanent underground water lines in the area needing watering.

stationary wave: a wave occurring in bays where the wave does not move but the water moves up and down.

stenophagous: an animal that eats a specialized diet.

stopping builder: person who constructs doors and walls in coal mine tunnels.

strip cropping: planting of alternating crops in parallel, narrow strips around the slope of a field, usually to help control erosion.

strip mine: the removal of coal near the surface by first removing the soil covering it.

subsoil thickness: in a soil profile, the distance measured from the bottom edge of the topsoil to the top edge of the parent material.

supratidal: the area of the ocean between low and high tide.

surface fire: a fire burning the surface litter of a forest floor.

surface irrigation: irrigation of crops by building a series of large and small ditches to transport the water.

surface water: water on the earth's surface, such as in our lakes, ponds, and streams.

suspension: solid particles held temporarily or permanently in a liquid; solids held (suspended) in a liquid but not in solution.

sustained yield: a method of using a natural resource in such a way that it can be used at a constant rate or for periodically repeated harvests indefinitely.

swamp: land area covered continuously or nearly continuously by standing water with trees or shrubs growing in the water. The water tends to be very stagnant and is usually dark and non-translucent.

T

taconite: a low grade of iron ore.

tar sands: sand containing a low-grade oil.

Taylor Grazing Act of 1934: the federal law that authorized the U.S. government to establish and enforce regulations regarding the grazing of public lands. It has eventually led to the management of millions of acres of public lands by the Bureau of Land Management.

temperature stratification: the temperature layers of the ocean.

Ten Commandments of Gun Safety: a set of sensible rules to ensure the safe use of firearms.

terrace: a physical structure constructed across the slope of a hill to prevent the free downhill movement of surface runoff.

territory: the area an animal will defend, usually to its death.

tertiary treatment: the chemical processing of sewage wastewater. This system involves removing from the water phosphates and nitrates, which could cause a nutrient-rich situation if the sewage effluent is dumped into a stream. This system is more costly than the secondary system and is used only when the water is going to be reused.

texture: with regard to soil, texture refers to the relative percentages of sand (largest), silt (mid-sized), and clay (smallest) particles in the soil. A course-textured soil is mostly sand and is gritty to the touch. A fine-textured soil is mostly clay and feels silky to the touch, becoming slick or sticky when wet.

thermal pollution: pollution by the release of heated water.

thermocline: the middle temperature zone in an ocean usually 10 to 100 meters deep.

thinning: the process of removing, normally harvesting, part of the trees in a stand to provide room for other trees to grow.

tidal power: using the force of the tides to run turbines in order to produce electricity.

tidal wave: the most common catastrophic wave known.

tides: specialized waves caused by gravitational attraction of the sun and moon on the earth.

TINSTAAFL paradox: *There Is No Such Thing As A Free Lunch.* The belief that for someone to get anything of value, something else of value must be given up either by that person or another person or group.

topsoil: the surface layer of soil, containing relatively high percentages of decomposed and partially decomposed organic matter.

topsoil thickness: the depth to which topsoil extends in the soil surface.

trail system: the system set up to provide outdoor recreation by the use of trails.

transformer: an organism that takes primary food sources from producers, combines them with other chemicals and energy forms, and converts them into more complex organic compounds, tissues, and foods. A dairy cow is a transformer when she manufactures milk from grass and water.

transpiration: the process of evaporation of water through the leaves of the tree.

tree height: the length of the tree trunk from the point where it will be cut to the end of the last usable section to be kept.

triangulation: the process of determining a location on the ground by measuring directions from two or more known points.

trunk: the "stem" of a tree, consisting of heartwood and sapwood.

turbidity: solid matter suspended in a liquid. In nature, water always contains some solid matter held in suspension. Thus all naturally occurring water has at least some degree of turbidity. When the concentration of solids in suspension becomes great enough to be visible, the water is aesthetically unpleasing. It may have color, odor, or taste. If the turbidity is more extreme, the water may appear to be muddy. Removal of turbidity is an important step in the processing of water for humans to use.

U

undercut: when felling a tree, the removal of a slice of wood from the side of the tree in the direction the tree is to fall.

undergrazed: a condition that occurs in a range area when too little grazing is done to keep the grass cover healthy.

urban pollution: pollution in our cities due to items such as road salts, street detergents, and sewer systems.

usable water: water that is readily accessible, may be freestanding (lake) or flowing (river).

V

vapor compression distillation: a distillation process in which salt water is turned to steam and then placed under pressure. This steam starts a chain reaction that heats other water, and when it condenses, it falls to a bottom tank to be collected as fresh water.

vegetative erosion control: use of plants to control erosion.

vernal pool: a special type of wetland that may only last for a few months each year. Like prairie potholes, vernal pools rely on periodic rainfall to form in the spring. They disappear in early summer.

vertebrate: an animal with a spinal column.

W

wastewater treatment plant operator: a person who oversees the waste treatment system.

water chute: a steep waterway lined with a material such as concrete, designed to handle large volumes of rapidly moving runoff.

water treatment plant operator: an employee who works in a water purification facility for a city, town, or other community. The water treatment plant operator operates the equipment that readies the water for human consumption.

watershed: that land draining into a pond.

waterway: a ditch or berm designed to conduct surface runoff safely to a desired location.

weathering: the process of breaking down rocks and minerals through the actions of weather, such as freezing/thawing or rain.

whalebone: tiny plates in the mouth of a baleen whale that strain food.

wilderness: an area that has not been changed by human activity.

wind erosion: soil loss resulting from the action of the wind.

windbreak: a structure, usually vegetative, across the prevailing wind direction designed to reduce surface wind speed.

Z

zone of aeration: empty spaces, filled with air, between soil mineral particles.

zoologist: a person who works with animals, both domestic and wild.